Nanomaterials for Environmental Purification and Energy Conversion

Nanomaterials for Environmental Purification and Energy Conversion

Special Issue Editors

Ewa Kowalska
Agata Markowska-Szczupak
Marcin Janczarek

MDPI • Basel • Beijing • Wuhan • Barcelona • Belgrade

Special Issue Editors

Ewa Kowalska
Hokkaido University
Japan

Agata Markowska-Szczupak
West Pomeranian University of Technology
Poland

Marcin Janczarek
Poznan University of Technology
Poland

Editorial Office
MDPI
St. Alban-Anlage 66
4052 Basel, Switzerland

This is a reprint of articles from the Special Issue published online in the open access journal *Catalysts* (ISSN 2073-4344) from 2017 to 2019 (available at: https://www.mdpi.com/journal/catalysts/special_issues/nano_envi_puri)

For citation purposes, cite each article independently as indicated on the article page online and as indicated below:

LastName, A.A.; LastName, B.B.; LastName, C.C. Article Title. *Journal Name* **Year**, *Article Number*, Page Range.

ISBN 978-3-03921-814-1 (Pbk)
ISBN 978-3-03921-815-8 (PDF)

Cover image courtesy of Kunlei Wang.

Contents

About the Special Issue Editors

Ewa Kowalska is a leader of Research Cluster for Plasmonic Photocatalysis in Institute for Catalysis (ICAT), Hokkaido University. She received her Ph.D. in chemical technology from Gdansk University of Technology, Poland in 2004. After completing Marie Skłodowska-Curie fellowships in France (Paris-sud University; 2002-2003) and in Germany (FAU University, Erlangen and Ulm University; 2009-2012) and postdoctoral fellowships in Japan (Hokkaido University; JSPS: 2005-2007 and GCOE: 2007-2009), she joined ICAT as an associate professor in 2012. Her current work focuses on heterogeneous photocatalysis, environmental protection, plasmonic nanomaterials, vis-responsive photocatalysts and anti-microbial properties of nanomaterials.

Agata Markowska-Szczupak was born in Szczecin in 1972, and graduated from Faculty of Environmental Science and Fisheries, University of Warmia and Mazury in Olsztyn. She completed a doctoral school at University of Agriculture in Szczecin and obtained a Ph.D. in plant genetics in 2001. She became an associate professor at Faculty of Chemical Technology and Engineering (West Pomeranian University of Technology in Szczecin) after obtaining habilitation (D.Sc.) in chemical engineering in 2014. From December 2018 she works as professor at the same Faculty.

Marcin Janczarek received his Ph.D. in chemical technology from Gdansk University Technology, Poland in 2005, where he mainly conducted his research until 2016. He also worked at University of Erlangen-Nurnberg, Germany (2003-2004) and Hokkaido University (2011-1013, 2016-2017). He obtained habilitation (D.Sc.) in chemical technology in 2019. Currently, he is an associate professor at Poznan University of Technology. His research interests include heterogeneous photocatalysis, materials engineering, and technologies for purification of water and air.

Editorial

Nanomaterials for Environmental Purification and Energy Conversion

Ewa Kowalska [1,*], Agata Markowska-Szczupak [2] and Marcin Janczarek [3]

[1] Institute for Catalysis (ICAT), Hokkaido University, N21 W10, Sapporo 001-0021, Japan
[2] Faculty of Chemical Technology and Engineering, West Pomeranian University of Technology in Szczecin, 70-322 Szczecin, Poland; Agata.Markowska@zut.edu.pl
[3] Institute of Chemical Technology and Engineering, Faculty of Chemical Technology, Poznan University of Technology, 60-965 Poznan, Poland; marcin.janczarek@put.poznan.pl
* Correspondence: kowalska@cat.hokudai.ac.jp; Tel.: +81-11-706-9130

Received: 10 October 2019; Accepted: 12 October 2019; Published: 14 October 2019

Nanomaterials, engineered structures of which a single unit is sized (in at least one dimension) between 1 to 100 nm, are probably the fastest growing market in the world. Although, nanotechnology is still a new discipline (proposed by Richard Feynman's talk "There's Plenty of Room at the Bottom" in 1959; and named by Norio Taniguchi in 1974), nanomaterials have already been commercialized for various purposes, including medicine, food, cosmetics, technology, and industry, as well as human and environmental health. A lot of studies on the preparation of more efficient, stable, and safe nanomaterials have been performed each year, as clearly shown by the growing number of published papers (Figure 1). Although, nanomaterials are extremely important for industrial and household purposes, it should be pointed out that properties of nanomaterials differ substantially from those of bulk materials of the same composition, resulting in high reactivity. Accordingly, possible undesirable effects might cause harmful interactions with the environment, living organisms, and humans and their parts (e.g., influence on structural integrity and functions of essential proteins, enzymes, and DNA), and thus have the potential to generate toxicity [1,2]. Therefore, possible negative impacts of novel materials (toxicity) must be also considered in the design of efficient, but also safe products. It is believed that most importantly, for future human development, nanomaterials/nanotechnology could be used to solve three of the top ten of humanity's problems (proposed by Prof. Smalley [3]), i.e., environment, water, and energy (which is critical for the rest of the problems). It is known that nanomaterials might be used for purification of water and air [4–6], wastewater treatment [7–10], microorganisms' inactivation [11–13], and energy conversion to the electricity and fuels [14–16].

Therefore, the special issue of *Catalysts* has been announced to discuss the progress of recent research on synthesis, properties, and applications of nanomaterials for environmental purification and energy conversion. This Special Issue was mainly dedicated as a platform for the contributions from The Symposium on Nanomaterials for Environmental Purification and Energy Conversion (SNEPEC), which was held in Sapporo, Japan in February 2018 (http://www.cat.hokudai.ac.jp/icat-nepec/). The contributions from those who could not attend SNEPEC were also welcomed. The Symposium covered a broad range of topics focusing on the exceptional role of catalytic nanomaterials in solving environmental and energy problems of modern societies. Accordingly, the Special Issue "Nanomaterials for Environmental Purification and Energy Conversion" is a collection of 17 papers, including 16 research papers and one review. Eleven papers present heterogeneous photocatalysis for efficient degradation of organic pollutants (phenol [17,18], 2-propanol [19], dyes [20–22], humic acid [23]), inactivation of microorganisms (*Escherichia coli*, *Staphylococcus epidermidis* [24], *Bacillus subtilis*, and *Clostridium* sp. [18]), H_2 evolution [25], and CO_2 reduction [26,27]. Six other papers focus on conventional catalysis ("dark" reactions), reporting efficient H_2 production [28,29], synthesis of ethanol and butanol [30], direct conversion of CO_2 and methanol to dimethyl carbonate [31], water purification [32], and advanced characterization of catalysts by X-ray absorption fine structure (EXAFS) spectroscopy [33]. The majority of studies have been performed with particulate catalysts (nanoparticles (NPs)), but organized nanostructures, such as nanotubes (TiO_2/Cu_xO_y [18], carbon nanotubes (CNTs) [30,32]) and nanowires (CeO_2 [31]) have also been used.

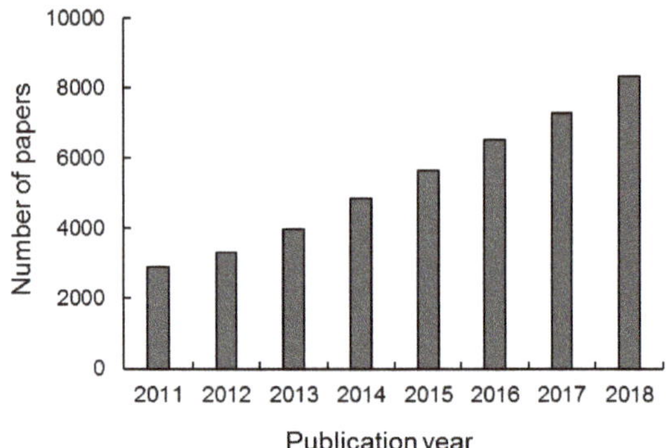

Figure 1. Number of papers published annually on nanomaterials (searched in Web of Science using "nanomaterials", 9 October 2019).

Fu et al. prepared CeO_2 nanowires by the advanced solvothermal method for direct catalytic synthesis of dimethyl carbonate from CO_2 and methanol [31]. They found that the surface reduction under H_2 atmosphere formed acidic/alkaline sites on the catalyst surface, and thus significantly improved catalytic activity, reducing the activation energy barrier from 74.7 to 41.9 kJ/mol. Complex catalytic studies were performed by Hafizi and Rahimpour for catalytic H_2 production [28]. The effect of alumina and Mg-Al spinel as the support for the formation of Fe_2O_3 catalyst was investigated. The $Fe_2O_3–MgAl_2O_4$ with narrowed mesopore-sized (2.3 nm) was successfully synthesized as an ultra-pure lattice oxygen transport medium. Furthermore, Daneshmand-Jahromi et al. analyzed the role of yttrium promoted Ni/SBA-16 as an oxygen carrier in steam methane reforming [29]. The reaction temperature, Y and Ni loading, steam/carbon molar ration, and lifetime of the oxygen carrier were investigated. The best catalytic activity was obtained for mesoporous silica (SBA-16) modified with 25 wt% Ni and 2.5 wt% Y, resulting in 99.83% CH_4 conversion and 85.34% H_2 production.

The synthesis of ethanol and butanol from synthesis gas on multiwalled carbon nanotubes (MWCNTs) functionalized with salicylic acid and impregnated with copper–cobalt catalyst was proposed by Zou et al. [30]. It was found that salicylic acid did not only functionalize carbon nanotubes, but also promoted the synthesis of ethanol and butanol, instead of methanol. Moreover, the surface properties of MWCNTs were crucial for efficient alcohol synthesis, i.e., the best activity was obtained with MWCNTs of 30 nm diameter. Carbon nanotubes (CNTs) were also used for efficient water purification by Li et al. [32]. The conductive cotton filter anodes were fabricated by a facile dying method to incorporate CNTs as filters. The developed filtration device achieved physical adsorption of organic compounds (ferrocyanide, methyl orange dye, and antibiotic tetracycline), and additionally, an application of external potential resulted in chemical oxidation of pollutants. The CNTs amount, total cell potential, and surfactant were key parameters affecting the electrochemical oxidation. It was proposed that the conductive cotton filter might be efficiently used for low-cost and energy-saving water purification. This very important research was reported by Wakisaka et al. who demonstrated the extended X-ray absorption fine structure (EXAFS) spectroscopy as an efficient technique to characterize Pt–Au fuel cell catalysts [33]. Previously, range-extended EXAFS was only achieved in high-energy resolution fluorescence detected XAFS (HERFD-XAFS). The presented results confirmed the feasibility of the range-extended EXAFS using the bent crystal Laue analyzer (BCLA) for fuel cells models containing Pt and Au.

Most papers discussed photocatalytic reactions on nanomaterials (heterogeneous photocatalysis). Titania (titanium(IV) oxide, TiO_2) is one of the most well-known and widely studied photocatalysts, due to its advantages, such as high activity, stability, low-cost, and nontoxicity (excluding toxicity of nanomaterials), as also confirmed in this issue (seven papers [17–19,21,24,26,27]). However, titania has two main shortcomings, i.e., recombination of charge carriers (typical for all semiconductors) and inactivity under visible light irradiation (due to wide bandgap). Therefore, various studies have been performed to improve photocatalytic performance of titania. The comprehensive review by Giovannetti et al. on recent advances in graphene-based TiO_2 nanocomposites for synthetic dye degradation shows the increasing potential of titania photocatalysts based on graphene matrix, in the field of extending the light absorption of TiO_2 from UV into the visible light range of radiation [21]. In this regard, the idea of titania modification is strongly present in the papers collected in the issue, e.g., titania surface modification with copper oxides (Cu_2O [19] and Cu_xO_y [18]), organic compounds (glucose [24] and urea [19]), polymers (polydopamine [27]), graphene [21], carbon/nitrogen [17], and titania doping (self-doped or hydrogenated) [34]. All kinds of modifications resulted in enhanced activity under either UV, visible light or solar radiation. Guan et al. prepared and characterized the colored core-shell structure of $TiO_2@TiO_{2-x}$ [26]. They reported visible light activity of this material towards CO_2 reduction under a simulated flue gas system. Visible light-induced photoreduction of CO_2 was also conducted with polydopamine-sensitized TiO_2 by Wang et al. [27]. The successful titania modifications with D-glucose was achieved by Markowska-Szczupak et al. [24], where the photocatalytic activities of suspended and immobilized photocatalysts were compared. In this study, it was shown for the first time that titania modification with monosaccharides could be efficient for water disinfection, and the immobilization of the photocatalyst on the concrete discs might be a prospective method for public water supplies and water storage tanks (as exemplified for a home aquarium in Figure 2a). The synergistic effect was observed by Janczarek et al. for titania bi-modification with urea (formed poly(amino-s-triazine) [35]) and Cu_2O [19]. Two types of possible mechanisms of visible light activity were proposed, i.e., the type II heterojunction and Z-scheme. An important issue was the morphological form of the photocatalytic material based on titania. Golabiewska et al. prepared highly active microspheres in the presence of ionic liquid and Kozak et al. designed TiO_2/Cu_xO_y nanotubes with visible light activity in the degradation of organic pollutants and bacteria inactivation. It was proposed that these nanotubes could be efficiently used for environmental purification under natural solar radiation, as shown on the journal cover (Figure 2b).

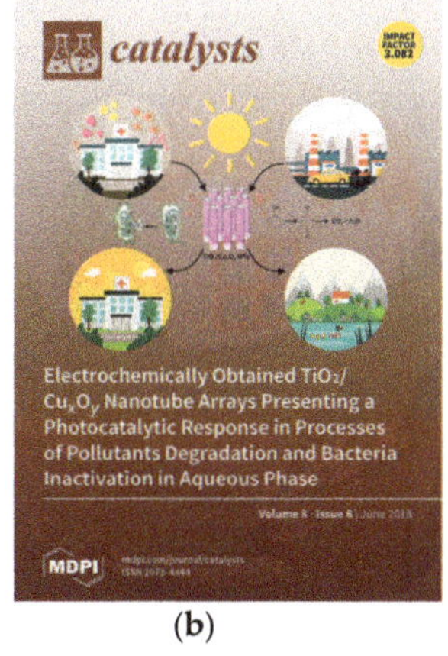

(a)   (b)

Figure 2. Journal covers of *Catalysts* showing possible applications of (**a**) modified titania with D-glucose for water tanks; volume 8, issue 8 (https://res.mdpi.com/data/covers/catalysts/big_cover-catalysts-v8-i8.png) [24] and (**b**) TiO_2/Cu_xO_y nanotube arrays for environmental purification; volume 8, issue 6 (https://res.mdpi.com/data/covers/catalysts/big_cover-catalysts-v8-i6.png) [18].

It should be pointed that also other semiconductors have been used successfully for environmental applications, such as ZnO [36], graphitic carbon nitride (g-C3N4) [37], WO_3 [10], $BiVO_4$ [38], and $SrTiO_3$ [39], and some of them exhibited higher activity than that of titania even under UV irradiation [40–42]. Accordingly, the photocatalytic activity of other semiconductors has also been discussed in this special issue, such as $ZnCr_2O_4$ [23], $TiOF_2$ (modified with NaOH) [20], SnO_2 (fluorine-doped SnO_2 (FTO), surface modified with Cu NPs) [25], and rectorite/Fe_3O_4/ZnO [22]. Wang et al. prepared rectorite/Fe_3O_4/ZnO composites with photocatalytic and magnetic properties enabling efficient photocatalyst separation after reaction [22]. Liu et al. obtained Cu/fluorine-doped tin oxide nanocomposites (Cu/FTO) dedicated to visible light-induced hydrogen production and photocurrent generation [25]. The high stability during recycling (24-h irradiation) should be considered as high advantage of this material. The nanosized $ZnCr_2O_4$ was synthesized by Dumitru et al. by thermolysis of a new Zn(II)–Cr(III) oxalate coordination compound [23]. The photocatalyst was much more efficient for humic acid degradation than simple photolysis (7%), reaching 60% degradation after 3 h of UV irradiation. Finally, $TiOF_2$ modified with NaOH of network structure was prepared via a modified low-temperature solvothermal method by Hou et al., and efficiently used for Rhodamine B degradation [20].

In conclusion, the significant role of catalytic nanomaterials in environmental remediation, energy production, and chemical synthesis systems has been discussed in the collected papers. We do believe that the SNEPEC symposium and this associated Special Issue have provided further insights to this area. We are looking forward to seeing how things will be progressed at the next SNEPEC symposium (July 2020).

Finally, we thank all authors for their valuable contributions, without which this special issue would not have been possible. We would like to express our sincerest thanks also to the editorial team of *Catalysts* for their kind support, advice, and fast responses.

Acknowledgments: Institute for Catalysis (ICAT), Hokkaido University is highly acknowledged for excellent support in organizing and hosting the Symposium on Nanomaterials for Environmental Purification and Energy Conversion (SNEPEC), 20–21 February 2018.

Conflicts of Interest: The authors declare no conflict of interest.

References

1. Nel, A.; Xia, T.; Mädler, L.; Li, N. Toxic Potential of Materials at the Nanolevel. *Science* **2006**, *311*, 622–627. [CrossRef] [PubMed]
2. Nel, A.E.; Mädler, L.; Velegol, D.; Xia, T.; Hoek, E.M.V.; Somasundaran, P.; Klaessig, F.; Castranova, V.; Thompson, M. Understanding biophysicochemical interactions at the nano–bio interface. *Nat. Mater.* **2009**, *8*, 543–557. [CrossRef] [PubMed]
3. Smalley, R.E. *Nanotechnology, Energy and People*; MIT Forum: River Oaks, TX, USA, 2003.
4. Hoffmann, M.R.; Martin, S.T.; Choi, W.; Bahnemann, D.W. Environmental Applications of Semiconductor Photocatalysis. *Chem. Rev.* **1995**, *95*, 69–96. [CrossRef]
5. Verbruggen, S.W.; Keulemans, M.; Goris, B.; Blommaerts, N.; Bals, S.; Martens, J.A.; Lenaerts, S. Plasmonic 'rainbow' photocatalyst with broadband solar light response for environmental applications. *Appl. Catal. B Environ.* **2016**, *188*, 147–153. [CrossRef]
6. Alfano, O.; Bahnemann, D.; Cassano, A.; Dillert, R.; Goslich, R. Photocatalysis in water environments using artificial and solar light. *Catal. Today* **2000**, *58*, 199–230. [CrossRef]
7. Herrmann, J.-M.; Disdier, J.; Pichat, P.; Malato, S.; Blanco, J. TiO_2-based solar photocatalytic detoxification of water containing organic pollutants. Case studies of 2,4-dichlorophenoxyaceticacid (2,4-D) and of benzofuran. *Appl. Catal. B Environ.* **1998**, *17*, 15–23. [CrossRef]
8. Wang, Z.-H.; Zhuang, Q.-X. Photocatalytic reduction of pollutant Hg(II) on doped WO_3 dispersion. *J. Photochem. Photobiol. A Chem.* **1993**, *75*, 105–111. [CrossRef]
9. Li Puma, G.L.; Khor, J.N.; Brucato, A. Modeling of an Annular Photocatalytic Reactor for Water Purification: Oxidation of Pesticides. *Environ. Sci. Technol.* **2004**, *38*, 3737–3745. [CrossRef]

10. Abe, R.; Takami, H.; Murakami, N.; Ohtani, B. Pristine Simple Oxides as Visible Light Driven Photocatalysts: Highly Efficient Decomposition of Organic Compounds over Platinum-Loaded Tungsten Oxide. *J. Am. Chem. Soc.* **2008**, *130*, 7780–7781. [CrossRef]

11. Guillard, C.; Bui, T.-H.; Felix, C.; Moules, V.; Lina, B.; Lejeune, P. Microbiological disinfection of water and air by photocatalysis. *Comptes Rendus Chim.* **2008**, *11*, 107–113. [CrossRef]

12. Markowska-Szczupak, A.; Wang, K.; Rokicka, P.; Endo, M.; Wei, Z.; Ohtani, B.; Morawski, A.W.; Kowalska, E. The effect of anatase and rutile crystallites isolated from titania P25 photocatalyst on growth of selected mould fungi. *J. Photochem. Photobiol. B Boil.* **2015**, *151*, 54–62. [CrossRef] [PubMed]

13. Markowska-Szczupak, A.; Ulfig, K.; Morawski, A. The application of titanium dioxide for deactivation of bioparticulates: An overview. *Catal. Today* **2011**, *169*, 249–257. [CrossRef]

14. Fujishima, A.; Honda, K. Electrochemical Photolysis of Water at a Semiconductor Electrode. *Nature* **1972**, *238*, 37–38. [CrossRef] [PubMed]

15. O'Regan, B.; Graetzel, M. A low-cost, high-efficiency solar cell based on dye-sensitized colloidal titanium dioxide films. *Nature* **1991**, *353*, 737–740. [CrossRef]

16. Abe, R. Recent progress on photocatalytic and photoelectrochemical water splitting under visible light irradiation. *J. Photochem. Photobiol. C Photochem. Rev.* **2010**, *11*, 179–209. [CrossRef]

17. Gołąbiewska, A.; Checa-Suárez, M.; Paszkiewicz-Gawron, M.; Lisowski, W.; Raczuk, E.; Klimczuk, T.; Polkowska, Ż.; Grabowska, E.; Zaleska-Medynska, A.; Łuczak, J. Highly Active TiO_2 Microspheres Formation in the Presence of Ethylammonium Nitrate Ionic Liquid. *Catalysts* **2018**, *8*, 279.

18. Kozak, M.; Mazierski, P.; Zebrowska, J.; Kobylański, M.; Klimczuk, T.; Lisowski, W.; Trykowski, G.; Nowaczyk, G.; Zaleska-Medynska, A. Electrochemically Obtained TiO_2/Cu_xO_y Nanotube Arrays Presenting a Photocatalytic Response in Processes of Pollutants Degradation and Bacteria Inactivation in Aqueous Phase. *Catalysts* **2018**, *8*, 237. [CrossRef]

19. Janczarek, M.; Wang, K.; Kowalska, E. Synergistic Effect of Cu_2O and Urea as Modifiers of TiO_2 for Enhanced Visible Light Activity. *Catalysts* **2018**, *8*, 240. [CrossRef]

20. Hou, C.T.; Liu, W.L.; Zhu, J.M. Synthesis of NaOH-Modified $TiOF_2$ and Its Enhanced Visible Light Photocatalytic Performance on RhB. *Catalysts* **2017**, *7*, 243. [CrossRef]

21. Giovannetti, R.; Rommozzi, E.; Zannotti, M.; D'Amato, C.A. Recent Advances in Graphene Based TiO_2 Nanocomposites ($GTiO_2$Ns) for Photocatalytic Degradation of Synthetic Dyes. *Catalysts* **2017**, *7*, 305. [CrossRef]

22. Wang, H.; Zhou, P.; Guo, R.; Wang, Y.; Zhan, H.; Yuan, Y. Synthesis of Rectorite/Fe_3O_4/ZnO Composites and Their Application for the Removal of Methylene Blue Dye. *Catalysts* **2018**, *8*, 107. [CrossRef]

23. Dumitru, R.; Manea, F.; Păcurariu, C.; Lupa, L.; Pop, A.; Cioablă, A.; Surdu, A.; Ianculescu, A. Synthesis, Characterization of Nanosized $ZnCr_2O_4$ and Its Photocatalytic Performance in the Degradation of Humic Acid from Drinking Water. *Catalysts* **2018**, *8*, 210. [CrossRef]

24. Markowska-Szczupak, A.; Rokicka, P.; Wang, K.; Endo, M.; Morawski, A.W.; Kowalska, E. Photocatalytic Water Disinfection under Solar Irradiation by d-Glucose-Modified Titania. *Catalysts* **2018**, *8*, 316. [CrossRef]

25. Liu, H.; Wang, A.; Sun, Q.; Wang, T.; Zeng, H. Cu Nanoparticles/Fluorine-Doped Tin Oxide (FTO) Nanocomposites for Photocatalytic H_2 Evolution under Visible Light Irradiation. *Catalysts* **2017**, *7*, 385. [CrossRef]

26. Guan, Y.; Xia, M.; Marchetti, A.; Wang, X.; Cao, W.; Guan, H.; Kong, X. Photocatalytic Reduction of CO_2 from Simulated Flue Gas with Colored Anatase. *Catalysts* **2018**, *8*, 78. [CrossRef]

27. Wang, T.; Xia, M.; Kong, X. The Pros and Cons of Polydopamine-Sensitized Titanium Oxide for the Photoreduction of CO_2. *Catalysts* **2018**, *8*, 215. [CrossRef]

28. Hafizi, A.; Rahimpour, M.R. Inhibiting Fe–Al Spinel Formation on a Narrowed Mesopore-Sized $MgAl_2O_4$ Support as a Novel Catalyst for H_2 Production in Chemical Looping Technology. *Catalysts* **2018**, *8*, 27. [CrossRef]

29. Daneshmand-Jahromi, S.; Rahimpour, M.R.; Meshksar, M.; Hafizi, A. Hydrogen Production from Cyclic Chemical Looping Steam Methane Reforming over Yttrium Promoted Ni/SBA-16 Oxygen Carrier. *Catalysts* **2017**, *7*, 286. [CrossRef]

30. Zou, J.; Wang, L.; Ji, P. Promoting the Synthesis of Ethanol and Butanol by Salicylic Acid. *Catalysts* **2017**, *7*, 295.

31. Fu, Z.; Yu, Y.; Li, Z.; Han, D.; Wang, S.; Xiao, M.; Meng, Y. Surface Reduced CeO_2 Nanowires for Direct Conversion of CO_2 and Methanol to Dimethyl Carbonate: Catalytic Performance and Role of Oxygen Vacancy. *Catalysts* **2018**, *8*, 164. [CrossRef]

32. Li, F.; Xia, Q.; Cheng, Q.; Huang, M.; Liu, Y. Conductive Cotton Filters for Affordable and Efficient Water Purification. *Catalysts* **2017**, *7*, 291. [CrossRef]
33. Wakisaka, Y.; Kido, D.; Uehara, H.; Yuan, Q.; Takakusagi, S.; Uemura, Y.; Yokoyama, T.; Wada, T.; Uo, M.; Sakata, T.; et al. A Demonstration of Pt L3-Edge EXAFS Free from Au L3-Edge Using Log–Spiral Bent Crystal Laue Analyzers. *Catalysts* **2018**, *8*, 204. [CrossRef]
34. Tsang, C.H.A.; Li, K.; Zeng, Y.; Zhao, W.; Zhang, T.; Zhan, Y.; Xie, R.; Leung, D.Y.; Huang, H. Titanium oxide based photocatalytic materials development and their role of in the air pollutants degradation: Overview and forecast. *Environ. Int.* **2019**, *125*, 200–228. [CrossRef] [PubMed]
35. Mitoraj, D.; Kisch, H. The Nature of Nitrogen-Modified Titanium Dioxide Photocatalysts Active in Visible Light. *Angew. Chem. Int. Ed.* **2008**, *47*, 9975–9978. [CrossRef] [PubMed]
36. Bandara, J.; Tennakone, K.; Jayatilaka, P. Composite tin and zinc oxide nanocrystalline particles for enhanced charge separation in sensitized degradation of dyes. *Chemosphere* **2002**, *49*, 439–445. [CrossRef]
37. Maeda, K.; Wang, X.; Nishihara, Y.; Lu, D.; Antonietti, M.; Domen, K. Photocatalytic Activities of Graphitic Carbon Nitride Powder for Water Reduction and Oxidation under Visible Light. *J. Phys. Chem. C* **2009**, *113*, 4940–4947. [CrossRef]
38. Kohtani, S.; Makino, S.; Kudo, A.; Tokumura, K.; Ishigaki, Y.; Matsunaga, T.; Nikaido, O.; Hayakawa, K.; Nakagaki, R. Photocatalytic degradation of 4-n-nonylphenol under irradiation from solar simulator: comparison between $BiVO_4$ and TiO_2 photocatalysts. *Chem. Lett.* **2002**, *7*, 660–661. [CrossRef]
39. Niishiro, R.; Kudo, A. Development of Visible-Light-Driven TiO_2 and $SrTiO_3$ Photocatalysts Doped with Metal Cations for H_2 or O_2 Evolution. *Solid State Phenom.* **2010**, *162*, 29–40. [CrossRef]
40. Shafaei, A.; Nikazar, M.; Arami, M. Photocatalytic degradation of terephthalic acid using titania and zinc oxide photocatalysts: Comparative study. *Desalination* **2010**, *252*, 8–16. [CrossRef]
41. Khodja, A.A.; Sehili, T.; Pilichowski, J.-F.; Boule, P. Photocatalytic degradation of 2-phenylphenol on TiO_2 and ZnO in aqueous suspensions. *J. Photochem. Photobiol. A Chem.* **2001**, *141*, 231–239. [CrossRef]
42. Lizama, C.; Freer, J.; Baeza, J.; Mansilla, H.D. Optimized photodegradation of Reactive Blue 19 on TiO_2 and ZnO suspensions. *Catal. Today* **2002**, *76*, 235–246. [CrossRef]

Article

Surface Reduced CeO$_2$ Nanowires for Direct Conversion of CO$_2$ and Methanol to Dimethyl Carbonate: Catalytic Performance and Role of Oxygen Vacancy

Zhongwei Fu [†], Yuehong Yu [†], Zhen Li, Dongmei Han, Shuanjin Wang, Min Xiao * and Yuezhong Meng *

The Key Laboratory of Low-Carbon Chemistry & Energy Conservation of Guangdong Province/State Key Laboratory of Optoelectronic Materials and Technologies, Sun Yat-sen University, No. 135, Xingang Xi Road, Guangzhou 510275, China; fuzhw@mail2.sysu.edu.cn (Z.F.); yuyueh@mail2.sysu.edu.cn (Y.Y.); lizh63@mail2.sysu.edu.cn (Z.L.); handongm@mail.sysu.edu.cn (D.H.); wangshj@mail.sysu.edu.cn (S.W.)
* Correspondence: stsxm@mail.sysu.edu.cn (M.X.); mengyzh@mail.sysu.edu.cn (Y.M.); Tel.: +86-20-8411-4113 (Y.M.)
† These authors have contributed equally to this work.

Received: 14 March 2018; Accepted: 4 April 2018; Published: 19 April 2018

Abstract: Ultralong 1D CeO$_2$ nanowires were synthesized via an advanced solvothermal method, surface reduced under H$_2$ atmosphere, and first applied in direct synthesis of dimethyl carbonate (DMC) from CO$_2$ and CH$_3$OH. The micro morphologies, physical parameters of nanowires were fully investigated by transmission electron microscopy (TEM), X-ray diffraction (XRD), N$_2$ adsorption, X-ray photoelectron spectrum (XPS), and temperature-programmed desorption of ammonia/carbon dioxide (NH$_3$-TPD/CO$_2$-TPD). The effects of surface oxygen vacancy and acidic/alkaline sites on the catalytic activity was explored. After reduction, the acidic/alkaline sites of CeO$_2$ nanowires can be dramatically improved and evidently raised the catalytic performance. CeO$_2$ nanowires reduced at 500 °C (CeO$_2$_NW_500) exhibited notably superior activity with DMC yield of 16.85 mmol gcat^{-1}. Furthermore, kinetic insights of initial rate were carried out and the apparent activation energy barrier of CeO$_2$_NW_500 catalyst was found to be 41.9 kJ/mol, much tiny than that of CeO$_2$_NW catalyst (74.7 KJ/mol).

Keywords: dimethyl carbonate; carbon dioxide; ceria nanowires; oxygen vacancy

---

1. Introduction

As an environmentally benign compound and unique intermediate of versatile chemical products, dimethyl carbonate (DMC) is widely applied in polymer industry and pharmaceutical as well as detergent, surfactant, and softener additives [1,2]. In addition, DMC is important raw material when serving as a non-toxic substitute for poisonous phosgene and dimethyl sulfate in sustainable chemistry of carbonylation, methylation, and polymer synthesis [3,4]. As an additive, DMC can improve the octane number and oxygen content of fuels, thereby enhancing its antiknock [1]. Furthermore, DMC can be used as a cleaning solvent in coating paints and the important composition of electrolyte [5]. Considering the wide applications, DMC is known as the "new cornerstone" for synthesis chemistry nowadays and lots of efforts have been made in finding appropriate routes to meet the demand of DMC industrial production since it is far from satisfaction until now. Several approaches including the methanolysis of phosgene [6], the oxidative carbonylation of methanol [7], the transesterification of alkene carbonates [8], and the alcoholysis of urea [3], have been developed, but it is still limited with strict operation conditions, highly toxicity, and corrosivity up to now.

Using carbon dioxide (CO_2) in DMC synthesis is particularly attractive since CO_2 is known as a recyclable and naturally abundant raw materials for the production of plentiful chemical reagents. Meanwhile, the emissions of CO_2 have significantly increased and contributed to global warming, thus the utilization of CO_2 has attracted more and more attention in the last decades [9,10]. In this regard, the direct synthesis of DMC from CO_2, and methanol (Scheme 1) is considered as one of the most attractive and effective methods since such an approach is environmentally benign not only for reduction of greenhouse gas emissions but also for development of a new carbon resource [11,12]. However, such a sustainable route also exists significant challenges due to facts including the highly thermodynamically stability of CO_2, as well as the kinetically inert and deactivation of catalysts induced by water formation in the reaction process [13–15].

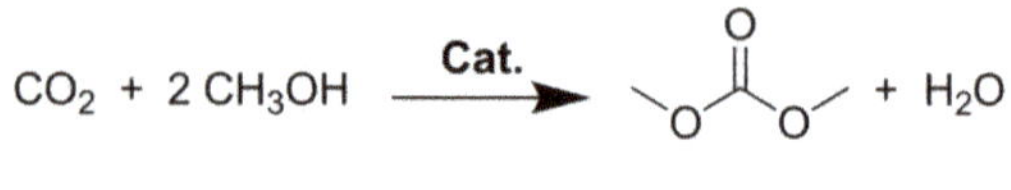

$$CO_2 + 2\ CH_3OH \xrightarrow{\text{Cat.}} \text{DMC} + H_2O$$

Scheme 1. Direct synthesis of DMC from CO_2 and methanol.

Several methods, such as adding co-reagents and dehydrants in the reaction systems, have been developed [16,17]. Furthermore, some new technologies, such as photo-assistant [14], electro-assistant [18], membrane separation [19], and supercritical CO_2 technology [20,21] have been introduced to boost the production of DMC in former reports. Even then, the reactions are preferred at strict conditions and the yield of DMC is relatively low. Though the efforts to these approaches are devoted today, the explorations of advanced heterogeneous catalysts are still regarded as the most effective route [22–24]. In particular, CeO_2 based catalysts have been transplanted in the direct synthesis of DMC and show much better catalytic activity as an excellent heterogeneous catalyst [25,26]. Plenty of references has employed CeO_2 as competent catalysts in DMC formation involving dehydration [27]. Furthermore, previously studies have revealed that the different crystal facets exposed on the surface of CeO_2 nanostructures were strongly controlled by its morphology, leading to differential physicochemical properties and further effecting the catalytic performance [28]. In this context, 1D structured CeO_2 nanorods catalyst demonstrated superior DMC yield (0.906 mmol DMC/mmol cat) from CO_2 and methanol when compared to CeO_2 nanocubes (0.582 mmol DMC/mmol cat) and CeO_2 nano-octahedrons (0.120 mmol DMC/mmol cat) [25,29]. However, major drawbacks of CeO_2 nanorods are the extremely low yield and high cost of hydrothermal method, preventing it from being used in practical applications [30]. In the meantime, the low aspect ratio of nanorods limits the specific surface area of catalysts, which would affect the catalytic performance further [25,26].

In this respect, we were especially interested in new trials for ultralong 1D CeO_2 nanostructure. Furthermore, oxygen deficiency of the CeO_2 based catalyst has been proved playing important roles in CO_2 and methanol activation in former research [26,31]. Thus, we conducted further research on surface reduced CeO_2 nanowires catalyst. Herein, CeO_2 nanowires with a diameter of 10 nm and an aspect ratio of more than 50 was successfully prepared by the refluxing approach established by Yu et al. [32] and then simply surface reduced under hydrogen atmospheres, followed by their application in the direct synthesis of DMC from CO_2 and methanol. Moreover, the influence of surface oxygen-deficiency and acid-basic sites were fully investigated. The catalytic recyclability was also detected. Finally, we conducted a detailed kinetic investigation for the direct formation of DMC in an autoclave reactor over catalysts.

2. Results and Discussion

2.1. Morphology and Microstructure of the Prepared Catalysts

CeO_2 nanowires catalyst was prepared using a solvothermal method in a mixed water/ethanol solvents (v/v = 1:1). Figure 1a,b show the morphology of the unreduced CeO_2 nanowires catalyst, exhibiting an intact nanowire structure with an average length of around 500 nm and a uniform diameter of less than 10 nm. After reduced with H_2, the nanowire structure was kept undestroyed, and the size of nanowires has made almost no change (Figure 1c). The crystal structures of CeO_2 nanowires catalysts were investigated by XRD, and the spectra are shown in Figure 2. For unreduced CeO_2 nanowires, the diffraction peaks of 2θ can be ascribed to the fluorite-structured CeO_2 (JCPDS 34-0394, 28.6° (111), 33.1° (200), 47.6° (220), and 56.4° (311)). After reduction under H_2 atmosphere as a function of temperature (450–700 °C), the spectra of nanowires remained almost unchanged, indicating that the crystalline stucture of the nanowires was not destroyed.

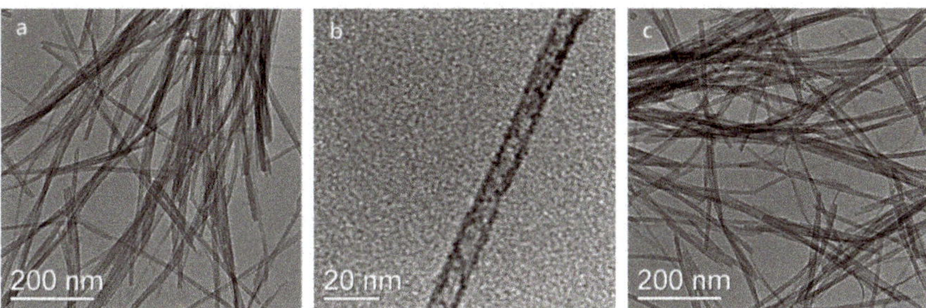

Figure 1. TEM images of (**a**,**b**) CeO_2_NW, and (**c**) CeO_2_NW_500.

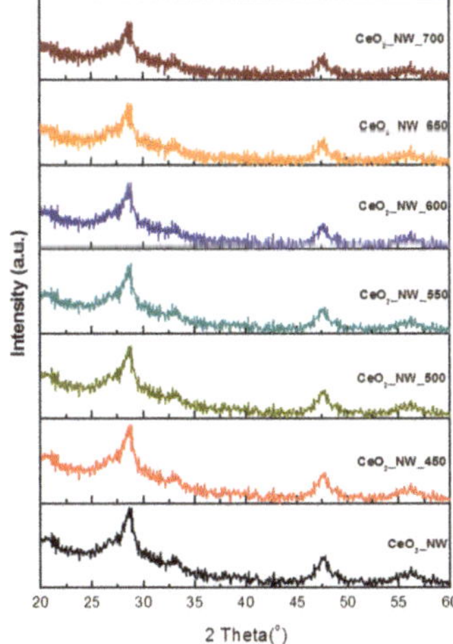

Figure 2. XRD patterns of CeO_2 nanowires reduced by H_2 as a function of temperature (450–700 °C).

Physical and chemical parameters of the as-prepared catalysts are summarized in Table 1. The specific surface area of the nanowires was acquired from BET method and seemed to decrease slightly from 116.33 m^2g^{-1} to 98.1 m^2g^{-1} upon increasing the reduction temperature up to 500 °C, indicating that the low-temperature reduction only can influence the specific surface area within tolerable extent. When elevating the reduction temperature, specific surface area of reduced CeO_2 nanowires drops abruptly, which inevitably lead to the covered up of efficient active sites and eventually cause the worse catalytic performance.

Further investigation on the surface acidic/alkaline properties of as-prepared catalysts was acquired by NH_3/CO_2-TPD. The amount of moderate acidic and alkaline sites is also summarized in Table 1. For CeO_2_NW_450 and CeO_2_NW_500, more plentiful moderately acidic and alkaline sites are generated with the elevating of reduction temperature, which is bond to benefit the DMC formation according to former research. [33] While for other nanowires reduced under higher temperature, the amount of moderate acidity and alkalinity lessens. It can be ascribe to the fast-declining specific surface area upon elevating the reduction temperature, and then result in the covered up of efficient active sites. CeO_2_NW_500 was determined to possess both the richest acidity and alkalify, which is mainly because of the enriching of oxygen vacancy on catalysts surface, further providing much richer active sites when compared to unreduced nanowires.

Table 1. Textural data of as-prepared catalysts basing on BET, XRD, and XPS investigation.

Catalysts	BET Surface Area (m^2g^{-1})	Sites Amount ($\mu mol\ g^{-1}$)		Oxygen Vacancy (%)
		Moderate Acidity	Moderate Alkalify	
CeO_2_NW	116.3	82	25	5.1
CeO_2_NW_450	104.9	190	52	14.2
CeO_2_NW_500	98.1	282	82	20.5
CeO_2_NW_550	70.4	235	65	22.7
CeO_2_NW_600	55.2	192	48	25.1
CeO_2_NW_650	42.5	171	47	27.3
CeO_2_NW_700	28.7	115	39	31.2

Surface chemical state of prepared catalysts was investigated by XPS. Based on the calculated result, the XPS result forecasts that the oxygen vacancy on the surface of as-prepared nanowires varies along the reduction temperature from 5.1% in CeO_2_NW to 31.2% in CeO_2_NW_700. Due to the Ce valent state partly shift from +4 to +3, reduction of CeO_2 nanowires leads to the formation of oxygen vacancy on the catalyst surface.

Based on the aforesaid result, a certain relationship between the surface active sites (mainly the moderate acidic and alkaline sites), specific surface area and surface oxygen vacancy was established. Both the moderate acidic and alkaline sites showed a linear relationship contrast specific surface area multiply surface oxygen vacancy (Figure 3). The combination of TPD, XPS, and BET reveals that the oxygen vacant structure of nanowires contributes to the formation of moderately acidic and basic sites, which is also influenced by the specific surface area.

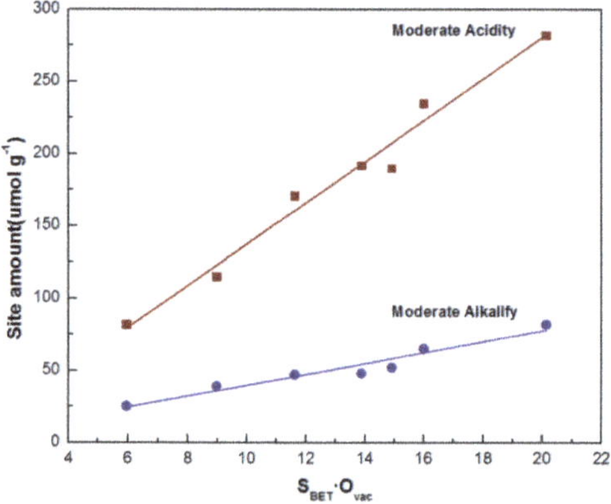

Figure 3. Liner relationship of moderate acidic/alkaline sites contrast specific surface area and surface oxygen vacancy.

2.2. Catalytic Performance

The effects of reduction temperature for nanowires on the catalytic activity were probed and the catalytic reaction was conducted in a stainless autoclave micro-reactor with high-speed stirring.

The DMC yield of as-prepared catalysts with different catalysts are demonstrated in Figure 4 and serves as the basis for original selection of reduction temperatures.

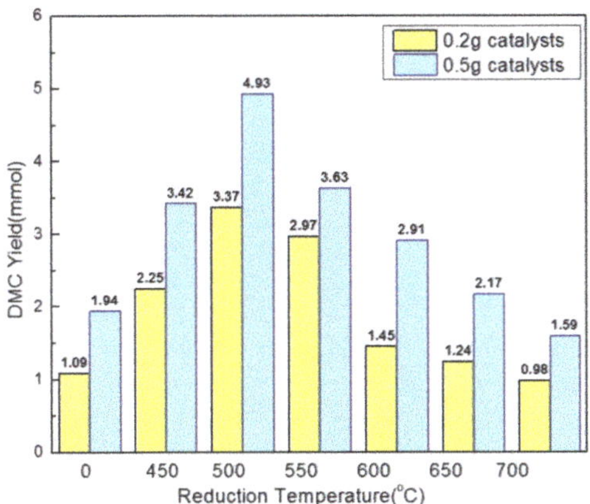

Figure 4. Effects of reduction temperature on the DMC yield over as-prepared nanowires. Reaction conditions: Methanol 500 mmol; catalysts 0.2 g or 0.5 g; CO_2 pressure 5 MPa; temperature 120 °C; reaction time 5 h.

Unreduced CeO_2 nanowires (CeO_2_NW) catalyst obtained much inferior DMC yield when compared with the surface reduced CeO_2 nanowires (CeO_2_NW_x). DMC yield enhanced with elevating the reduction temperature of CeO_2 nanowires, reached a maximum at 500 °C and then declined with further temperature rise. We observed the catalytic performance of all nanowires catalysts with loading amount of 0.2 g and 0.5 g respectively. The reaction found to reach saturated and catalytic performance was influenced by leveling effect when loading 0.5 g catalyst, while the catalyst was efficiently utilized at 0.2 g. Among the catalysts examined, CeO_2_NW_500 catalyst achieves excellent DMC yield of 16.85 mmol gcat^{-1}, superior than the catalytic activity of CeO_2_NW catalyst (5.45 mmol gcat^{-1}) under the same condition. Associating with the specific surface area of as-prepared catalysts, CeO_2_NW_x (x > 500) catalysts with tinier surface area proof inferior catalytic activity, illustrating that the catalytic performance is directly related to the specific surface area of as-prepared catalysts. Smaller specific surface area necessarily leads to the covering of efficient active sites and causes lower catalytic activity [34,35].

Further research on direct synthesis of DMC from CO_2 and methanol over CeO_2_NW_500 was conducted. The effects of different catalytic conditions was fully investigated. Figure 5 shows the DMC amount with different reaction time and reaction temperatures over CeO_2_NW_500 catalysts catalyst. The generation rate of the destination product DMC enhanced when elevated the catalytic temperature, while the final yield of DMC constantly decreased due to the limitations of thermodynamic and generation of side product. Under 140 °C, the yield of DMC reached the maximum value at 75 min, and then seemed to be almost unchanged. However, the formation amount of DMC even trends increasing after 5 h at 120 °C.

Further investigation for the recyclability of CeO_2_NW_500 was carried out and the used nanowires catalyst was thermal reduced under H_2 atmosphere before re-catalyze the direct synthesis of DMC under the same reaction conditions. BET specific surface area and catalytic performance of the recovered CeO_2_NW_500 catalysts are demonstrated in Figure 6. Both specific surface area and the catalytic performance were found mildly falling as the number of reuses accumulates, which is on account of the slight surface collapse during the retreatment of the catalysts. Anyway, CeO_2_NW_500 shows favorable stability for the direct formation of DMC from CO_2 and methanol.

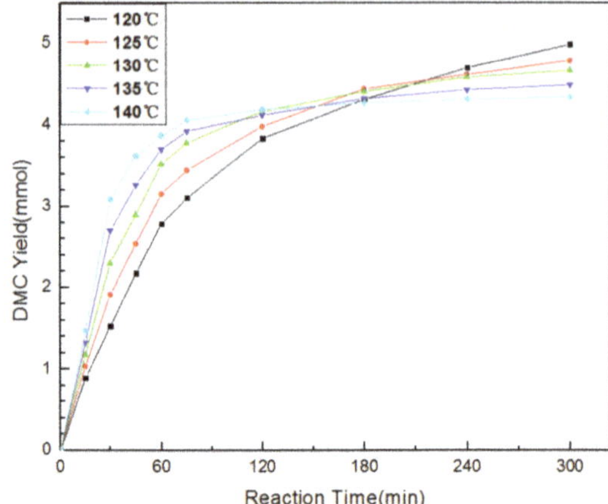

Figure 5. Effects of reaction temperature of nanowires on the catalytic performance for DMC formation. Reaction conditions: Methanol 500 mmol; CeO$_2$_NW_500 catalysts 0.5 g; CO$_2$ pressure 5 MPa.

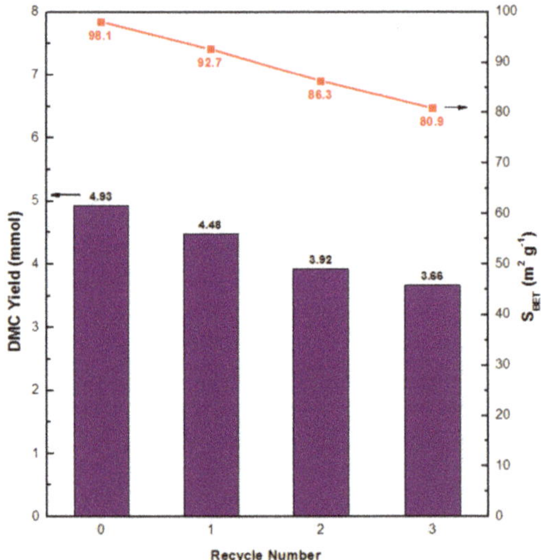

Figure 6. Recyclability study of CeO$_2$_NW_500 catalyst for the direct synthesis of DMC from CO$_2$ and methanol. Reaction conditions: Methanol 500 mmol; catalysts 0.5 g; CO$_2$ pressure 5 MPa; temperature 120 °C; reaction time 5 h.

2.3. Kinetic Analysis

Initial rate kinetic insights in the direct synthesis of DMC over CeO$_2$_NW_500 catalyst were conducted. Based on the result of Figure 5, the yield data within 60 min was selected as the initial rate region. In addition, similar initial reaction was carried out on CeO$_2$_NW and compared with that of CeO$_2$_NW_500. The linear fitting of Arrhenius plot in Figure 7 gives a slope at −5.04 (±0.41), indicating the apparent activation energy at 41.9 ± 3.4 kJ/mol for CeO$_2$_NW_500 catalyst, which is lower than CeO$_2$_NW catalyst (74.7 kJ/mol). It suggests that the surface reduction of CeO$_2$ nanowires has reduced the activation energy barriers and improved the catalytic performance by enriching the surface active sites.

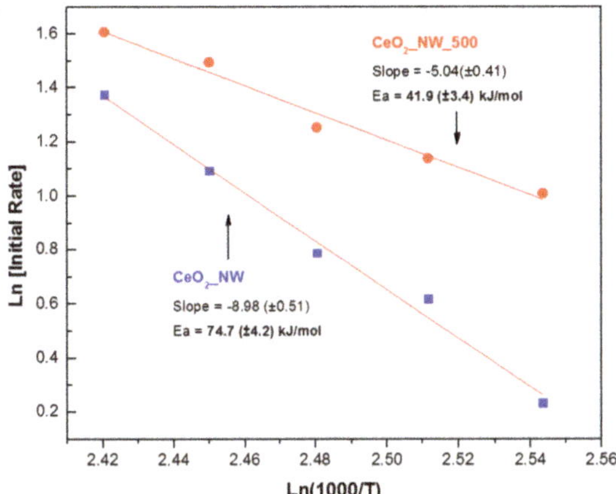

Figure 7. Arrhenius plot for direct synthesis of DMC over CeO_2_NW_500 and CeO_2_NW catalyst.

Furthermore, based on the former proposed mechanism for the direct synthesis of DMC, surface adsorption and activation of CO_2 and methanol occurs on the alkaline sites and acidic sites, respectively [36]. As a consequence, the inferior catalytic performance of unreduced CeO_2_NW catalyst in this study should mainly ascribe to much poorer surface acidic and alkaline sites. While for CeO_2_NW_x catalysts, more abundant moderately acidic and alkaline sites generate along with the surface reduction process, thus resulting in more favorable catalytic performance than CeO_2_NW. Figure 8 shows the Ln–Ln curves of initial rate for each catalysts contrast the concentration of moderately acidic/alkaline sites. There is a positive liner relationship of these parameters, suggesting the initial rates of this reaction influenced by the activation of both CO_2 and methanol. This result corresponds to the deduction of Langmuir–Hinshelwood mechanism [37,38].

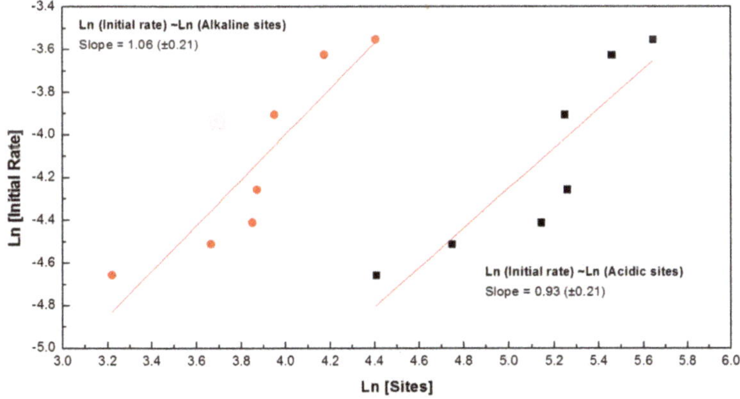

Figure 8. Kinetics study of the initial rate of DMC production contrast acidic/alkaline sites. Reaction conditions: Methanol 500 mmol; catalysts 0.2 g; CO_2 pressure 5 MPa; temperature 120 °C; reaction time 60 min.

3. Materials and Methods

3.1. Materials

Cerium (III) nitrate hexahydrate $Ce(NO_3)_3 \cdot 6H_2O$, methanol, and dimethyl carbonate (DMC) were purchased from Aladdin Co., Ltd. (Shanghai, China). Ammonium hydroxide $NH_3 \cdot H_2O$ (25 wt %) and ethanol was purchased from Guangdong Chemical Reagent Factory (Guangzhou, China). All the reactants were of analytical purity and used without any further treatment.

High purity CO_2 (>99.9999%) and H_2 (>99.999%) were obtained from Guangqi Gas Co., Ltd. (Guangzhou, China).

3.2. Catalysts Preparation

Ceria nanowires catalyst was prepared using a solvothermal method in a mixed water/ethanol solvents [32]. Briefly, stoichiometric $Ce(NO_3)_3 \cdot 6H_2O$ was dissolved in a flash with water/ethanol mixed solution (v/v = 1:1), following by oil bath heating up to 140 °C. Then the $NH_3 \cdot H_2O$ was added into the flash and the reaction mixture was refluxed for 12 h under stirring. After cooling to room temperature, the resulting mixture was separated by centrifugation. Afterward, the solid product was bathed with a mixture of ethanol and water (v/v = 1:1) for several times. After that, the pre-synthesized nanowires were freeze-dried in a lyophilizer (Four-Ring Science Instrument Plant Beijing Corporation, Beijing, China) at a vacuity of 3 mbar and a frigorific temperature of −40 °C. Finally, as-prepared nanowires catalyst were thermal reduced under H_2 atmosphere for 4 h. The samples were named after CeO_2_NW and $CeO_2\_NW\_x$, in which x represented the reduced temperature.

3.3. Catalyst Characterization

Micromorphology measurement was carried out on a transmission electron microscope (TEM, JSM-2010HR, JEOL Ltd., Tokyo, Japan) at a high voltage of 200 kV. Samples were ultrasonic dispersed into ethanol absolute, and then dropwise loaded onto the micro copper grid, followed by drying under air condition at room temperature.

Powder X-ray diffraction (XRD) was measured on a XRD diffractometer (Dmax 2200, Rigaku Ltd., Tokyo, Japan) at a scan rate of 5°/min. High voltage of 40 kV and current of 30 mA were employed in this measurement and Cu Kα radiation target (λ = 0.154178 nm) was used.

N_2 adsorption characterization was acquired on a nitrogen adsorption apparatus (ASAP-2020, Micrometrics Ltd., Cumming, GA, USA) and the specific surface area was calculated through the Brunauer–Emmett–Teller method from the adsorption results. Samples were pre-treated under nitrogen atmosphere at 200 °C for 2 h. After cooling, N_2 at a flow rate of 110 mL/min was adsorbed on the samples surface in a U tube surrounded by liquid nitrogen.

Temperature programmed desorption (TPD) was conducted on a chemical adsorption apparatus (Chem-BET 3000, Quantachrome Ltd., Boynton Beach, FL, USA). Firstly, samples were pre-treated under nitrogen atmosphere at 200 °C for 1 h. Then, a mixture standard gas of $10\%CO_2/90\%N_2$ or $10\%NH_3/90\%N_2$ saturated with the samples for 30 min at a flow rate of 60 mL/min in a U tube. After that, surface physical adsorption of CO_2 or NH_3 was dislodged by bathing with $30\%N_2/70\%$ He standard gas for 2 h at a flow rate of 50 mL/min. Then, the samples were thermal treated under N_2/He $30\%N_2/70\%$ He from room temperature up to 600 °C with 8 °C/min heating rate. Finally, the total desorption of NH_3/CO_2 was determined through back-titration method. HCl/NaOH (0.01 mol/L) was employed as an adsorbent for NH_3/CO_2, NaOH/HCl (0.01 mol/L) was served as titrant together with a mixed indicator reagent, which consisted of bromocresol green ethanol solution (1%, 3 equivalent volumes) and methyl red ethanol solution (2%, 1 equivalent volume) [39].

X-ray photoelectron spectrum (XPS) was acquired on an X-ray photoelectron spectrometer (ESCALAB250, Thermo Fisher Scientific Ltd., Waltham, MA, USA) with a scan survey of 1100-0 eV binding energy range. Monochromatized Al-Ka source at 1486.6 eV and 150 w was applied in the characterization with a voltage of 15 kV. Surface oxygen vacancy can be roughly calculated according to the equation

$$O_{Vac}\,(\%)\,=\,\frac{C(Ce)\,-\,0.5\,C(O)}{C(Ce)} \tag{1}$$

3.4. Catalytic Performance Measurement

Direct synthesis of DMC from CO_2 and methanol was carried out in a stainless steel autoclave with a volume of 50 mL and high-velocity stirring. As-prepared catalyst and a certain amount of absolute methanol were added into the reactor, following be purging CO_2 for several times to evacuate the air inside and obtained the strict oxygen-free and water-free condition. Reaction pressure of CO_2 was set at 5 MPa and the reaction was conducted at 120 °C for 5 h if no otherwise specified. The final products were measured and quantified by a gas chromatograph (GC-7900II, Techcomp Ltd., Beijing, China) equipped with a flame ionization detector (FID) after filtrating with PES membrane with a pore size of 0.45 um.

4. Conclusions

Ultralong 1D CeO_2 nanowires were synthesized via an advanced solvothermal method, surface reduced under H_2 atmosphere, and firstly applied in direct synthesis of dimethyl carbonate (DMC) from CO_2 and CH_3OH. The influences of reduction temperatures for the nanowires and different operating conditions for the catalysis reactivity were fully explored. The catalysis reactivity of ceria nanowires was founded to be greatly improved after surface reduction by generating more surface acidic-alkaline sites. Among the catalysts investigated, $CeO_2\_NW\_500$ obtains the most favorable catalytic activity for DMC formation than CeO_2_NW and all of the other $CeO_2\_NW\_x$ catalysts. Under optimal reaction conditions, $CeO_2\_NW\_500$ catalyst achieves the best catalysis reactivity with DMC yield of 16.85 mmol gcat^{-1} in an autoclave reactor. Based on the approach of initial rates method, the kinetic insight were conducted for the direct synthesis of DMC over $CeO_2\_NW\_500$ catalyst and the activation energy barrier is determined to be 41.9 kJ/mol, tinier than 74.7 kJ/mol for unreduced CeO_2 nanowires. Moreover, a certain relationship between the initial rate and the surface acidity/alkalify was found, which is identical to the deduction of former proposed Langmuir–Hinshelwood mechanism where the initial rates of this reaction are influenced by the activation of both CO_2 and methanol.

Acknowledgments: The authors would like to thank the National Natural Science Foundation of China (grant no. 21376276, 21643002), Guangdong Province Sci. & Tech. Bureau (grant no. 2017B090901003, 2016B010114004, 2016A050503001), and Fundamental Research Funds for the Central Universities (171gjc37) for financial support of this work.

Author Contributions: Zhongwei Fu, Yuezhong Meng, Min Xiao, Dongmei Han, and Shuanjin Wang conceived and designed the experiments; Zhongwei Fu performed the experiments; Zhongwei Fu and Yuehong Yu analyzed the data; Zhen Li and Yuehong Yu contributed analysis tools; Zhongwei Fu wrote the paper.

Conflicts of Interest: The authors declare no conflict of interest.

References

1. And, M.A.P.; Christopher, L.M. Review of dimethyl carbonate (DMC) manufacture and its characteristics as a fuel additive. *Energy Fuels* **1997**, *11*, 2–29.
2. Ono, Y. Catalysis in the production and reactions of dimethyl carbonate, an environmentally benign building block. *Appl. Catal. A Gen.* **1997**, *155*, 133–166. [CrossRef]
3. Santos, B.A.V.; Silva, V.M.T.M.; Loureiro, J.M.; Rodrigues, A.E. Review for the direct synthesis of dimethyl carbonate. *Chem. Rev.* **2015**, *1*, 214–229. [CrossRef]
4. Keller, N.; Rebmann, G.; Keller, V. Catalysts, mechanisms and industrial processes for the dimethylcarbonate synthesis. *J. Mol. Catal. A Chem.* **2010**, *317*, 1–18. [CrossRef]
5. Tundo, P.; Selva, M. The chemistry of dimethyl carbonate. *Acc. Chem. Res.* **2002**, *35*, 706. [CrossRef] [PubMed]

6. Chaturvedi, D.; Mishra, N.; Mishra, V. Various approaches for the synthesis of organic carbamates. *Curr. Org. Synth.* **2007**, *4*, 308–320. [CrossRef]

7. King, S.T. Reaction mechanism of oxidative carbonylation of methanol to dimethyl carbonate in cu–y zeolite. *J. Catal.* **1996**, *161*, 530–538. [CrossRef]

8. Bhanage, B.M.; Fujita, S.; Ikushima, Y.; Torii, K.; Arai, M. Synthesis of dimethyl carbonate and glycols from carbon dioxide, epoxides and methanol using heterogeneous mg containing smectite catalysts: Effect of reaction variables on activity and selectivity performance. *Green Chem.* **2003**, *5*, 71–75. [CrossRef]

9. Zhao, G.; Huang, X.; Wang, X.; Wang, X. Progress in catalyst exploration for heterogeneous CO_2 reduction and utilization: A critical review. *J. Mater. Chem. A* **2017**, *41*, 21625–21649. [CrossRef]

10. Olajire, A.A. Recent advances in the synthesis of covalent organic frameworks for CO_2 capture. *J. CO_2 Util.* **2017**, *17*, 137–161. [CrossRef]

11. Fu, Z.; Meng, Y. *Research Progress in the Phosgene-Free and Direct Synthesis of Dimethyl Carbonate From CO_2 and Methanol*; Springer International Publishing: Berlin/Heidelberg, Germany, 2016.

12. Garciaherrero, I.; Cuéllarfranca, R.M.; Alvarezguerra, M.; Irabien, A.; Azapagic, A. Environmental assessment of dimethyl carbonate production: Comparison of a novel electrosynthesis route utilizing CO_2 with a commercial oxidative carbonylation process. *ACS Sustain. Chem. Eng.* **2016**, *4*, 2088–2097. [CrossRef]

13. Fang, S.; Fujimoto, K. Direct synthesis of dimethyl carbonate from carbon dioxide and methanol catalyzed by base. *Appl. Catal. A Gen.* **1996**, *142*, L1–L3. [CrossRef]

14. Wang, X.J.; Xiao, M.; Wang, S.J.; Lu, Y.X.; Meng, Y.Z. Direct synthesis of dimethyl carbonate from carbon dioxide and methanol using supported copper (Ni, V, O) catalyst with photo-assistance. *J. Mol. Catal. A Chem.* **2007**, *278*, 92–96. [CrossRef]

15. Han, D.; Wang, S.; Meng, Y.; Chen, Y.; Lu, Y.; Xiao, M. Porous diatomite-immobilized cu–ni bimetallic nanocatalysts for direct synthesis of dimethyl carbonate. *J. Nanomater.* **2012**, *2012*, 8.

16. Eta, V.; Mäkiarvela, P.; Murzin, D.Y.; Salmi, T.; Mikkola, J.P. Synthesis of dimethyl carbonate from methanol and carbon dioxide : The effect of dehydration. *Ind. Eng. Chem. Res.* **2009**, *49*, 9609–9617. [CrossRef]

17. Lin, C.M.; Shen, D.P.; You, Y.U.; Ming-Xian, X.U. Catalysts and dehydration agents of one-pot synthesis of dimethyl carbonate. *J. Zhejiang Univ. Technol.* **2013**, *13*, 329–331.

18. Zhou, Y.; Fu, Z.; Wang, S.; Xiao, M.; Han, D.; Meng, Y. Electrochemical synthesis of dimethyl carbonate from CO_2 and methanol over carbonaceous material supported dbu in a capacitor-like cell reactor. *RSC Adv.* **2016**, *6*, 40010–40016. [CrossRef]

19. Mengers, H.J. Membrane reactors for the direct conversion of CO_2 to dimethyl carbonate. *Clin. Nephrol.* **2015**, *33*, 1653–1659.

20. Ballivet-Tkatchenko, D.; Ligabue, R.A.; Plasseraud, L. Synthesis of dimethyl carbonate in supercritical carbon dioxide. *Braz. J. Chem. Eng.* **2006**, *23*, 111–116. [CrossRef]

21. Hong, S.T.; Park, H.S.; Lim, J.S.; Lee, Y.W.; Anpo, M.; Kim, J.D. Synthesis of dimethyl carbonate from methanol and supercritical carbon dioxide. *Res. Chem. Intermed.* **2006**, *32*, 737–747. [CrossRef]

22. Bian, J.; Xiao, M.; Wang, S.; Lu, Y.; Meng, Y. Direct synthesis of dmc from CH_3 oh and CO_2 over v -doped cu–ni/ac catalysts. *Catal. Commun.* **2009**, *10*, 1142–1145. [CrossRef]

23. Chen, H.; Wang, S.; Xiao, M.; Han, D.; Yixin, L.U.; Meng, Y. Direct synthesis of dimethyl carbonate from coand choh using 0.4 nm molecular sieve supported cu-ni bimetal catalyst. *Chin. J. Chem. Eng.* **2012**, *20*, 906–913. [CrossRef]

24. Zhang, M.; Xiao, M.; Wang, S.; Han, D.; Lu, Y.; Meng, Y. Cerium oxide-based catalysts made by template-precipitation for the dimethyl carbonate synthesis from carbon dioxide and methanol. *J. Clean. Prod.* **2015**, *103*, 847–853. [CrossRef]

25. Wang, S.; Zhao, L.; Wang, W.; Zhao, Y.; Zhang, G.; Ma, X.; Gong, J. Morphology control of ceria nanocrystals for catalytic conversion of CO_2 with methanol. *Nanoscale* **2013**, *5*, 5582–5588. [CrossRef] [PubMed]

26. Fu, Z.; Zhong, Y.; Yu, Y.; Long, L.; Xiao, M.; Han, D.; Wang, S.; Meng, Y. TiO_2-doped CeO_2 nanorod catalyst for direct conversion of CO_2 and CH_3OH to dimethyl carbonate: Catalytic performance and kinetic study. *ACS Omega* **2018**, *3*, 198–207. [CrossRef]

27. Honda, M.; Sonehara, S.; Yasuda, H.; Nakagawa, Y.; Tomishige, K. Heterogeneous CeO_2 catalyst for the one-pot synthesis of organic carbamates from amines, CO_2 and alcohols. *Green Chem.* **2011**, *13*, 3406–3413. [CrossRef]

28. Sun, C.; Li, H.; Zhang, H.; Wang, Z.; Chen, L. Controlled synthesis of ceo2 nanorods by a solvothermal method. *Nanotechnology* **2005**, *16*, 1454. [CrossRef]

29. Zhao, S.Y.; Wang, S.P.; Zhao, Y.J.; Ma, X.B. An in situ infrared study of dimethyl carbonate synthesis from carbon dioxide and methanol over well-shaped CeO_2. *Chin. Chem. Lett.* **2017**, *28*, 65–69. [CrossRef]

30. Vantomme, A.; Yuan, Z.Y.; Du, G.; Su, B.L. Surfactant-assisted large-scale preparation of crystalline CeO_2 nanorods. *Langmuir ACS J. Surf. Colloids* **2005**, *21*, 1132. [CrossRef] [PubMed]

31. Watanabe, S.; Ma, X.; Song, C. Characterization of structural and surface properties of nanocrystalline TiO_2-CeO_2 mixed oxides by XRD, XPS, TPR, and TPD. *J. Phys. Chem.C* **2009**, *113*, 14249–14257. [CrossRef]

32. Yu, X.F.; Liu, J.W.; Cong, H.P.; Xue, L.; Arshad, M.N.; Albar, H.A.; Sobahi, T.R.; Gao, Q.; Yu, S.H. Template- and surfactant-free synthesis of ultrathin CeO_2 nanowires in a mixed solvent and their superior adsorption capability for water treatment. *Chem. Sci.* **2015**, *6*, 2511. [CrossRef] [PubMed]

33. Li, A.; Pu, Y.; Li, F.; Luo, J.; Zhao, N.; Xiao, F. Synthesis of dimethyl carbonate from methanol and CO_2 over fe–zr mixed oxides. *J. CO_2 Util.* **2017**, *19*, 33–39. [CrossRef]

34. Li, H.; Jiao, X.; Li, L.; Zhao, N.; Xiao, F.; Wei, W.; Sun, Y.; Zhang, B. Synthesis of glycerol carbonate by direct carbonylation of glycerol with CO_2 over solid catalysts derived from zn/al/la and zn/al/la/m (m = li, mg and zr) hydrotalcites. *Catal. Sci. Technol.* **2015**, *5*, 989–1005. [CrossRef]

35. Cosimo, J.I.D.; DıEz, V.K.; Xu, M.; Iglesia, E.; ApesteguıA, C.R. Structure and surface and catalytic properties of mg-al basic oxides. *J. Catal.* **1998**, *178*, 499–510. [CrossRef]

36. Jung, K.T.; Bell, A.T. An in situ infrared study of dimethyl carbonate synthesis from carbon dioxide and methanol over zirconia. *J. Catal.* **2001**, *204*, 339–347. [CrossRef]

37. Tomishige, K.; Ikeda, Y.; Sakaihori, T.; Fujimoto, K. Catalytic properties and structure of zirconia catalysts for direct synthesis of dimethyl carbonate from methanol and carbon dioxide. *J. Catal.* **2000**, *192*, 355–362. [CrossRef]

38. Marin, C.M.; Li, L.; Bhalkikar, A.; Doyle, J.E.; Zeng, X.C.; Cheung, C.L. Kinetic and mechanistic investigations of the direct synthesis of dimethyl carbonate from carbon dioxide over ceria nanorod catalysts. *J. Catal.* **2016**, *340*, 295–301. [CrossRef]

39. Jin, D.; Jing, G.; Hou, Z.; Yan, G.; Lu, X.; Zhu, Y.; Zheng, X. Microwave assisted in situ synthesis of usy-encapsulated heteropoly acid (hpw-usy) catalysts. *Appl. Catal. A Gen.* **2009**, *352*, 259–264. [CrossRef]

 *catalysts*

Article

Inhibiting Fe–Al Spinel Formation on a Narrowed Mesopore-Sized MgAl$_2$O$_4$ Support as a Novel Catalyst for H$_2$ Production in Chemical Looping Technology

Ali Hafizi and Mohammad Reza Rahimpour *

Department of Chemical Engineering, Shiraz University, Shiraz 71345, Iran; hafizi@shirazu.ac.ir
* Correspondence: rahimpor@shirazu.ac.ir

Received: 13 October 2017; Accepted: 29 December 2017; Published: 15 January 2018

Abstract: In this paper, the structure of Al$_2$O$_3$ is modified with magnesium to synthesize MgAl$_2$O$_4$ as an oxygen carrier (OC) support. The surface properties and structural stability of the modified support are improved by the incorporation of magnesium in the structure of the support and additionally by narrowing the pore size distribution (about 2.3 nm). Then, iron oxide is impregnated on both an Al$_2$O$_3$ support and a MgAl$_2$O$_4$ support as the oxygen transfer active site. The XRD results showed the formation of solely Fe$_2$O$_3$ on the MgAl$_2$O$_4$ support, while both Fe$_2$O$_3$ and Fe$_3$O$_4$ are detected in the synthesized Fe$_2$O$_3$-Al$_2$O$_3$ structure. The synthesized samples are investigated in chemical looping cycles, including CO reduction (as one of the most important side reactions of chemical looping reforming), at different temperatures (300–500 °C) and oxidation with steam at 700 °C for hydrogen production. The obtained results showed the inhibition of Fe–Al spinel formation in the structure of the Fe$_2$O$_3$-MgAl$_2$O$_4$ OC. In addition, H$_2$ with a purity higher than 98% is achievable in oxidation of the OC with steam. In addition, the activity and crystalline change of the Fe$_2$O$_3$-MgAl$_2$O$_4$ OC is investigated after 20 reduction-oxidation cycles.

Keywords: chemical looping; oxygen carrier; hydrogen production; narrow pore size distribution; Fe$_2$O$_3$ dispersion; texture modification

---

1. Introduction

The production of hydrogen (as an energy carrier) from primary resources, such as methane and water, is industrially developed [1–4]. Hydrogen with high purity can be applied in different applications, such as fuel cells. Nevertheless, most hydrogen production processes need further purification with the purpose of preventing electrode poisoning [5]. One-step pure hydrogen production has attracted attention during the past few years [6,7]. The chemical looping technique (CLT) is known as a novel process for hydrogen production from different sources. Chemical looping processes are based on the transportation of lattice oxygen from an oxidizing solid environment to a reducing environment in two interconnected reactors called the "fuel reactor" and the "oxidation reactor" as indicated in Figure 1 [8,9]. The principal reactions involved in the chemical looping steam methane reforming (CL-SMR) process are as follows [10,11]:

Fuel reactor:

$$CH_4 + Me_xO_y \rightarrow CO + 2H_2 + Me_xO_{y-1} \tag{1}$$

Oxidation reactor:

$$H_2O + Me_xO_{y-1} \rightarrow H_2 + Me_xO_y \tag{2}$$

The in-situ oxidation of carbon monoxide is one of the most important side reactions to progress the reaction network toward pure hydrogen. Carbon monoxide can be used as a reducing agent and the steam is used as an oxidant as indicated in the following reactions:

$$CO + Me_xO_y \rightarrow CO_2 + Me_xO_{y-1} \tag{3}$$

Therefore, hydrogen with high purity could be obtained in both the oxidation and reduction periods of chemical looping cycles, which is depicted in Figure 1.

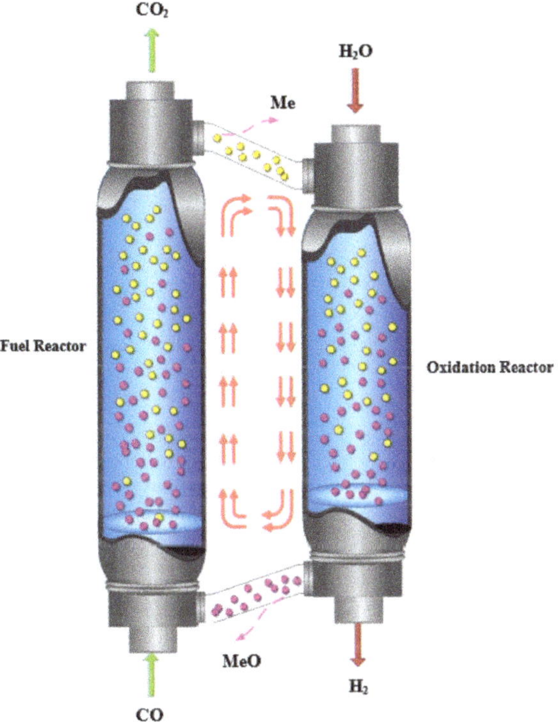

Figure 1. Conceptual scheme of chemical looping reforming technology.

The selection of an appropriate oxygen carrier (OC) has a significant impact on the reaction performance of this process. Recently, the application of metal oxides as oxygen transfer media have attracted much attention [11–13]. The lattice oxygen released from these composites has the advantage of being pure in addition to transferring lattice oxygen intelligently. The selection of a suitable OC seems to be of great importance in large-scale use. The oxygen carrier must have sufficient oxygen transport capacity, high reactivity, easy oxygen release, sorption properties, agglomeration resistance, a low cost, no environmental impact, and high chemical and mechanical stability [14,15]. One of the major challenges in the large-scale usage of oxygen transfer materials is the development of a stable OC in continuous alternating cycles of reduction and oxidation without deactivation. Fe_2O_3 is known as one of the best candidates among different metal oxides because of its low cost, high thermal resistance, and its thermodynamic tendency toward H_2 production [11]. In an iron-based OC, the oxygen is mainly supplied from the reduction of hematite to magnetite ($Fe_2O_3 \rightarrow Fe_3O_4 \rightarrow Fe_{0.947}O$) in the presence of oxygen demandant for the first cycle. The first step of the reaction occurs faster than other iron oxide reduction reactions. The reoxidation reaction ($Fe_{0.947}O + H_2O \rightarrow H_2 + Fe_3O_4$) occurs by passing steam over the OC bed at high temperatures in order to regenerate the reacted OC. The oxidation of the reduced oxygen carrier ($Fe_{0.947}O$) with steam for the production of hydrogen is thermodynamically limited. Thus, magnetite (Fe_3O_4) could be formed after oxidization with steam [16,17].

The active metal oxide should be supported on an appropriate material that can effectively increase the surface area and availability of active metal oxides, improves the OC's structural properties, and subsequently increases its reactivity [18]. Al_2O_3 is known as one of the best catalysts and oxygen carrier supports owing to its high stability (chemical and mechanical) and considerably higher melting temperature [18]. However, the imperfect formation of iron-alumina spinel ($FeAl_2O_4$) is an important phenomenon that occurs at high temperatures. Furthermore, the transformation of primary formed iron oxide particles to Fe_3O_4 on the support surface is the other defect of aluminum oxide as the OC support [11]. The formation of $FeAl_2O_4$ in the OC structure could deactivate the iron oxides and slow down the rate and capacity of oxygen transference.

Despite these well-known problems, many efforts have been made to enhance the activity and stability of an iron-based oxygen carrier. There are several ways currently being applied to reach this purpose, such as applying various types of supports with enhanced structural and textural properties [19–21], and the application of different promoters and inhibitors are the most important efforts to solve these drawbacks [22–27].

With the purpose of suppressing the carbon deposition and improving its catalytic activity, a wide diversity of promoters, such as lanthanum, cerium, magnesium, and calcium, have been incorporated into the structure of the support [28–30]. Li et al. [28] assessed ceria-promoted Ni/SBA-15 catalysts for ethanol steam reforming. The results showed that the incorporation of CeO_2 could effectively control the size of Ni particles via strong metal–support interaction. Hafizi et al. [11] examined a Ca-promoted Fe/Al_2O_3 oxygen carrier in the chemical looping reforming of methane during the redox cycles. The obtained results showed the interaction of the calcium promoter and the alumina support for the improvement of an iron-based oxygen carrier with an inhibition of coke deposition. Ce-SBA-15-supported nickel catalysts were synthesized by Wang et al. [31] and applied for the dry reforming of methane. The results revealed that the cerium incorporated into the skeleton of the SBA-15 promoted the dispersion of nano-sized Ni particles and prohibited carbon formation.

The addition of magnesium to Al_2O_3 could effectively prevent the formation of Fe–Al spinel and improve the stability of the support at higher reaction temperatures by forming $MgAl_2O_4$ [8,31]. Furthermore, a rearrangement of the support's crystalline structure and a regulation of its pore size in a narrow distribution range with magnesium addition could effectively increase the OC's activity and stability. On the other hand, it could effectively control the coke deposition during reforming processes due to the nature of the Mg promoter.

This study investigates the effect of alumina and Mg–Al spinel as the support for the formation of Fe_2O_3. In addition, a narrowing of the oxygen carrier's pore size in the range of 2–3 nm is expected in order to control the coke deposition. The activity and feasibility of synthesized OC in the chemical looping process, including CO oxidation and hydrogen production in alternating cycles, were investigated at different temperatures. The crystalline phase transformation in the reduction-oxidation process is examined on the successfully synthesized Fe_2O_3-$MgAl_2O_4$ as an ultra-pure lattice oxygen transport medium.

2. Results and Discussion

2.1. Sample Characterization

The XRD patterns of Al_2O_3, Fe_2O_3-Al_2O_3, and Fe_2O_3-$MgAl_2O_4$ are shown in Figure 2a. The patterns were identified by the typical peaks of iron oxides, alumina, magnesium oxide, and spinel phases considering the JCPDS data bank. The XRD results suggest that the $MgAl_2O_4$ support could help the formation of pure Fe_2O_3, while alumina disperses iron on its surface in both the Fe^{2+} and Fe^{3+} forms.

The Fe_2O_3 could be clearly assigned at 2θ = 24.3°, 33.4°, 35.8°, 49.7°, 62.7°, 64.3°, and 89° in the XRD patterns of Fe_2O_3-Al_2O_3 and Fe_2O_3-$MgAl_2O_4$ according to JCPDS No. 01-084-0308. The formation of $MgAl_2O_4$ spinel is clear in the Mg-modified OC, and significantly inhibits the formation of iron alumina spinel. The diffraction peak at 37.6° for the (110) plane, 46° for the (111) plane, and 66.4° for the (211) plane of Al_2O_3 shifts about 0.4–0.8° to left in the Mg-promoted OC, representing the presence of $MgAl_2O_4$ spinel (JCPDS No. 01-073-2210) shown in Figure 2a. The four peaks located at 19.25°, 36.9°, 42.5°, and 66.3° match well with $MgAl_2O_4$. Furthermore, Mg or MgO species are not detected in the modified OC, which suggests the complete transformation of Mg to its spinel form. The XRD pattern of Fe_2O_3-$MgAl_2O_4$ after 20 reduction-oxidation cycles showed no iron–alumina spinel ($FeAl_2O_4$) formation (Figure 2b) at high reaction temperatures during the process (JCPDS No. 01-086-2320). However, our previous study on Fe_2O_3-Al_2O_3 or Fe_2O_3-CaO-Al_2O_3 oxygen transfer materials showed the formation of $FeAl_2O_4$ or $CaFe_2O_4$ in the OC structure after using it in the chemical looping process [11].

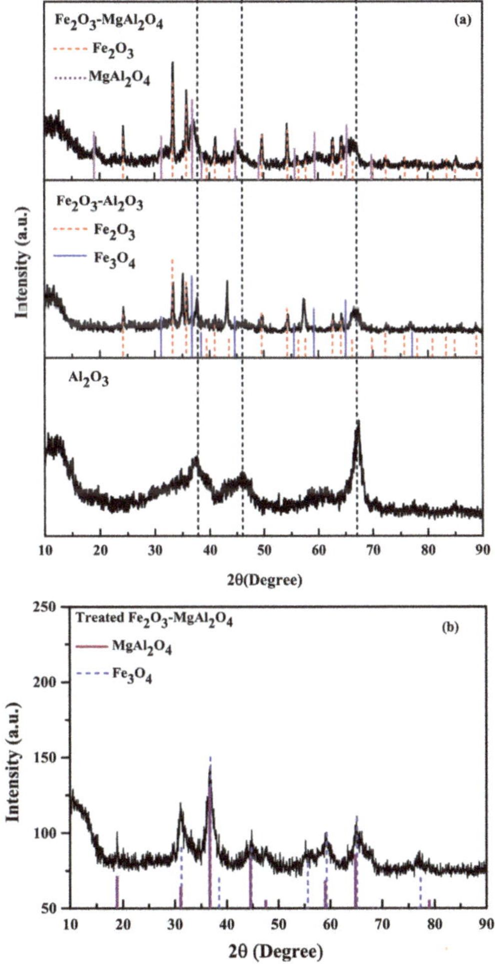

Figure 2. XRD patterns of (**a**) Al_2O_3, Fe_2O_3-Al_2O_3, and Fe_2O_3-$MgAl_2O_4$; (**b**) treated Fe_2O_3-$MgAl_2O_4$ in 20 redox cycles analyzed after the oxidation reaction.

Table 1 presents the physical properties, including Brunauer–Emmett–Teller (BET) surface area, pore volume average, pore size distribution, and energy dispersive X-ray Spectroscopy (EDX) results, of the synthesized OC samples. The BET surface area of the Fe_2O_3-Al_2O_3 is 174.3 m^2 g^{-1}, while the surface area of the Fe_2O_3-$MgAl_2O_4$ oxygen carrier shows a reduction to 113.1 m^2 g^{-1}. This reduction is due to the change of alumina structure and sequential calcinations that can result in pore blockage, sintering, and structural destruction. The reduction in pore volume from 0.416 to 0.389 cm^3 g r^{-1} can verify the reduction in surface area of the Mg-modified sample.

Table 1. Structural properties of the fresh and used samples.

Sample	Fe [a] (wt %)	Mg [a] (wt %)	Al [a] (wt %)	C [a] (wt %)	O [a] (wt %)	BET Surface Area (m^2 g^{-1})	Average Pore Size (nm)	Pore Volume (cm^3 g^{-1})
Fresh Fe_2O_3-Al_2O_3	17.72	-	39.82	-	42.46	174.3	4.3	0.416
Fresh Fe_2O_3-$MgAl_2O_4$	16.42	10.81	21.94	-	50.83	113.1	2.3	0.389
Used Fe_2O_3-Al_2O_3	16.97	-	40.01	5.83	37.29	-	-	-
Used Fe_2O_3-$MgAl_2O_4$	14.86	9.25	33.46	1.20	41.23	-	-	-

[a] Measured by EDX. BET: Brunauer–Emmett–Teller.

Figure 3 shows the Barrett–Joyner–Halenda (BJH) pore size distribution of both the Fe_2O_3-Al_2O_3 and Fe_2O_3-$MgAl_2O_4$ oxygen carriers. Since the average pore size distribution of the Fe_2O_3-Al_2O_3 is 4.3 nm, the distribution peak shows a wide range of pores (about 2–20 nm). On the other hand, the sharp peak of the Fe_2O_3-$MgAl_2O_4$ indicated in this figure confirms a narrow pore size distribution centered at 2.3 nm. Consequently, modifying the OC alumina support by Mg could effectively regulate the pore size distribution in a narrow range.

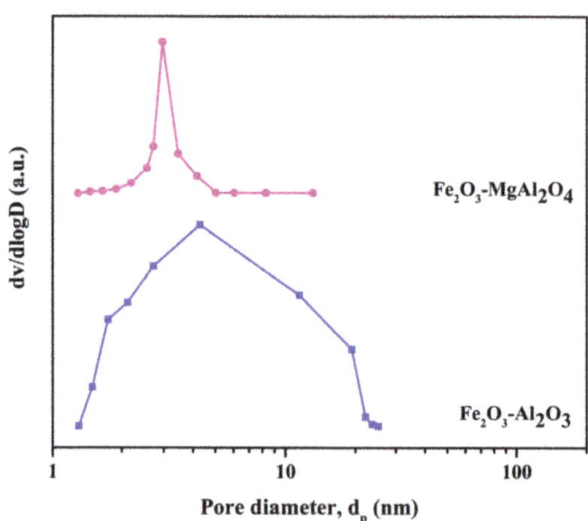

Figure 3. Pore size distributions of Fe_2O_3-Al_2O_3 and Fe_2O_3-$MgAl_2O_4$.

Field emission scanning electron microscopy (FESEM) micrographs of the prepared Al_2O_3, Fe_2O_3-Al_2O_3, and Fe_2O_3-$MgAl_2O_4$ samples are presented in Figure 4a–c. The synthesized blank alumina has a uniform aggregated surface as indicated in Figure 4a, and the distribution of Fe_2O_3-Al_2O_3 particles is indicated in Figure 4b. On the other hand, the Fe_2O_3-$MgAl_2O_4$ OC has approximately uniformly dispersed aggregated Fe_2O_3 particles (Figure 4c). A comparison of the FESEM graphs of both OC samples indicates that the structure of the Fe_2O_3-Al_2O_3 seems to have higher porosity than that of the Fe_2O_3-$MgAl_2O_4$. The FESEM-EDX results of the used Fe_2O_3-$MgAl_2O_4$

OC shown in Figure 4d and Table 1 revealed coke deposition of about 1.2 wt %. In addition, the slight aggregation of OC nanoparticles that is demonstrated in this figure could be due to the high oxidation temperature. The TEM image of the prepared Fe_2O_3-$MgAl_2O_4$ OC is shown in Figure 4e to examine the dispersion and size of iron oxide particles on the surface of the Mg–Al spinel. Nevertheless, the image shows the agglomeration of iron oxide particles with a diameter range of about 40–50 nm.

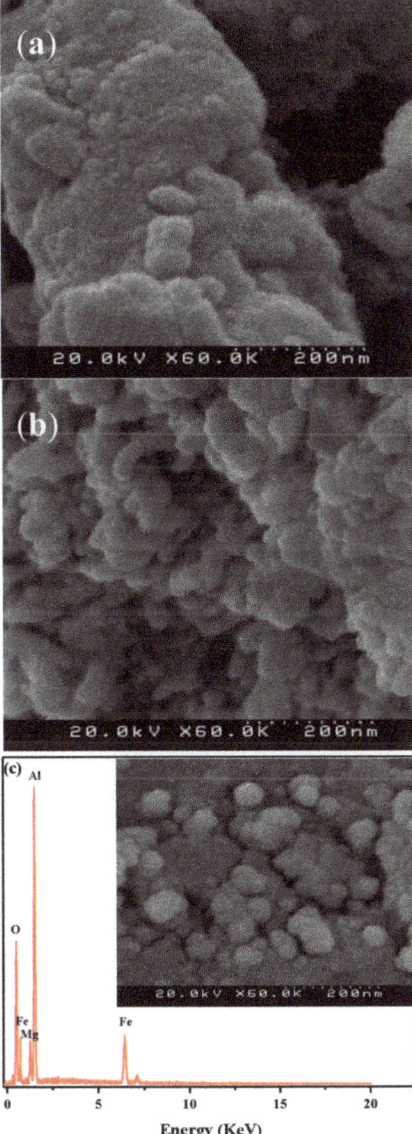

Figure 4. *Cont.*

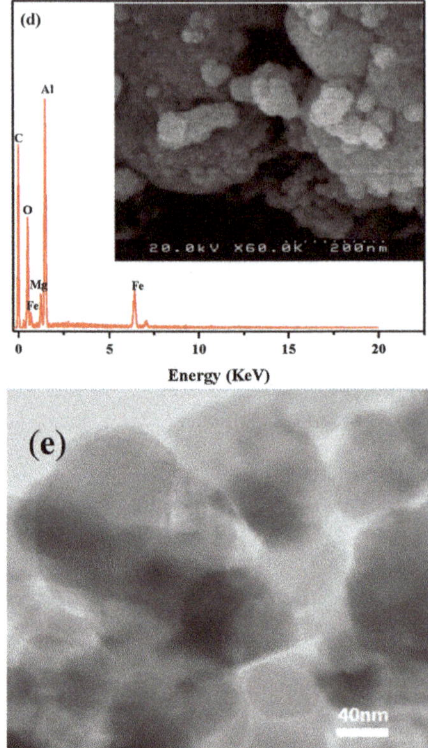

Figure 4. FESEM and EDX of blank Al_2O_3 (**a**); fresh Fe_2O_3-Al_2O_3 (**b**); fresh Fe_2O_3-$MgAl_2O_4$ (**c**); and used Fe_2O_3-$MgAl_2O_4$ (**d**) and TEM of Fe_2O_3-$MgAl_2O_4$ (**e**).

2.2. Activity Results

The catalytic activity of not-modified and modified OC was evaluated in chemical looping CO oxidation, where CO is oxidized to CO_2 in the reduction period and H_2 is produced in the oxidation section. Figure 5 represents the effect of reduction temperature (300–500 °C) on average CO conversion. In addition, the purity of produced hydrogen in the oxidation section at 700 °C for each relevant reduction temperature is shown in this figure. The results show the positive effect of reduction temperature on CO conversion for both samples. For instance, the CO conversion is increased from about 45.3% to 98.3% using the Fe_2O_3-Al_2O_3 oxygen carrier. On the other hand, the conversion of carbon monoxide is 48.2%, 75.7%, 91.3%, 95.2%, and 96.6% at 300, 350, 400, 450, and 500 °C, respectively. The improvement in CO conversion with temperature for both samples is related to the surface activation of iron oxide for oxygen transfer to the reducing agent at elevated temperatures. However, the related H_2 purity in the oxidation section is decreased with a further increase in reduction temperature. In addition to improving the activity of the OC, increasing the reduction temperature facilitates coke formation on its surface. CO disproportionation is an important side reaction that leads to the formation of coke through the following equation [32,33]:

$$2CO \rightarrow C + CO_2. \tag{4}$$

The coke is oxidized in the oxidation section with steam to CO and CO_2 that reduce the purity of produced hydrogen. For example, in the oxidation step related to the reduction temperature of 400 °C, hydrogen with about 98.5% purity is produced, while oxidation of OC treated in lower reduction temperatures results in the H_2 purity of 100% (Figure 5b). In addition, the CO conversion of about 91.3% is achieved at 400 °C in the reduction period.

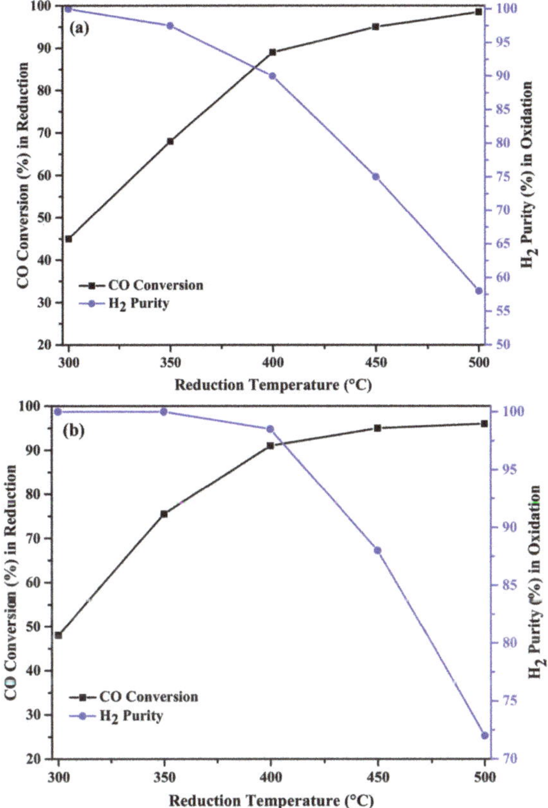

Figure 5. Carbon monoxide conversion at different reduction temperatures and hydrogen purity in the oxidation section using (**a**) the Fe_2O_3-Al_2O_3 oxygen carrier and (**b**) the Fe_2O_3-$MgAl_2O_4$ oxygen carrier.

With the aim of investigating the effect of an Mg promoter on the stability of the synthesized oxygen carriers, the performance of Fe_2O_3-Al_2O_3 and Fe_2O_3-$MgAl_2O_4$ was tested through 20 consecutive reduction-oxidation cycles. Therefore, the variation of CO conversion in the reduction period at 400 °C and the hydrogen purity in the oxidation step over the cycles were determined as shown in Figure 6a,b. The results for the Fe_2O_3-Al_2O_3 oxygen carrier revealed a reduction in CO conversion from 89.9% to 83.5% during 20 redox cycles (about 7%). In addition, the hydrogen purity in the oxidation step is reduced during the cycles (Figure 6a). These behaviors could be mainly due to the deposition of coke on the surface of the oxygen carrier in the reduction step that reduces the purity of produced hydrogen in the oxidation section. However, the crystalline change and destruction in the structure of the oxygen carrier at a high redox temperature has a negative effect on the durability of the sample. On the other hand, the Mg-promoted oxygen carrier (Fe_2O_3-$MgAl_2O_4$) showed a 0.7% reduction in activity through the first five cycles and remained approximately constant over the rest of cycles. The hydrogen purity in the oxidation section is firstly 97.7 and increased to 98.8% in the fifth cycle and remained constant. The improvement in H_2 purity seems to be due to the coke inhibition in the reduction cycles. This phenomenon is related to the reduction of emitted gases produced from the catalyst's decoking simultaneous with hydrogen production in the reaction between metal and water.

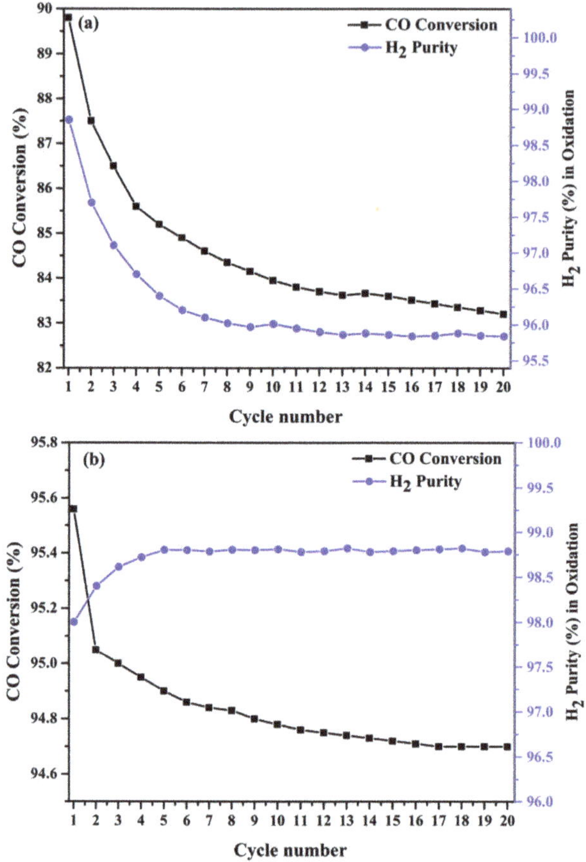

Figure 6. Life time of (**a**) Fe_2O_3-Al_2O_3 and (**b**) Fe_2O_3-$MgAl_2O_4$ oxygen carriers during 20 redox cycles.

In order to better investigate the durability results of the two samples, the CO conversion and hydrogen purity of the first, second, fifth, tenth, and twentieth cycles at 400 °C are demonstrated in Figures 7 and 8. The reduction in CO conversion during each cycle could be related to the reduction in the lattice oxygen of surface active sites with reduction time and coke deposition on the surface. The reduction in CO conversion in the tenth and twentieth cycles compared to the previous cycles shows a continuous reduction in the activity of the oxygen carrier (Figure 7a). On the other hand, the activity of the Mg-promoted sample in Figure 8a revealed that the conversion remained approximately constant after cycle 5. The purity of the produced hydrogen in the oxidation step of both samples increased during the step time to reach 100% (Figures 7 and 8b). The increase in hydrogen purity could be due to the coke decomposition during the process. The higher purity of the produced hydrogen using the Fe_2O_3-$MgAl_2O_4$ oxygen carrier compared to that using the Fe_2O_3-Al_2O_3 oxygen carrier is due to the lower coke deposition.

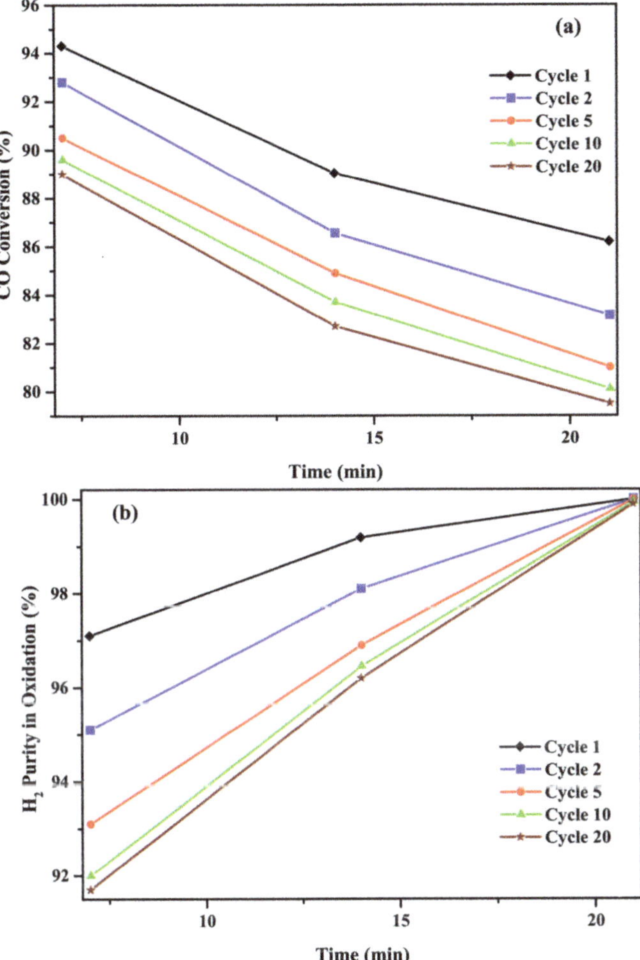

Figure 7. (**a**) The CO conversion and (**b**) hydrogen purity in the oxidation step with time for the Fe$_2$O$_3$-Al$_2$O$_3$ oxygen carrier at 400 °C (in the reduction step) for different cycles.

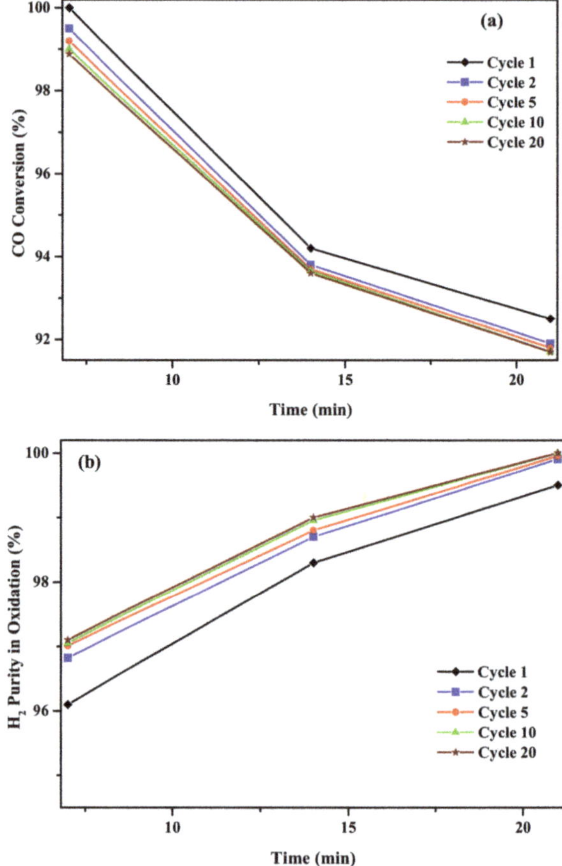

Figure 8. (**a**) The CO conversion and (**b**) hydrogen purity in the oxidation step with time for the Fe_2O_3-$MgAl_2O_4$ oxygen carrier at 400 °C (in the reduction step) for different cycles.

3. Experimental Methods

3.1. Oxygen Carrier Preparation

The OC support affects the performance of oxygen transference, and both alumina and magnesia–alumina spinels were applied as the support for iron oxide. The Fe_2O_3-$MgAl_2O_4$ oxygen transfer material was synthesized using a two-step sequential wetness impregnation method. $MgAl_2O_4$ spinel was prepared by adding a solution of magnesium acetate (0.2 M) to previously synthesized alumina (Al_2O_3 was synthesized by the precipitation method). Prior to impregnation, the slurry was sonicated in a bath-type ultrasound at 30 °C for better incorporation of magnesium precursor into the pores' alumina structure, followed by vacuum impregnation at 80 °C in a rotary evaporator. After that, the sample was dried at 70 °C in a vacuum oven for 10 h. The resulting paste was calcined at 600 °C for 2 h in flowing air. After that, the prepared $MgAl_2O_4$ was used as the support for the preparation of Fe_2O_3-Al_2O_3 or Fe_2O_3-$MgAl_2O_4$ samples. For this purpose, a 0.5 M solution of $Fe(NO_3)_2 \cdot 6H_2O$ (Merck KGaA, Darmstadt, Germany) was added to the as-prepared $MgAl_2O_4$ or Al_2O_3 supports with continuous sonication in a bath ultrasound for 15 min. Subsequently, the impregnated oxygen carrier was dried at 80 °C for 12 h in a vacuum oven followed by calcination at 800 °C for 2 h.

3.2. Oxygen Carrier Characterization

Textual properties, including specific surface area, pore volume, and average pore diameter of the as-prepared samples, were calculated according to the Brunauer–Emmett–Teller (BET) method by N_2 adsorption-desorption isotherms. The specific surface area of the oxygen carriers was measured by BET tests using a (ASAP-2020, Micromeritics, Norcross, GA, USA) gas adsorption apparatus. The Barrett–Joyner–Halenda (BJH) method was implemented to calculate the pore size distributions according to the adsorption branches of the isotherms. The freshly calcined samples were degassed with nitrogen at 200 °C for 3 h prior to the BET test.

The morphology of the synthesized samples was analyzed by an energy dispersive X-ray Spectroscopy (EDX)-equipped field emission scanning electron microscope using a HITACHI S-4160 apparatus (Hitachi, Ltd., Chiyoda, Tokyo, Japan). In addition, a Philips CM30 operated at 300 kV was applied for transmission electron microscopy (TEM) (Philips, Andover, MA, USA). The phase composition and crystallinity of the oxygen transfer materials were analyzed by means of X-ray powder diffraction (XRD; Bruker, D8 Advance, Karlsruhe, Germany) using Cu Kα radiation and operated at 40 kV and 40 mA. The support and oxygen carriers were scanned with 0.05°/s resolution to collect the spectra 2θ between 10° and 90° at ambient temperature.

3.3. Process Activity

In order to evaluate the structural resistance and crystalline transformation of $MgAl_2O_4$-supported Fe_2O_3 as an oxygen carrier, a series of cyclic reduction-oxidation tests were performed at different reduction temperatures (300–500 °C). Hereinafter, the term "cycle" means a reduction cycle involving the reduction of the oxygen carrier with carbon monoxide, followed by oxidation with steam. In the reduction step, carbon monoxide diluted in Ar was fed to the reactor for 21 min and the effluent gas was analyzed continuously. Afterwards, the reduced OC was treated with steam for 21 min in order to reoxidize the sample and hydrogen production simultaneously. Each step in a cycle is dissociated with pure Ar for 20 min in order to drive all of the remaining gases out of the reactor. After 20 reduction-oxidation cycles, the residue oxygen transfer material was examined by the XRD technique.

The activity tests of the chemical looping process were conducted in a fixed-bed reactor with an inner diameter of 16 mm and height of 1000 mm filled with 1 g of finely powdered oxygen carrier with a 100–200 mesh size (Figure 9). The gas streams, including argon (Ar, 100 mL·min^{-1}) as carrier gas and carbon monoxide (CO, 100 mL·min^{-1}) as reactant, were passed through two distinct mass flow controllers (MFC-Unit instruments, model UFC 1661, Yorba Linda, CA, USA) before mixing in the gas mixer. Deionized water was injected into the evaporator with an HPLC pump (Gilson, 307, Middleton, WI, USA) to generate steam. The generated steam was mixed with Ar in a heat-traced pipeline. The products of the reactor were passed through a condenser in order to liquefy the steam and separate them from the gaseous stream. Finally, the products and the unreacted gas streams were analyzed using an online Bruker 450 gas chromatograph (GC-Bruker, 450 series, Billerica, MA, USA) system. The data were collected after one complete cycle in order to ensure the structural stability of the oxygen carriers at a high process temperature.

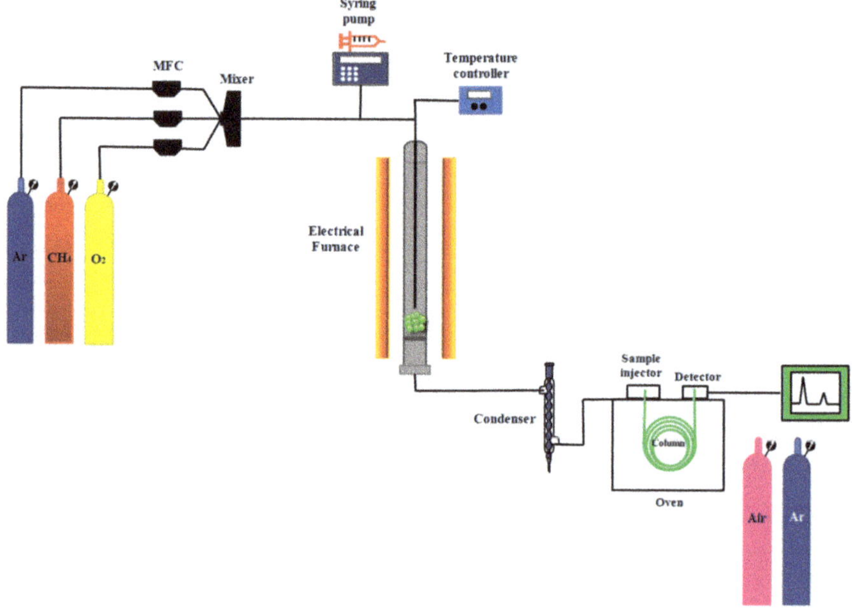

Figure 9. Reactor system for chemical looping hydrogen production.

CO conversion in the reduction section (X_{CO}) was calculated for evaluating the activity of catalyst-sorbents in the chemical looping hydrogen production process according to the following equations.

$$X_{CO} = \frac{\left(F_{CO_{in}}\right) - \left(F_{CO_{out}}\right)}{\left(F_{CO_{in}}\right)} * 100 \qquad (5)$$

where $F_{CO_{in}}$ and $F_{CO_{out}}$ represent the inlet and outlet molar flow rates of carbon monoxide (mol·min^{-1}), respectively.

4. Conclusions

In summary, iron oxide was impregnated on a magnesium-modified alumina support for the transference of pure lattice oxygen. The synthesized $MgAl_2O_4$ with a narrow pore size distribution centered at 2.3 nm revealed a better performance on the formation of Fe_2O_3 on the support's surface in comparison with Al_2O_3. However, the higher lattice oxygen transfer capacity could be attained using Fe_2O_3-$MgAl_2O_4$ as an oxygen carrier, which is due to the inhibition of Fe–Al spinel formation. It means that more iron oxides are exposed to transfer lattice oxygen to feed carbon monoxides (for Fe_2O_3-$MgAl_2O_4$ OC) in the fuel reactor compared to Fe_2O_3-Al_2O_3 with more iron atoms associated in spinel of Fe–Al. In addition, the characterization results of the Fe_2O_3-$MgAl_2O_4$ sample showed the agglomeration of particles with a diameter of about 40–50 nm. The modified OC was applied in chemical looping CO oxidation cycles. The obtained results revealed that increasing the reduction temperature to higher than 400 °C accelerates coke formation on the OC's surface. The conversion of deposited coke in the oxidation section reduced the purity of produced hydrogen in the oxidation section. A CO conversion of 91.3% and hydrogen purity of about 98.5% were achieved in the reduction (400 °C) and oxidation periods, respectively.

Author Contributions: A.H. conceived and designed the experiments, performed the experiments, performed the catalyst synthesis, and analyzed the data; M.R.R. contributed reagents/materials/analysis tools and participated in the analysis and interpretation of the characterization results; and A.H. and M.R.R. wrote the paper.

Conflicts of Interest: The authors declare no conflict of interest.

References

1. Sarshar, Z.; Kleitz, F.; Kaliaguine, S. Novel oxygen carriers for chemical looping combustion: La$_{1-x}$CexBO$_3$ (B = Co, Mn) perovskites synthesized by reactive grinding and nanocasting. *Energy Environ. Sci.* **2012**, *4*, 4258–4269. [CrossRef]
2. Hafizi, A.; Rahimpour, M.R.; Hassanajili, S. Hydrogen production via chemical looping steam methane reforming process: Effect of cerium and calcium promoters on the performance of Fe$_2$O$_3$/Al$_2$O$_3$ oxygen carrier. *Appl. Energy* **2016**, *165*, 685–694. [CrossRef]
3. Mohsenzadeh, A.; Richards, T.; Bolton, K. DFT study of the water gas shift reaction on Ni(111), Ni(100) and Ni(110) surfaces. *Surf. Sci.* **2016**, *644*, 53–63. [CrossRef]
4. Tang, Q.-L.; Chen, Z.-X.; He, X. A theoretical study of the water gas shift reaction mechanism on Cu(111) model system. *Surf. Sci.* **2009**, *603*, 2138–2144. [CrossRef]
5. Alihoseinzadeh, A.; Khodadadi, A.A.; Mortazavi, Y. Enhanced catalytic performance of Au/CuO–ZnO catalysts containing low CuO content for preferential oxidation of carbon monoxide in hydrogen-rich streams for PEMFC. *Int. J. Hydrogen Energy* **2014**, *39*, 2056–2066. [CrossRef]
6. Tosti, S.; Zerbo, M.; Basile, A.; Calabrò, V.; Borgognoni, F.; Santucci, A. Pd-based membrane reactors for producing ultra-pure hydrogen: Oxidative reforming of bio-ethanol. *Int. J. Hydrogen Energy* **2013**, *38*, 701–707. [CrossRef]
7. Akbari-Emadabadi, S.; Rahimpour, M.R.; Hafizi, A.; Keshavarz, P. Promotion of Ca-Co bifunctional catalyst/sorbent with yttrium for hydrogen production in modified chemical looping steam methane reforming. *Catalysts* **2017**, *7*, 270. [CrossRef]
8. Liu, W.; Ismail, M.; Dunstan, M.T.; Hu, W.; Zhang, Z. Fennell PS Inhibiting the interaction between FeO and Al$_2$O$_3$ during chemical looping production of hydrogen. *RSC Adv.* **2015**, *5*, 1759–1771. [CrossRef]
9. Akbari-Emadabadi, S.; Rahimpour, M.R.; Hafizi, A.; Keshavarz, P. Production of hydrogen-rich syngas using Zr modified Ca-Co bifunctional catalyst-sorbent in chemical looping steam methane reforming. *Appl. Energy* **2017**, *206*, 51–62. [CrossRef]
10. Alirezaei, I.; Hafizi, A.; Rahimpour, M.R. Syngas production in chemical looping reforming process over ZrO$_2$ promoted Mn-based catalyst. *J. CO$_2$ Util.* **2017**, *23*, 105–116. [CrossRef]
11. Hafizi, A.; Rahimpour, M.R.; Hassanajili, S. Calcium promoted Fe/Al$_2$O$_3$ oxygen carrier for hydrogen production via cyclic chemical looping steam methane reforming process. *Int. J. Hydrogen Energy* **2015**, *40*, 16159–16168. [CrossRef]
12. Zhu, X.; Sun, L.; Zheng, Y.; Wang, H.; Wei, Y.; Li, K. CeO$_2$ modified Fe$_2$O$_3$ for the chemical hydrogen storage and production via cyclic water splitting. *Int. J. Hydrogen Energy* **2014**, *39*, 13381–13388. [CrossRef]
13. Alirezaei, I.; Hafizi, A.; Rahimpour, M.R.; Raeissi, S. Application of zirconium modified Cu-based oxygen carrier in chemical looping reforming. *J. CO$_2$ Util.* **2016**, *14*, 112–121. [CrossRef]
14. Ortiz, M.; de Diego, L.F.; Abad, A.; García-Labiano, F.; Gayán, P.; Adánez, J. Catalytic Activity of Ni-Based Oxygen-Carriers for Steam Methane Reforming in Chemical-Looping Processes. *Energy Fuels* **2012**, *26*, 791–800. [CrossRef]
15. Forutan, H.; Karimi, E.; Hafizi, A.; Rahimpour, M.R.; Keshavarz, P. Expert representation chemical looping reforming: A comparative study of Fe, Mn, Co and Cu as oxygen carriers supported on Al$_2$O$_3$. *J. Ind. Eng. Chem.* **2015**, *21*, 900–911. [CrossRef]
16. Hacker, V.; Faleschini, G.; Fuchs, H.; Fankhauser, R.; Simader, G.; Ghaemi, M.; Spreitz, B.; Friedrich, M. Usage of biomass gas for fuel cells by the SIR process. *J. Power Sources* **1998**, *71*, 226–230. [CrossRef]
17. Hacker, V.; Fankhauser, R.; Faleschini, G.; Fuchs, H.; Friedrich, K.; Muhr, M.; Kordesch, K. Hydrogen production by steam–iron process. *J. Power Sources* **2000**, *86*, 531–535. [CrossRef]
18. Hafizi, A.; Rahimpour, M.R.; Hassanajili, S. High purity hydrogen production via sorption enhanced chemical looping reforming: Application of 22Fe$_2$O$_3$/MgAl$_2$O$_4$ and 22Fe$_2$O$_3$/Al$_2$O$_3$ as oxygen carriers and cerium promoted CaO as CO$_2$ sorbent. *Appl. Energy* **2016**, *169*, 629–641. [CrossRef]

19. Li, F.; Sun, Z.; Luo, S.; Fan, L.S. Ionic diffusion in the oxidation of iron—Effect of support and its implications to chemical looping applications. *Energy Environ. Sci.* **2011**, *4*, 876–880. [CrossRef]
20. Xu, B.-Q.; Wei, J.-M.; Wang, H.-Y.; Sun, K.-Q.; Zhu, Q.-M. Nano-MgO: Novel preparation and application as support of Ni catalyst for CO_2 reforming of methane. *Catal. Today* **2001**, *68*, 217–225. [CrossRef]
21. Kambolis, A.; Matralis, H.; Trovarelli, A.; Papadopoulou, C. Ni/CeO_2-ZrO_2 catalysts for the dry reforming of methane. *Appl. Catal. A Gen.* **2010**, *377*, 16–26. [CrossRef]
22. Hafizi, A.; Rahimpour, M.R.; Hassanajili, S. Hydrogen production by chemical looping steam reforming of methane over Mg promoted iron oxygen carrier: Optimization using design of experiments. *J. Taiwan Inst. Chem. Eng.* **2016**, *62*, 140–149. [CrossRef]
23. Rezaei, M.; Alavi, S.; Sahebdelfar, S.; Bai, P.; Liu, X.; Yan, Z.-F. CO_2 reforming of CH_4 over nanocrystalline zirconia-supported nickel catalysts. *Appl. Catal. B Environ.* **2008**, *77*, 346–354. [CrossRef]
24. Wang, N.; Yu, X.; Shen, K.; Chu, W.; Qian, W. Synthesis, characterization and catalytic performance of MgO-coated Ni/SBA-15 catalysts for methane dry reforming to syngas and hydrogen. *Int. J. Hydrogen Energy* **2013**, *38*, 9718–9731. [CrossRef]
25. Laosiripojana, N.; Sutthisripok, W.; Assabumrungrat, S. Synthesis gas production from dry reforming of methane over CeO_2 doped Ni/Al_2O_3: Influence of the doping ceria on the resistance toward carbon formation. *Chem. Eng. J.* **2005**, *112*, 13–22. [CrossRef]
26. Huang, L.; Xie, J.; Chen, R.; Chu, D.; Chu, W.; Hsu, A.T. Effect of iron on durability of nickel-based catalysts in auto-thermal reforming of ethanol for hydrogen production. *Int. J. Hydrogen Energy* **2008**, *33*, 7448–7456. [CrossRef]
27. Hafizi, A.; Jafari, M.; Rahimpour, M.R.; Hassanajili, S. Experimental investigation of sorption enhanced chemical looping reforming for high purity hydrogen production using CeO_2–CaO CO_2 sorbent and 15Fe–5Ca/Al_2O_3 oxygen carrier. *J. Taiwan Inst. Chem. Eng.* **2016**, *65*, 185–196. [CrossRef]
28. Li, D.; Zeng, L.; Li, X.; Wang, X.; Ma, H.; Assabumrungrat, S.; Gong, J. Ceria-promoted Ni/SBA-15 catalysts for ethanol steam reforming with enhanced activity and resistance to deactivation. *Appl. Catal. B Environ.* **2015**, *176*, 532–541. [CrossRef]
29. Jabbour, K.; Massiani, P.; Davidson, A.; Casale, S.; Hassan, N. Ordered mesoporous "one-pot" synthesized Ni-Mg(Ca)-Al_2O_3 as effective and remarkably stable catalysts for combined steam and dry reforming of methane (CSDRM). *Appl. Catal. B Environ.* **2017**, *201*, 527–542. [CrossRef]
30. Jiang, B.; Dou, B.; Wang, K.; Song, Y.; Chen, H.; Zhang, C.; Xu, Y.; Li, M. Hydrogen production from chemical looping steam reforming of glycerol by Ni based Al-MCM-41 oxygen carriers in a fixed-bed reactor. *Fuel* **2016**, *183*, 170–176. [CrossRef]
31. Ren, B.; Zhang, L.; Tang, C.; Dong, L.; Li, J. Construction of hierarchical $MgAl_2O_4$ spinel as catalytic supports. *Mater. Lett.* **2015**, *159*, 204–206. [CrossRef]
32. Damyanova, S.; Pawelec, B.; Arishtirova, K.; Fierro, J. Ni-based catalysts for reforming of methane with CO_2. *Int. J. Hydrogen Energy* **2012**, *37*, 15966–15975. [CrossRef]
33. Wang, P.; Tanabe, E.; Ito, K.; Jia, J.; Morioka, H.; Shishido, T.; Takehira, K. Filamentous carbon prepared by the catalytic pyrolysis of CH_4 on Ni/SiO_2. *Appl. Catal. A Gen.* **2002**, *231*, 35–44. [CrossRef]

 *catalysts*

Article

Hydrogen Production from Cyclic Chemical Looping Steam Methane Reforming over Yttrium Promoted Ni/SBA-16 Oxygen Carrier

Sanaz Daneshmand-Jahromi, Mohammad Reza Rahimpour *, Maryam Meshksar and Ali Hafizi

Department of Chemical Engineering, Shiraz University, Shiraz 71345, Iran;
Sanaz.daneshmand@yahoo.com (S.D.-J.); Meshksar.maryam@gmail.com (M.M.); hafizi@shirazu.ac.ir (A.H.)
* Correspondence: rahimpor@shirazu.ac.ir

Academic Editors: Ewa Kowalska, Marcin Janczarek and Agata Markowska-Szczupak
Received: 3 September 2017; Accepted: 22 September 2017; Published: 25 September 2017

Abstract: In this work, the modification of Ni/SBA-16 oxygen carrier (OC) with yttrium promoter is investigated. The yttrium promoted Ni-based oxygen carrier was synthesized via co-impregnation method and applied in chemical looping steam methane reforming (CL-SMR) process, which is used for the production of clean energy carrier. The reaction temperature (500–750 °C), Y loading (2.5–7.4 wt. %), steam/carbon molar ratio (1–5), Ni loading (10–30 wt. %) and life time of OCs over 16 cycles at 650 °C were studied to investigate and optimize the structure of OC and process temperature with maximizing average methane conversion and hydrogen production yield. The synthesized OCs were characterized by multiples techniques. The results of X-ray powder diffraction (XRD) and energy dispersive X-ray spectroscopy (EDX) of reacted OCs showed that the presence of Y particles on the surface of OCs reduces the coke formation. The smaller NiO species were found for the yttrium promoted OC and therefore the distribution of Ni particles was improved. The reduction-oxidation (redox) results revealed that 25Ni-2.5Y/SBA-16 OC has the highest catalytic activity of about 99.83% average CH_4 conversion and 85.34% H_2 production yield at reduction temperature of 650 °C with the steam to carbon molar ratio of 2.

Keywords: chemical looping reforming of methane; yttrium promoted oxygen carrier; SBA-16; hydrogen production

---

1. Introduction

Fast depletion of conventional fossil fuel sources and increasing concerns over the global warming phenomenon due to the emissions of greenhouse gases especially carbon dioxide, initiated various researchers for the production of clean energies [1–6]. H_2 has been widely identified as a favorable clean energy carrier because of its non-polluting nature and high specific energy density (120.7 kJ/g) [7–9]. Recently, hydrogen can be produced by applying novel techniques with lower cost [9–11]. Reforming of fossil fuels, photo-catalytic water splitting, electrolysis and biomass gasification are hydrogen production technologies [9,12–15]. Steam methane reforming (SMR) is the most commonly used process in industry for the generation of H_2 [16,17]. However, SMR is energy intensive process and needs high level of capital investment that is not economically [18,19]. As an alternative method, chemical looping steam methane reforming (CL-SMR) process was proposed in order to overcome these drawbacks [20,21]. In this process the necessity of the gas separation is eliminated since the produced gas is not diluted with N_2 [22,23]. A typical CL-SMR scheme consists of two interconnected reactors where there is no direct mixing of fuel and air as indicated in Figure 1. Methane is partially oxidized to syngas (H_2 and CO) in the fuel reactor, while the metal oxide (Me_xO_y) used as an oxygen carrier (OC) is reduced to Me_xO_{y-1}. The principle reactions that are involved in the fuel reactor are as follows:

$$CH_4 + Me_xO_y \rightarrow Me_xO_{y-s} + sCO + 2H_2 \tag{1}$$

$$CO + H_2O \leftrightarrow H_2 + CO_2 \tag{2}$$

Then air is applied to re-oxidized the reduced OC at high temperature in the air reactor through the following reaction:

$$2Me_xO_{y-s} + sO_2 \rightarrow Me_xO_y \tag{3}$$

Selection and development of an appropriate oxygen carrier is one of the most essential issues in CL-SMR process. The OC needs to possess adequate stability over multiple reduction-oxidation (redox) cycles. It also should have high methane conversion and selectivity to syngas, negligible coke deposition, high oxygen transfer rate and considerable strength to agglomeration, attrition, fragmentation and other chemical and mechanical degeneration types [2,19,22,24].

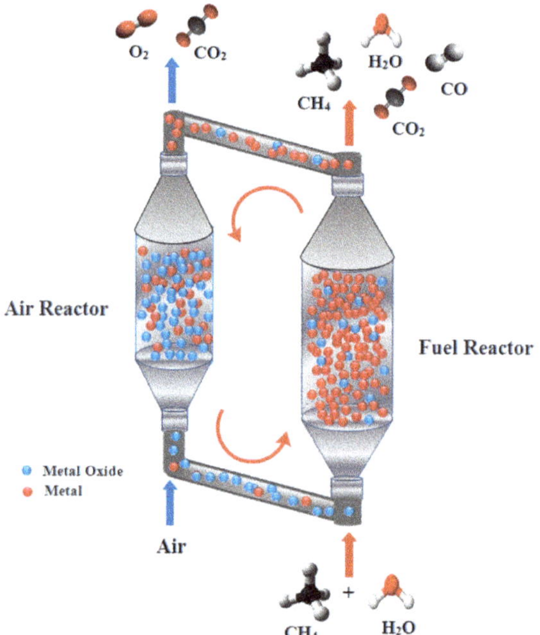

Figure 1. The schematic of CL-SMR reactor for synthesis gas production.

Among the various oxygen carriers, nickel-based OCs are the most promising and attractive candidate as active phase for CL-SMR process because of their high activity and selectivity toward H_2 production. Also, they have wide availability and low cost compared to noble metal materials [25,26]. However, Ni-based systems generally suffer from severe deactivation caused by coke deposition and/or thermal sintering of the metallic phases due to their low tammann temperature (863 K) [27–29]. Therefore, some approaches have been applied in order to enhance the catalytic performance and durability of nickel-based materials in reforming reactions.

The first strategy is using a support with appropriate structural and textural properties that improves Ni dispersion over the carrier surface and inhibits the agglomeration of Ni nanoparticles [30,31]. Previous experiments have shown that the size of nickel particles have remarkable influence on its catalytic activity. Ni particles with sizes of several nanometers could supply more active Ni surface and could excellently reduce the graphite coke deposition [32–34]. Dispersing Ni particles on suitable supports such as porous silica can control the size of metallic Ni particles [35–39].

Silica-based mesoporous materials (e.g., SBA-n, MCM-48, MCM-41) have attracted much attention in many areas of material science and technology such as separations, adsorptions and catalysis [40,41]. They have shown good characteristics because of their high specific surface area, controllable pore size and pore volume. MCM-41 and SBA-15 have a uniform two-dimensional hexagonal ordered mesopore channel structure. However, the weak interaction between Ni nanoparticles and these kinds of supports leads nickel metallic particles to diffuse out of the mesoporous silica channels at elevated temperature [42]. Therefore, using three-dimensional cubic SBA-16 as a support of Ni-based material is an alternative way in order to enhance metal-support interaction. Santa Barbara Amorphous 16 (SBA-16) silica, has been known as an outstanding support for Ni-based materials due to its thick pore walls, excellent hydrothermal and thermal stability, high specific surface area and uniform pore size distribution [43].

Another popular strategy in order to avoid carbon formation and/or stabilizing Ni nanoparticles in the channels of two dimensional mesoporous silica is the utilization of promotional oxides such as Ce_2O_3, La_2O_3, CaO, MgO and Y_2O_3 [44–48]. Li et al. [46] prepared Ce promoted Ni/SiO_2 catalyst using co-impregnation method and applied for producing syngas in the combined partial oxidation of methane with CO_2 reforming. The results revealed that nickel-based catalyst has better performance in the presence of cerium promoter. Qian et al. [47] assessed Ni/SBA-15 catalysts with La promoter for dry reforming reaction. They have proved that highly dispersed La led to an increase of the CO_2 conversion.

It was reported that the presence of Y_2O_3 particles reduces the metal particle size and prevents the coke formation due to high surface oxygen mobility. Shi et al. [49] have been investigated the performance of Y_2O_3 promoted Pd/Al_2O_3 in dry (CO_2) reforming of CH_4. They concluded that the addition of yttrium to Pd/Al_2O_3 suppressed carbon formation and maintained Pd particle size below 10 nm. Furthermore, the oxygen species mobility increased in the presence of Y_2O_3 promoter. Li et al. [50] applied sol gel method to synthesize 0–9%Y-NiO/SBA-15 catalysts and tested them in dry reforming of CH_4 process. The obtained results showed that the stronger interaction between support and metallic nickel particles is created as a result of Y_2O_3 promoter addition. Also, 9%Y-NiO/SBA-15 showed low coke deposition and remarkable catalytic activity through dry methane reforming reaction.

Based on the above consideration, three-dimensional cubic SBA-16 material was applied as a support in order to synthesize Ni-based oxygen carriers in this research, and yttrium promoter was introduced to enhance the catalytic activity in CL-SMR process. Besides, some characteristics were employed for investigating the effect of yttrium addition on the catalytic and structural properties of $NiO-Y_2O_3$/SBA-16 oxygen carrier.

2. Results and Discussion

2.1. Sample Characterization

2.1.1. X-ray Powder Diffraction (XRD) Study

Figure 2 displays the wide angle XRD diffraction patterns of fresh 25Ni/SBA-16 and 25Ni-2.5Y/SBA-16 oxygen carriers. Five well-resolved peaks at $2\theta = 37.3°, 43.3°, 62.9°, 75.4°$ and $79.4°$, ascribable respectively to (111), (200), (220), (311) and (222) crystal planes of cubic nickel oxide (cod No. 01-078-0643), are observed for both samples. The crystallite sizes of the nickel particles were calculated using the Debye Scherrer's equation [51]. The results showed that the size of NiO crystallite of 25Ni-2.5Y/SBA-16 oxygen carrier is smaller than that of 25Ni/SBA-16 (Table 1). This suggests that the presence of Y_2O_3 led to the formation of NiO species with smaller size, which caused better dispersion of NiO particles on the surface of SBA-16 support. According to 01-083-0927 reference pattern code, Y_2O_3 is observed in 25Ni-2.5Y/SBA-16 sample at peaks of $2\theta = 20.5°, 33.8°, 48.5°, 54.7°$ and $85.9°$.

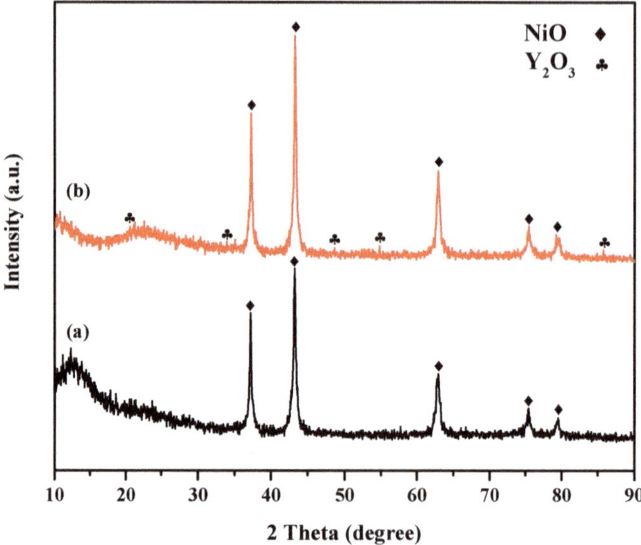

Figure 2. XRD patterns of (a) 25Ni/SBA-16 and (b) 25Ni-2.5Y/SBA-16.

Table 1. Structure properties of the prepared samples.

Samples	BET Surface Area (m^2/g)	Pore Diameter (nm)	Pore Volume (cm^3/g)	Crystal Size [a] (nm)
SBA-16	743.93	3.30	0.38	-
25Ni/SBA-16	321.99	3.68	0.25	42.91
25Ni-2.5Y/SBA-16	363.09	3.59	0.26	34.34

[a] Crystal size of NiO calculated using the Debye Scherrer equation (d = 0.89λ/β cos θ).

2.1.2. N$_2$ Adsorption-Desorption Isotherms

Nitrogen-physisorption analyses of pure SBA-16, 25Ni/SBA-16 and 25Ni-2.5Y/SBA-16 are shown in Figure 3a. As reported by IUPAC classification, all these prepared samples demonstrated type IV isotherm patterns with a H2 hysteresis loop at a range of about P/P$_0$ = 0.4–0.7, which represents the mesoporous hexagonal materials with cage like structure. This confirms that the mesoporous structure of SBA-16 is well preserved after impregnation of nickel active sites and yttrium promoters. However, the amount of adsorbed nitrogen is decreased upon impregnation of nickel and yttrium particles and therefore some porosity loss would be expected. The pore size distribution of pure SBA-16, 25Ni/SBA-16 and 25Ni-2.5Y/SBA-16 samples are obtained by BJH method and depicted in Figure 3b. All materials present sharp and narrow pore size distribution peaks, indicative of uniform mesopore structure. The mesopores of SBA-16 and 25Ni-2.5Y/SBA-16 samples possess a pore size of about 3.5 nm, while it is about 3.9 nm for 25Ni/SBA-16 oxygen carrier [52,53].

The major characteristics obtained by the nitrogen adsorption experiments are exhibited in Table 1. It can be clearly observed that after the addition of nickel active sites and yttrium promoter to the surface of SBA-16, the pore volume and specific surface area decreased significantly. These declines are due to the partial pore blocking caused by NiO and Y$_2$O$_3$ species posited on the surface of the SBA-16 support [54,55]. However, the pore volume and specific surface area of 25Ni-2.5Y/SBA-16 sample are higher than 25Ni/SBA-16 oxygen carrier, which showed that yttrium promoter developed the surface area of 25Ni/SBA-16 oxygen carrier and therefore improved the dispersion of Ni nanoparticles on the SBA-16 support [56,57]. Also, the addition of yttrium oxide to the oxygen carrier disperses fine NiO particles through the structure of SBA-16 support and inhibits the agglomeration of Ni active sites and pore blockage. Thus, the Brunauer-Emmett-Teller (BET) surface area of yttrium promoted oxygen carrier is higher than that of non-promoted 25Ni/SBA-16 oxygen carrier.

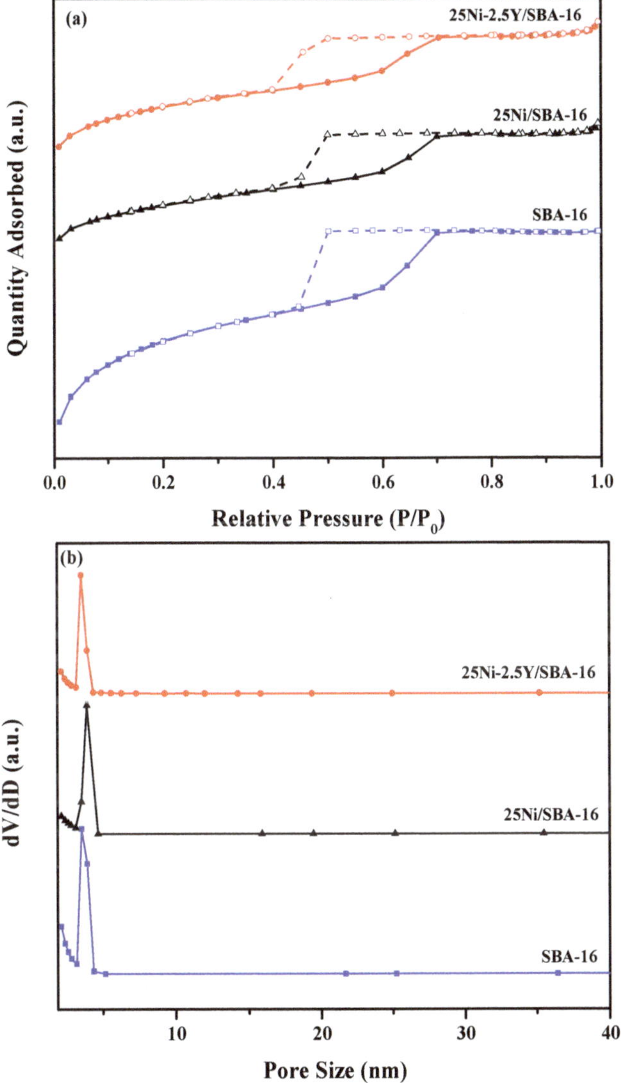

Figure 3. (a) N_2 adsorption/desorption isotherms and (b) Pore size distribution of: SBA-16, 25Ni/SBA-16 and 25Ni-2.5Y/SBA-16.

2.1.3. Field Emission Scanning Electron Microscopy (FESEM) Analysis

The FESEM analysis was performed to evaluate the surface morphology of some samples. The FESEM micrographs of 25Ni-2.47Y/SBA-16 and 25Ni/SBA-16 oxygen carriers before CL-SMR process are dedicated in Figure 4. As exhibited in this figure, SBA-16 possesses 3-dimentional cubic and solid mesoporous structure.

Figure 4. FESEM images of fresh (**a**) 25Ni/SBA-16 and (**b**) 25Ni-2.5Y/SBA-16 oxygen carriers.

2.2. Effect of Yttrium Weight Percentage and Temperature on the Catalytic Activity

To investigate the catalytic activity of different yttrium promoted Ni-based oxygen carriers, the samples with different Y percentages are synthesized and tested in CL-SMR process. The yttrium weight percentage was changed from 0 to 7.4 wt. % in different reaction temperatures (500–750 °C) with the steam to carbon molar ratio of 1.5. As depicted in Figure 5a, the methane conversion increases with the rise of temperature, reflecting the endothermic nature of steam methane reforming process. Therefore, high temperature is favorable for increasing the methane conversion [58,59]. The yttrium free oxygen carrier (20Ni/SBA-16) showed the lowest methane conversion in all the temperature range studied. The performance of this oxygen carrier is greatly improved by the addition of yttrium to the support especially at lower reaction temperatures. As discussed previously in the result of XRD, the promotion of oxygen carrier with yttrium led to the better dispersion of nickel nanoparticles. Thus, the oxygen transfer rate is improved in these samples. For instance, by adding 2.5 wt. % yttrium to the oxygen carrier, the average methane conversion is increased from 56.37% to 93.13% at the reduction temperature of 550 °C. It can be clearly observed that 20Ni-2.5Y/SBA-16 oxygen carrier had the highest catalytic activity and further increase of yttrium loading has negative effect on methane conversion. In addition, as indicated in Figure 5b, hydrogen production yield is increased by raising the reaction temperature. The average hydrogen yield of 20Ni-2.5Y/SBA-16 oxygen carrier is increased from 69.13% to 85.66% by increasing the temperature from 500 to 750 °C. The highest average hydrogen yield is achieved by 20Ni-2.5Y/SBA-16 oxygen carrier. Consequently, 20Ni-2.5Y/SBA-16 has better proficiency and higher activity according to the results of methane conversion and average hydrogen yield.

In addition to high activity and hydrogen yield, it is important to reduce the coke deposition as the main effect of adding yttrium promoter. Thus, the energy dispersive X-ray spectroscopy (EDX) analysis is carried out for the used samples with different yttrium contents.

The EDX analysis of used 20Ni-2.5Y/SBA-16, 20Ni-3.7Y/SBA-16 and 20Ni-7.4Y/SBA-16 oxygen carriers are presented in Figure 6a–c. As shown in this figure, carbon (C) element is detected in addition to Si, O, Ni and Y elements on the surface of all these used OCs. For 20Ni-7.4Y/SBA-16 OC, the coke deposited became less because of higher amount of Y_2O_3 particles on the surface of OC. As demonstrated in Figure 6b,c, the amount of Y element for these two OCs is much lower than that the theoretical value, indicating the coverage of Y_2O_3 species by NiO nanoparticles on the surface of SBA-16. As a result, yttrium particles that are covered by NiO particles cannot inhibit the coke by releasing oxygen easily. Therefore, in spite of higher amount of yttrium in the synthesized mixture of 20Ni-3.7/SBA-16 OC compared with 20Ni-2.5Y/SBA-16, the coke deposited is approximately equal due to the similar amount of Y_2O_3 particles on the surface of the support. Although the carbon deposited on the 20Ni-2.5Y/SBA-16 OC was approximately higher than other OCs, the catalytic activity of this oxygen carrier was much better especially at lower temperature (Figure 5). Therefore, the yttrium weight percentage of 2.5 is the best yttrium loading on the surface of oxygen carrier.

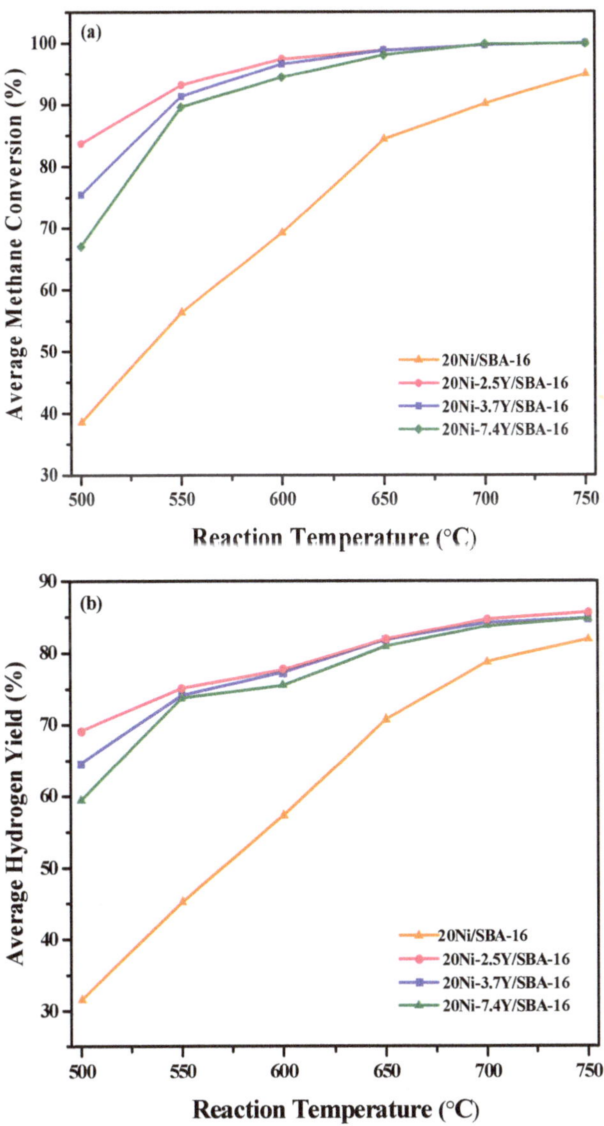

Figure 5. The effect of Y weight percentage and different reduction temperature on (**a**) Average methane conversion and (**b**) Average hydrogen yield of 20Ni-yY/SBA-16 oxygen carriers.

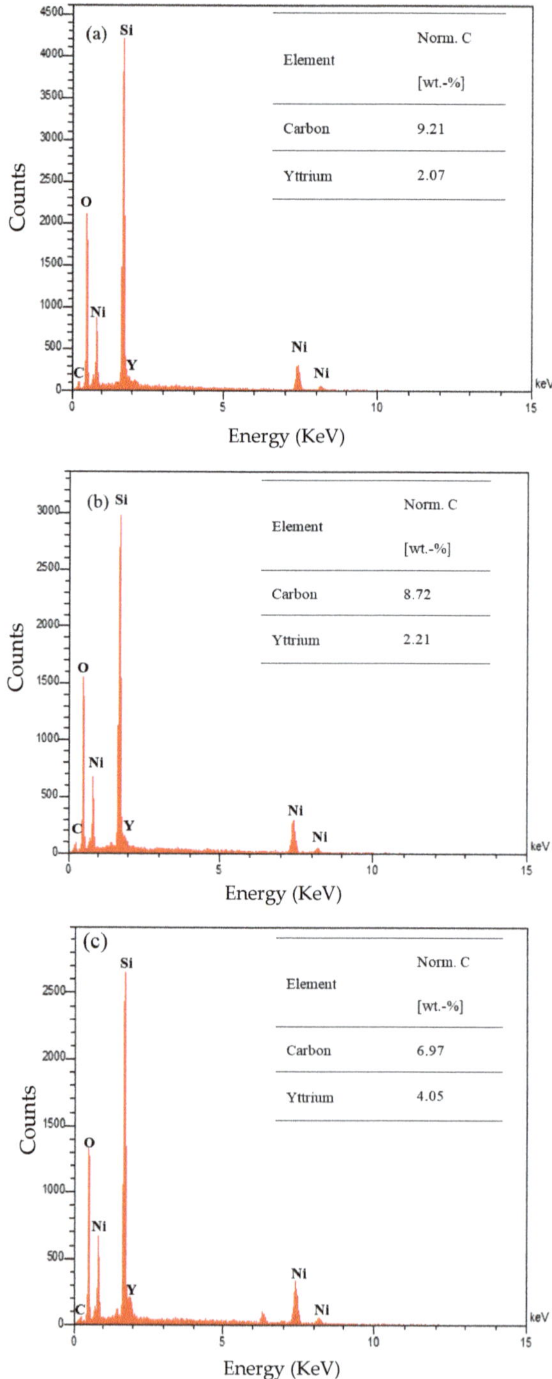

Figure 6. EDX of used (**a**) 20Ni-2.5Y/SBA-16; (**b**) 20Ni-3.7Y/SBA-16 and (**c**) 20Ni-7.4Y/SBA-16 oxygen carriers after 16 redox cycles at 650 °C.

2.3. Life Time Investigation of Different Yttrium Weight Percentage

To investigate the influence of different yttrium weight percentage on the life time of oxygen carrier, the synthesized samples were tested during 16 oxidation/reduction cycles at the temperature of 650 °C with the steam to methane molar ratio of 1.5. Figure 7 demonstrates the average methane conversion and average hydrogen yield of 20Ni/SBA-16, 20Ni-2.5Y/SBA-16, 20Ni-3.7Y/SBA-16 and 20Ni-7.4Y/SBA-16 oxygen carriers. As shown in Figure 7, the average methane conversion and the average hydrogen yield of promoted oxygen carriers are higher than yttrium free sample during the cycles. It confirms the result obtained in the characterization section, which indicated the better distribution of nickel particles on the surface of support. In addition, it shows the higher oxygen mobility in the presence of yttrium promoter. The results revealed that the average CH_4 conversion and average H_2 yield of 20Ni-2.5Y/SBA-16 oxygen carrier are higher than other promoted oxygen carriers in most of successive redox cycles. It is noteworthy to say that the methane conversion of 99.21% and the average hydrogen yield of 84.87% were achieved using 20Ni-2.5Y/SBA-16 oxygen carrier. Therefore, according to Figure 7a,b, 20Ni-2.5Y/SBA-16 oxygen carrier has highest stability compared with other oxygen carriers and also as demonstrated before, this oxygen carrier has higher activity and better proficiency. Therefore, the yttrium weight percentage of 2.5 is detected as the optimum yttrium loading.

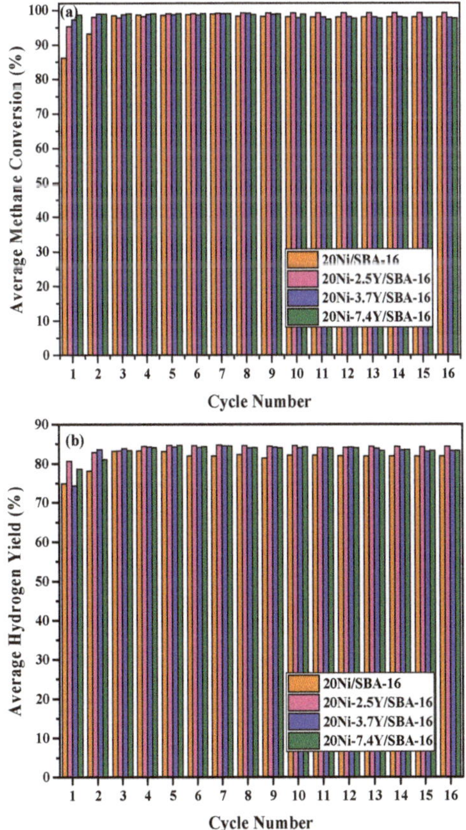

Figure 7. The Life time of 20Ni-yY/SBA-16 oxygen carriers at 650 °C: (**a**) Average methane conversion and (**b**) Average hydrogen yield.

2.4. Effect of Steam/Carbon Molar Ratio on the Catalytic Performance

Figure 8 shows the average methane conversion and hydrogen yield at various steam to methane molar ratio (1, 1.5, 2, 3, 4 and 5) at 650 °C using 20Ni-2.5Y/SBA-16 oxygen carrier. The S/C molar ratio is a significant parameter to determine reaction pathway of CH_4 and distribution of product in the CL-SMR process. Using large excess steam is undesirable, since high operating expenditures are required [60,61]. As demonstrated in this figure, by increasing the H_2O/CH_4 ratio to 2, the CH_4 conversion and hydrogen yield were increased to 99.44% and 85.48%, respectively and further increase of H_2O/CH_4 ratio, caused slight negative effect on the activity of OC due to the pore blockage of oxygen carriers at higher steam to carbon molar ratios [62,63]. Therefore, the optimum average CH_4 conversion and H_2 yield were obtained at steam to carbon molar ratio of 2.

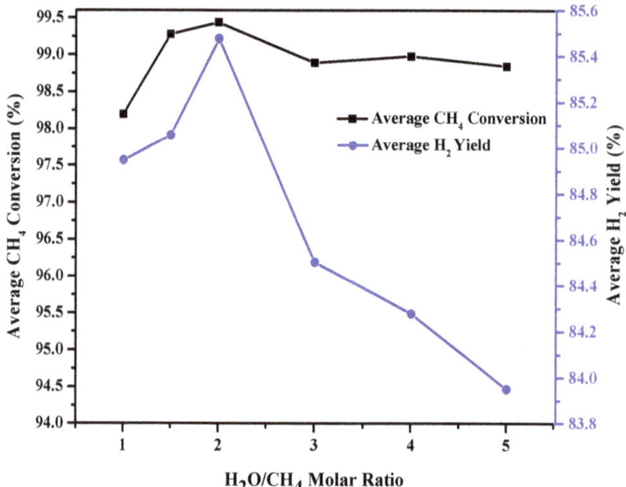

Figure 8. Average methane conversion and hydrogen yield of 20Ni-2.5Y/SBA-16 oxygen carrier at 650 °C for different H_2O/CH_4 molar ratio.

2.5. Effect of Ni Loading Percentage and Temperature on the Catalytic Activity

In order to detect the optimum nickel weight percentage, xNi-2.5Y/SBA-16 (x = 10, 15, 20, 25, 30 wt. %) oxygen carriers were synthesized and investigated in CL-SMR process in various reduction temperatures (500–750 °C) with steam to carbon molar ratio of 2. The methane conversion during the cycle time and average hydrogen yield in different temperature are plotted in Figure 9. Because Ni loading percentage has great effect on the catalyst activity, its variation with time at different temperatures is plotted in Figure 9. As demonstrated in this figure, 25Ni-2.5Y/SBA-16 OC revealed the highest catalytic activity and average H_2 yield in all temperature range. The average methane conversion of 25Ni-2.5Y/SBA-16 OC is about 91.09% at 500 °C (Figure 9a) and increased to 100% at 700 °C (Figure 9b). In addition, the average hydrogen yield increased from about 78.05% to 85.33% with temperature rise from 500 to 650 °C using this OC. Actually, the performance of OCs with different nickel loadings displayed a maximum with raising the nickel content up to 25% and further increase of nickel loading percentage results in the reduction of CH_4 conversion. The improvement of methane conversion with Ni loading percentage could be due to the more available lattice oxygen by increasing the NiO molecules. Afterwards the reduction in the activity is related to the agglomeration tendency of nickel particles at higher Ni weight percentage [48]. It can be concluded from Figure 9 that at higher reduction temperature the difference between catalytic activity and hydrogen production yield of xNi-2.5Y/SBA-16 (x = 10, 15, 20, 25, 30 wt. %) oxygen carriers are less significant.

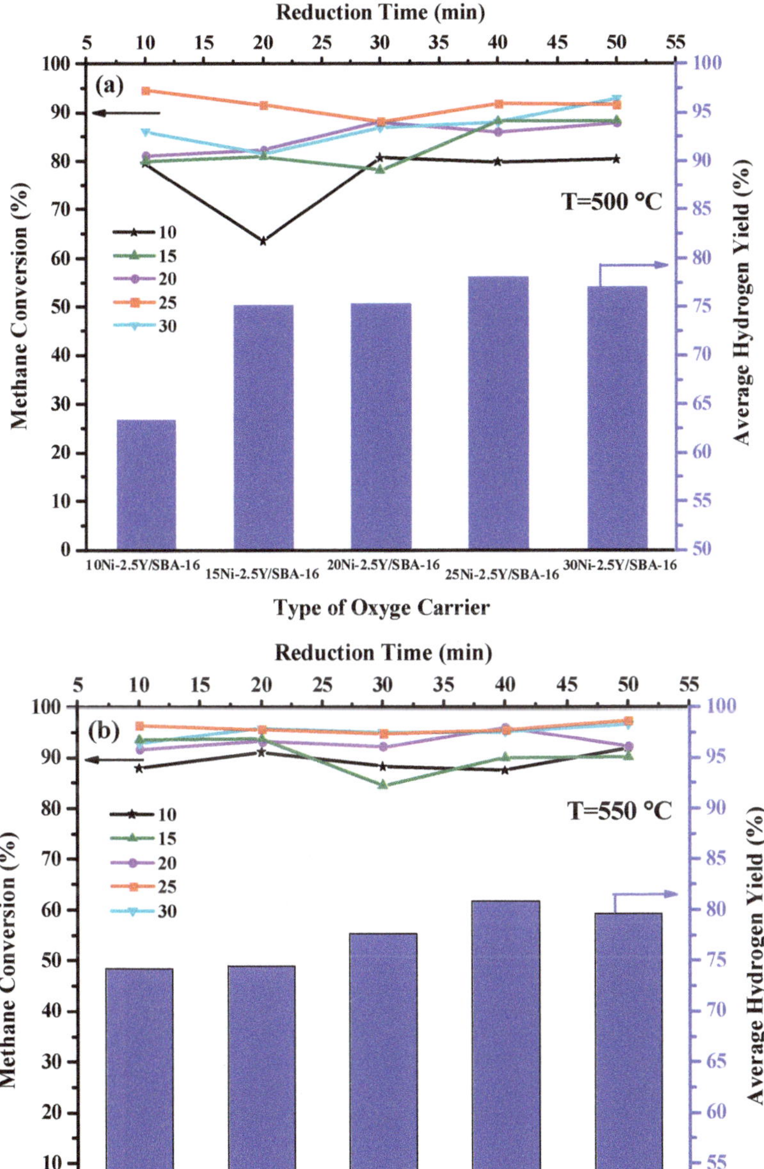

Figure 9. *Cont.*

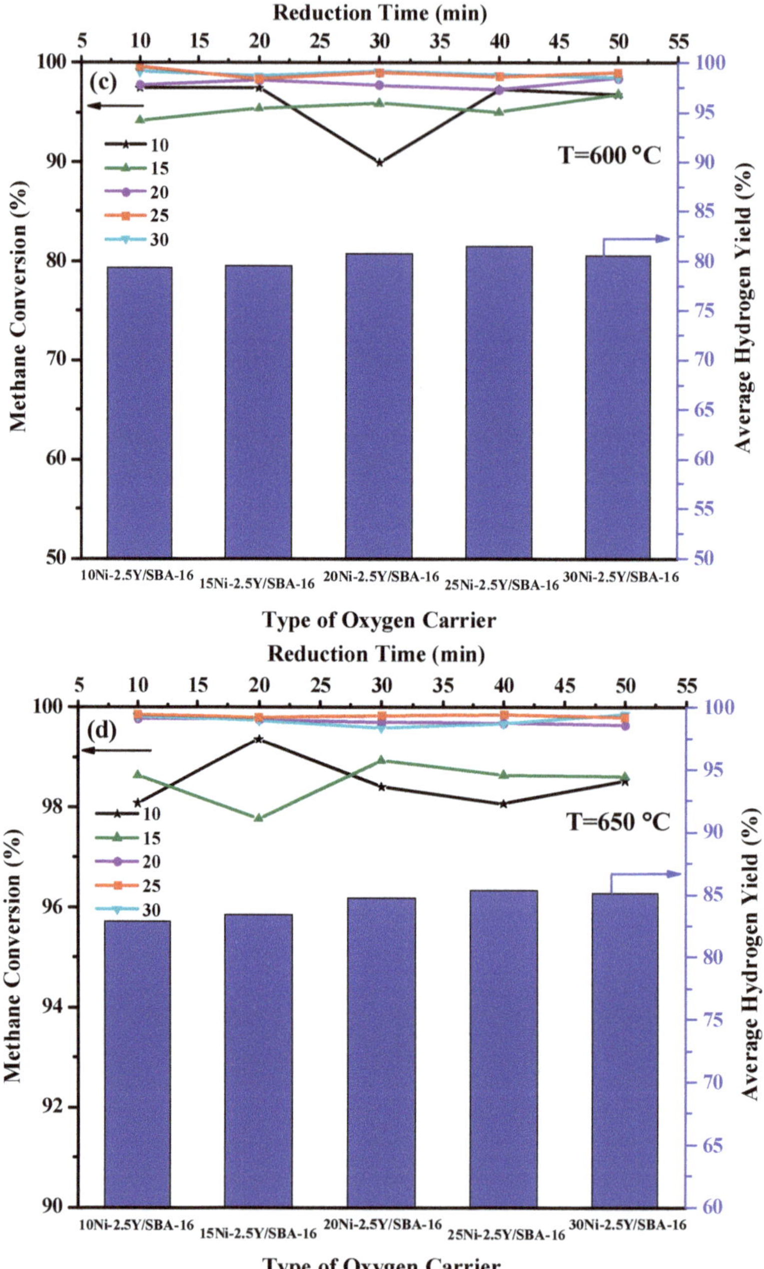

Figure 9. *Cont.*

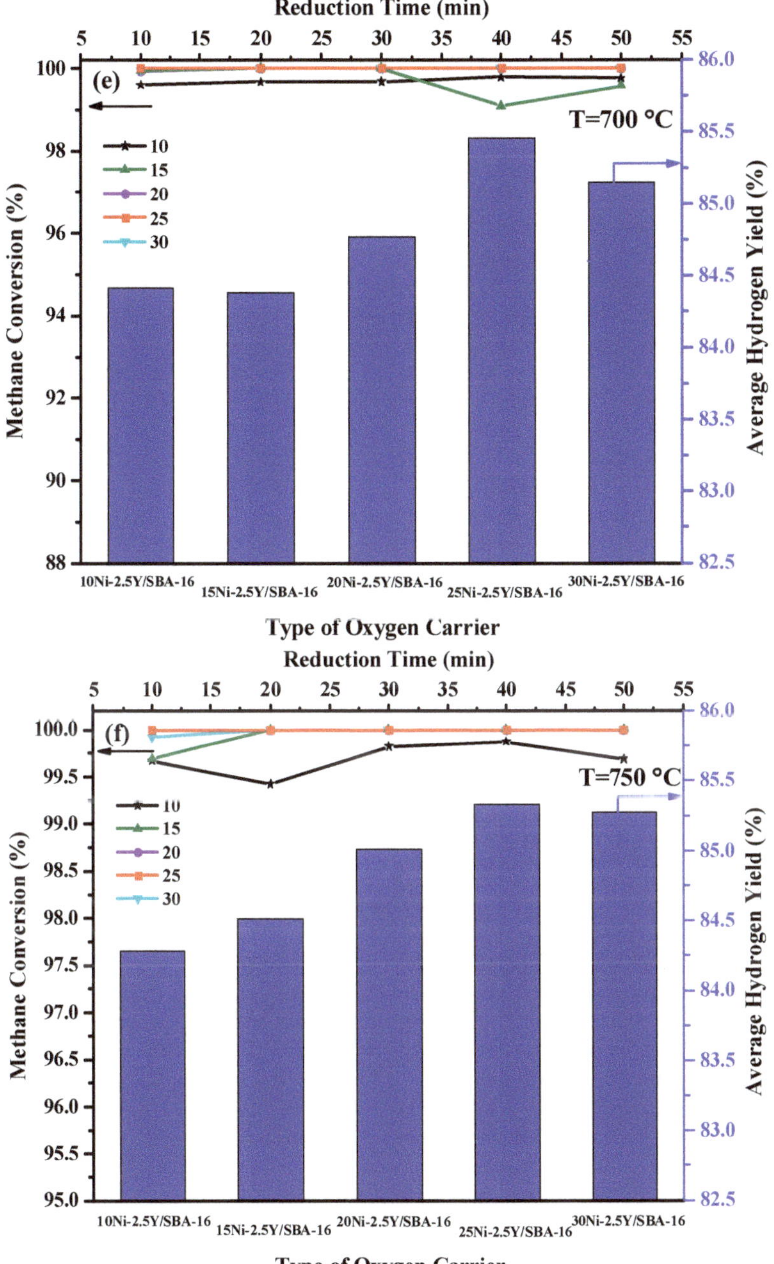

Figure 9. The effect of Ni weight percentage on the methane conversion and average hydrogen yield of xNi-2.5Y/SBA-16 oxygen carriers at reduction temperature of (**a**) 500 °C; (**b**) 550 °C; (**c**) 600 °C; (**d**) 650 °C; (**e**) 700 °C and (**f**) 750 °C.

2.6. Life Time of 25Ni/SBA-16 and 25Ni-2.5Y/SBA-16 Oxygen Carriers

In order to investigate the effect of yttrium promoter on the durability of samples, 25Ni/SBA-16 and 25Ni-2.5Y/SBA-16 oxygen carriers were examined over 16 oxidation-reduction cycles at 650 °C with steam to carbon molar ration of 2. As represented in Figure 10, the average CH_4 conversion and H_2 production yield of promoted oxygen carrier is higher than that of yttrium free OC during cycles. A slight variation of methane conversion observed during the cycles and no sensible activity loss was detected over the 16 redox cycles for 25Ni-2.5Y/SBA-16 OC. The highest CH_4 conversion of about 99% and H_2 yield of about 85% were achieved during the cycles at 650 °C using this OC. This can be due to better distribution of nickel nanoparticles on the surface of oxygen carrier in yttrium promoted OC according to XRD result. Therefore, oxygen transfer rate is improved in the presence of yttrium promoter. As shown in this figure, the average methane conversion and hydrogen production yield of non-promoted oxygen carrier increased during first six cycles. Then they decreased with slight slope in the next three cycles and after that remained constant. Coke formation on the surface of 25Ni/SBA-16 and/or agglomeration of Ni particles can be the result of this reduction in the catalytic activity [19].

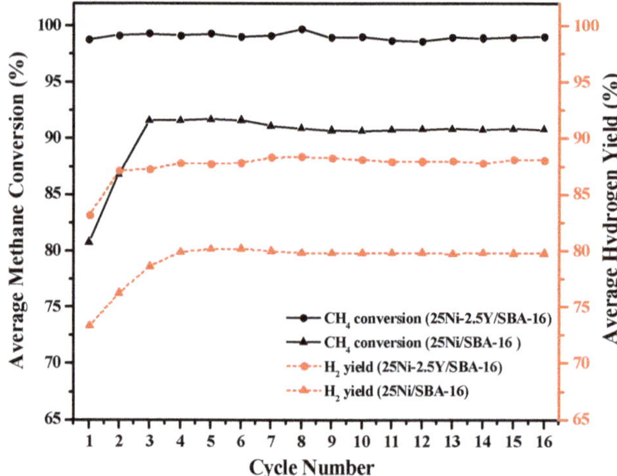

Figure 10. Life time of 25Ni/SBA-16 and 25Ni-2.5Y/SBA-16 oxygen carriers during 16 cycles at 650 °C.

Table 2 shows the BET surface area, total pore volume and average pore size of 25Ni-2.5Y/SBA-16 and 25Ni/SBA-16 oxygen carriers after 16 redox cycles. Specific surface area and pore volume of both samples were decreased significantly due to sintering and blocking the pores during CL-SMR process at high temperature of 650 °C. In addition, coke formation on the surface of oxygen carrier in reduction section can effectively decrease the specific surface area of samples [1]. It is noteworthy that the presence of yttrium on the surface of support leads to lower reduction in pore volume and specific surface area of 25Ni-2.5Y/SBA-16 compared to 25Ni/SBA-16 OC.

Table 2. Structure characteristics of the spent 25Ni/SBA-16 and 25Ni-2.5Y/SBA-16 oxygen carriers after 16 redox cycles at 650 °C.

Oxygen Carrier	BET Surface Area (m^2/g)	Pore Diameter (nm)	Pore Volume (cm^3/g)
25Ni/SBA-16	58.26	4.78	0.08
25Ni-2.5Y/SBA-16	144.04	3.99	0.15

The XRD patterns of 25Ni/SBA-16 and 25Ni-2.5Y/SBA-16 oxygen carriers after 16 redox cycles are depicted in Figure 11. Both samples show the diffraction peaks of nickel at 2θ value of 44.5°, 51.8° and 76.4° which respectively correspond to (111), (200) and (220) planes. This indicates that nickel oxide was reduced to metallic Ni during CL-SMR process. Y_2O_3 is observed in 25Ni-2.5Y/SBA-16 sample at peak of 2θ = 79.7°. The carbon diffraction peak observed at 44.5° for both samples was less pronounced in the case of 25Ni-2.5Y/SBA-16 oxygen carrier. It was reported that the presence of yttrium oxide enhanced the oxygen vacancies on the surface of support and promoted the mobility of oxygen so the carbon can be removed more easily over the yttrium promoted oxygen carrier [62]. Also, the peak at 2θ = 43.8° can be assigned to carbon for 25Ni/SBA-16 sample (cod. No. 00-050-1084). The formation of $NiCO_3$ spinel in yttrium free oxygen carrier (25Ni/SBA-16) is evident at the diffraction peak of 43.4° according to 00-012-0771 reference pattern code.

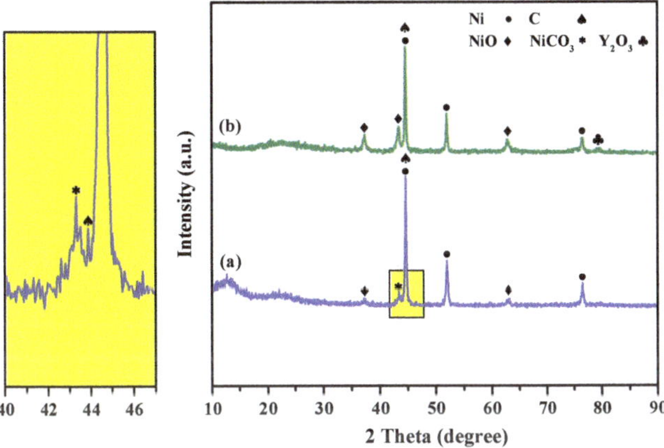

Figure 11. XRD patterns of (**a**) 25Ni/SBA-16 and (**b**) 25Ni-2.5Y/SBA-16 oxygen carriers.

Figure 12 shows the EDX analysis and FESEM images of spent 25Ni/SBA-16 and 25Ni-2.5Y/SBA-16 OCs in 16 redox cycles. As mentioned in this figure, C, O, Ni, Si and Y atoms are presented in these OCs and the C element which represents the coke deposited on the SBA-16 support, was reduced from 10.61% to 4.93% on the 25Ni-2.5Y/SBA-16 OC. This can be due to the presence of Y_2O_3 species that have the ability to form oxycarbonate for oxidizing the surface carbon [64]. The FESEM images indicate the formation of Ni active sites with appropriate distribution on the surface of yttrium promoted OC which enhanced the catalytic activity.

The results presented in this paper demonstrated a better performance of Ni/SBA-16 OC with yttrium promoter compared with other oxygen carriers used previously in CL-SMR process. The average methane conversion of about 99.8% was achieved at 650 °C using 25Ni-2.5Y/SBA-16 oxygen carrier in the present work, while Belhadi et al. [65] obtained the methane conversion of about 72.0% and 88.0% at 700 °C using nickel based ZrO_2 and La_2O_3 supports, respectively. As a comparison with Ni-based catalysts in SMR process, Wan et al. [66] achieved the methane conversion of 99.5% at 800 °C using $Ce_xZr_{1-x}O_2$ promoted Ni/SBA-15 catalyst. Furthermore, Rakass et al. [67] tested unsupported nickel powder catalyst for SMR and the methane conversion of 98.0% was achieved at 700 °C.

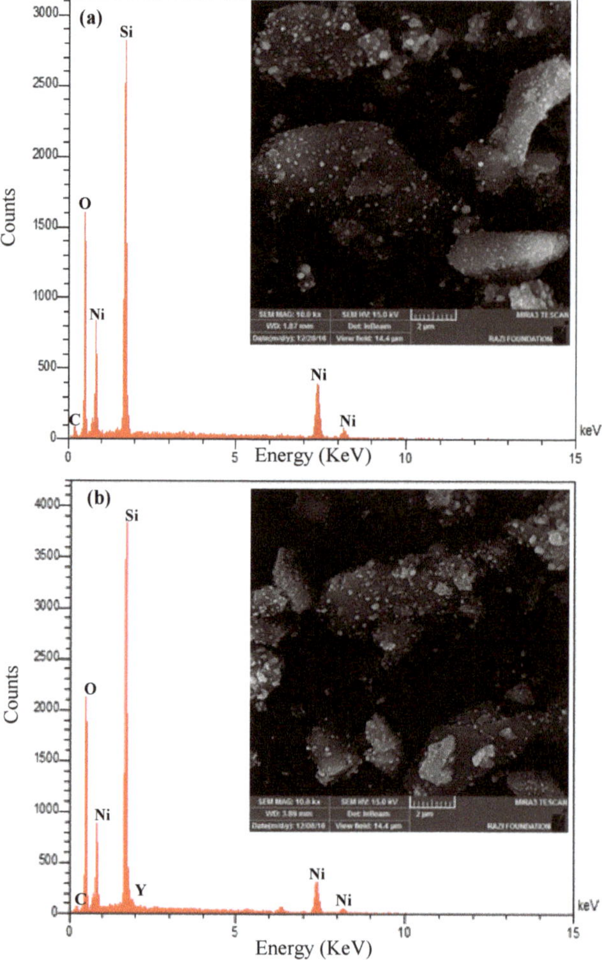

Figure 12. EDX and FESEM analysis of used (**a**) 25Ni/SBA-16 and (**b**) 25Ni-2.5Y/SBA-16 oxygen carriers.

3. Experimental Methods

3.1. Oxygen Carrier Preparation

SBA-16 support was synthesized using triblock copolymer Pluronic F127 ($EO_{106}PO_{70}EO_{106}$, Mw = 12,600, Aldrich, St. Louis, MO, USA) as the structure-directing agent and tetraethyl orthosilicate (TEOS, Merck, Kenilworth, NJ, USA) as the silica source by means of hydrothermal method. In a typical synthesis, 1.15 g F127 was dissolved in HCl (37 wt. %) and de-ionized water under vigorous stirring at 40 °C. Then, TEOS and butanol were added dropwise to the above solution. After stirring the mixture for 27 h at 40 °C, it was transferred to the Teflon-lined autoclave and heated at 100 °C for two days. The solid product was recovered, washed with de-ionized water and ethanol for several times and dried at 100 °C for one day. Calcination was occurred at 550 °C for 6 h with a heating rate of 4 °C/min to remove the organic compounds of template and the formation of mesoporous structure of cubic SBA-16. It was used as a supporting material for Ni-Y/SBA-16 oxygen carrier.

Final stage of preparation procedure was performed by co-impregnation of SBA-16 support with nickel nitrate and yttrium nitrate solutions. For this purpose, the separated solution of $Y(NO_3)_3 \cdot 6H_2O$ (Merck) and $Ni(NO_3)_2 \cdot 6H_2O$ (Merck) were added simultaneously to the support at 40 °C. The solution was stirred for 5 h at 40 °C in order to better diffusion of nickel and yttrium precursors into the pores of SBA-16. The impregnated OCs were dried at 100 °C for 15 h and then calcined at 650 °C for 3 h. The solid product designated as xNi-yY/SBA-16 where x = 10, 15, 20, 25 & 30 wt. % and y = 0, 2.5, 3.7 & 7.4 wt. %.

3.2. Oxygen Carrier Characterization

The synthesized OCs were characterized before and after reduction section in order to study their structural and catalytic properties via different techniques. The XRD patterns were collected on a powder Bruker D8 Advance Germany instrument equipped with Cu Kα source at 40 kV and 40 mA in the range of 2θ = 10°–90° with a step size of 0.05°. In order to measure the specific surface area, BET method was applied by N_2 adsorption/desorption isotherms using ASPA-2020 Instrument (Norcross, GA, USA). All the synthesized samples were degassed at 250 °C with nitrogen. The pore size distribution and the cumulative pore volume were achieved by the Barret-Joyner-Halenda (BJH) method from the desorption branches of N_2 isotherms. FESEM images were recorded on a HITAGHI S-4160 system equipped with an EDX spectroscopy.

3.3. Process Activity

Gas-solid reactions consist of redox multi-cycles were done in a cylindrical stainless steel reactor which has an inner diameter of 16 mm with 1000 mm height. The fixed-bed reactor was inserted into a vertical electrical furnace and a k-type thermocouple posited at the center of the OC bed to monitor the temperature of the catalytic bed during the process. In each activity test, 1 g of freshly synthesized powdered OC (mesh size: 100–200 μm) was packed on a thin porous layer in the middle of the reformer reactor. Deionized water was injected to the evaporator with syringe pump to generate steam. In the reduction step, the reactant gas streams (CH_4 as reactant and Ar as carrier gas) were controlled through two distinct mass flow controllers (MFCs) and mixed with steam before the reactor entrance. The reduction section was carried out by changing temperature from 500 to 750 °C at atmospheric pressure for 50 min. The feed mixture entered the bed at the reaction temperature and reacted with lattice oxygen of solid OCs. In the oxidation step, the stream of 20 vol % O_2 diluted in Ar with the total flow rate of 124 mL min^{-1} was fed into the reactor to reoxidize the oxygen carriers and removing the deposited carbon. The oxidation and reduction periods were dissociated by purging Ar for 3 min. The products of the reactor were passed through a condenser in order to liquefy the steam from gaseous product. Finally, the product and unreacted gas streams were analyzed using an online Bruker 450 gas chromatograph (GC) system every ten minutes. This means that the outlet gas was analyzed five times for each temperature. It is noteworthy to say that the first sample injection was carried out 10 min after the beginning of the reaction. Thus, the short transient state period is passed and it could be in the steady state section. Figure 13 indicates the schematic of the designated reactor applied for investigating the performance of oxygen carriers. Methane conversion (X_{CH_4}) and hydrogen production yield (y_{H_2}) were calculated as follows:

$$X_{CH_4} = \frac{(moles\ of\ CH_4)_{in} - (moles\ of\ CH_4)_{out}}{(moles\ of\ CH_4)_{in}} \times 100 \tag{4}$$

$$y_{H_2} = \frac{(moles\ of\ H_2)_{out}}{2 * (moles\ of\ CH_4)_{in}} \times 100 \tag{5}$$

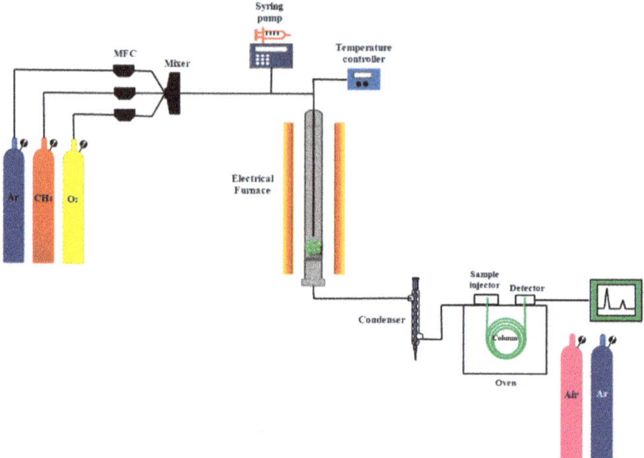

Figure 13. Reactor system for CL-SMR.

4. Conclusions

Structural characterization and catalytic activity of yttrium promoted Ni-based oxygen carriers were examined in chemical looping steam methane reforming process to produce synthesis gas. The oxygen carriers were synthesized via co-impregnation method and the effect of Ni loading (10–30 wt. %), Y weight percentage (0–7.4 wt. %), reaction temperature (500–750 °C), steam to methane molar ratio (1–5) and life time of oxygen carriers over 16 redox cycles were successfully investigated. The reduction temperature revealed significant effect on CH_4 conversion and H_2 production yield of all prepared samples. The reaction temperature of 650 °C and H_2O/CH_4 molar ratio of 2 were the optimum reduction condition using 25Ni-60Y/SBA-16 oxygen carrier to achieve 99.83% methane conversion and 85.34% hydrogen yield. The characterization results showed that better dispersion of Ni active sites, higher specific surface area and lower coke deposited were achieved using yttrium as a promoter on the surface of oxygen carrier. Thus, the catalytic activity and long-term stability of OCs were improved after the addition of yttrium promoter on SBA-16 support.

Author Contributions: S.D.-J., M.M. and A.H. conceived and designed the experiments; S.D.-J. and M.M. performed the experiments; S.D.-J. and M.M. performed catalyst synthesis; S.D.-J., M.M. and A.H. analyzed the data; M.R.R. contributed reagents/materials/analysis tools; M.R.R. participated in the analysis and interpretation of characterization results; and S.D.-J. and M.M. wrote the paper.

Conflicts of Interest: The authors declare no conflict of interest.

References

1. Hafizi, A.; Rahimpour, M.R.; Hassanajili, S. Calcium promoted Fe/Al_2O_3 oxygen carrier for hydrogen production via cyclic chemical looping steam methane reforming process. *Int. J. Hydrogen Energy* **2015**, *40*, 16159–16168. [CrossRef]
2. Hafizi, A.; Rahimpour, M.R.; Hassanajili, S. Hydrogen production by chemical looping steam reforming of methane over Mg promoted iron oxygen carrier: Optimization using design of experiments. *J. Taiwan Inst. Chem. Eng.* **2016**, *62*, 140–1409. [CrossRef]
3. Karimi, E.; Forutan, H.; Saidi, M.; Rahimpour, M.R.; Shariati, A. Experimental study of chemical-looping reforming in a fixed-bed reactor: Performance investigation of different oxygen carriers on Al_2O_3 and TiO_2 support. *Energy Fuels* **2014**, *28*, 2811–2820. [CrossRef]
4. Frusteri, F.; Freni, S.; Spadaro, L.; Chiodo, V.; Bonura, G.; Donato, S.; Cavallaro, S. H_2 production for MC fuel cell by steam reforming of ethanol over MgO supported Pd, Rh, Ni and Co catalysts. *Catal. Commun.* **2004**, *5*, 611–615. [CrossRef]

5. Wang, Y.; Chen, M.; Liang, T.; Yang, Z.; Yang, J.; Liu, S. Hydrogen generation from catalytic steam reforming of acetic acid by Ni/attapulgite catalysts. *Catalysts* **2016**, *6*, 172. [CrossRef]

6. Du, Y.-L.; Wu, X.; Cheng, Q.; Huang, Y.-L.; Huang, W. Development of Ni-based catalysts derived from hydrotalcite-like compounds precursors for synthesis gas production via methane or ethanol reforming. *Catalysts* **2017**, *7*, 70. [CrossRef]

7. Midilli, A.; Ay, M.; Dincer, I.; Rosen, M. On hydrogen and hydrogen energy strategies: I: Current status and needs. *Renew. Sustain. Energy Rev.* **2005**, *9*, 255–271. [CrossRef]

8. Momirlan, M.; Veziroglu, T.N. The properties of hydrogen as fuel tomorrow in sustainable energy system for a cleaner planet. *Int. J. Hydrogen Energy* **2005**, *30*, 795–802. [CrossRef]

9. Holladay, J.D.; Hu, J.; King, D.L.; Wang, Y. An overview of hydrogen production technologies. *Catal. Today* **2009**, *139*, 244–260. [CrossRef]

10. Abbasi, M.; Farniaei, M.; Rahimpour, M.R.; Shariati, A. Enhancement of Hydrogen Production and Carbon Dioxide Capturing in a Novel Methane Steam Reformer Coupled with Chemical Looping Combustion and Assisted by Hydrogen Perm-Selective Membranes. *Energy Fuels* **2013**, *27*, 5359–5372. [CrossRef]

11. Akbari-Emadabadi, S.; Rahimpour, M.R.; Hafizi, A.; Keshavarz, P. Production of hydrogen-rich syngas using Zr modified Ca-Co bifunctional catalyst-sorbent in chemical looping steam methane reforming. *Appl. Energy* **2017**, *206*, 51–62. [CrossRef]

12. Schädel, B.T.; Duisberg, M.; Deutschmann, O. Steam reforming of methane, ethane, propane, butane, and natural gas over a rhodium-based catalyst. *Catal. Today* **2009**, *142*, 42–51. [CrossRef]

13. Basagiannis, A.; Verykios, X. Catalytic steam reforming of acetic acid for hydrogen production. *Int. J. Hydrogen Energy* **2007**, *32*, 3343–3355. [CrossRef]

14. Armor, J. Catalysis and the hydrogen economy. *Catal. Lett.* **2005**, *101*, 131–135. [CrossRef]

15. Chaubey, R.; Sahu, S.; James, O.O.; Maity, S. A review on development of industrial processes and emerging techniques for production of hydrogen from renewable and sustainable sources. *Renew. Sustain. Energy Rev.* **2013**, *23*, 443–462. [CrossRef]

16. Forutan, H.; Karimi, E.; Hafizi, A.; Rahimpour, M.R.; Keshavarz, P. Expert representation chemical looping reforming: A comparative study of Fe, Mn, Co and Cu as oxygen carriers supported on Al_2O_3. *J. Ind. Eng. Chem.* **2015**, *21*, 900–911. [CrossRef]

17. Rydén, M.; Arjmand, M. Continuous hydrogen production via the steam–iron reaction by chemical looping in a circulating fluidized-bed reactor. *Int. J. Hydrogen Energy* **2012**, *37*, 4843–4854. [CrossRef]

18. Zhu, X.; Wei, Y.; Wang, H.; Li, K. Ce-Fe oxygen carriers for chemical-looping steam methane reforming. *Int. J. Hydrogen Energy* **2013**, *38*, 4492–4501. [CrossRef]

19. Hafizi, A.; Rahimpour, M.R.; Hassanajili, S. Hydrogen production via chemical looping steam methane reforming process: Effect of cerium and calcium promoters on the performance of Fe_2O_3/Al_2O_3 oxygen carrier. *Appl. Energy* **2016**, *165*, 685–694. [CrossRef]

20. Moghtaderi, B. Review of the recent chemical looping process developments for novel energy and fuel applications. *Energy Fuels* **2011**, *26*, 15–40. [CrossRef]

21. Solunke, R.D.; Veser, G.T. Hydrogen production via chemical looping steam reforming in a periodically operated fixed-bed reactor. *Ind. Eng. Chem. Res.* **2010**, *49*, 11037–11044. [CrossRef]

22. Hafizi, A.; Rahimpour, M.R.; Hassanajili, S. High purity hydrogen production via sorption enhanced chemical looping reforming: Application of $22Fe_2O_3/MgAl_2O_4$ and $22Fe_2O_3/Al_2O_3$ as oxygen carriers and cerium promoted CaO as CO_2 sorbent. *Appl. Energy* **2016**, *169*, 629–641. [CrossRef]

23. Rahimpour, M.R.; Hesami, M.; Saidi, M.; Jahanmiri, A.; Farniaei, M.; Abbasi, M. Methane Steam Reforming Thermally Coupled with Fuel Combustion: Application of Chemical Looping Concept as a Novel Technology. *Energy Fuels* **2013**, *27*, 2351–2362. [CrossRef]

24. Alirezaei, I.; Hafizi, A.; Rahimpour, M.R.; Raeissi, S. Application of zirconium modified Cu-based oxygen carrier in chemical looping reforming. *J. CO_2 Util.* **2016**, *14*, 112–121. [CrossRef]

25. Zafar, Q.; Mattisson, T.; Gevert, B. Redox investigation of some oxides of transition-state metals Ni, Cu, Fe, and Mn supported on SiO_2 and $MgAl_2O_4$. *Energy Fuels* **2006**, *20*, 34–44. [CrossRef]

26. Adanez, J.; Abad, A.; Garcia-Labiano, F.; Gayan, P.; Luis, F. Progress in chemical-looping combustion and reforming technologies. *Prog. Energy Comb. Sci.* **2012**, *38*, 215–282. [CrossRef]

27. Trimm, D.L. Catalysts for the control of coking during steam reforming. *Catal. Today* **1999**, *49*, 3–10. [CrossRef]

28. Van Dillen, A.J.; Terörde, R.J.; Lensveld, D.J.; Geus, J.W.; De Jong, K.P. Synthesis of supported catalysts by impregnation and drying using aqueous chelated metal complexes. *J. Catal.* **2003**, *216*, 257–264. [CrossRef]

29. Zhong, X.; Xie, W.; Wang, N.; Duan, Y.; Shang, R.; Huang, L. Dolomite-Derived Ni-Based Catalysts with Fe Modification for Hydrogen Production via Auto-Thermal Reforming of Acetic Acid. *Catalysts* **2016**, *6*, 85. [CrossRef]

30. Xu, B.-Q.; Wei, J.-M.; Wang, H.-Y.; Sun, K.-Q.; Zhu, Q.-M. Nano-MgO: Novel preparation and application as support of Ni catalyst for CO_2 reforming of methane. *Catal. Today* **2001**, *68*, 217–225. [CrossRef]

31. Kambolis, A.; Matralis, H.; Trovarelli, A.; Papadopoulou, C. Ni/CeO_2-ZrO_2 catalysts for the dry reforming of methane. *Appl. Catal. A Gen.* **2010**, *377*, 16–26. [CrossRef]

32. Kim, J.-H.; Suh, D.J.; Park, T.-J.; Kim, K.-L. Effect of metal particle size on coking during CO_2 reforming of CH_4 over Ni–alumina aerogel catalysts. *Appl. Catal. A Gen.* **2000**, *197*, 191–200. [CrossRef]

33. Bengaard, H.S.; Nørskov, J.K.; Sehested, J.; Clausen, B.; Nielsen, L.; Molenbroek, A.; Rostrup-Nielsen, J. Steam reforming and graphite formation on Ni catalysts. *J. Catal.* **2002**, *209*, 365–384. [CrossRef]

34. Bradford, M.; Vannice, M. CO_2 reforming of CH_4. *Catal. Rev.* **1999**, *41*, 1–42. [CrossRef]

35. Takahashi, R.; Sato, S.; Sodesawa, T.; Tomiyama, S. CO_2-reforming of methane over Ni/SiO_2 catalyst prepared by homogeneous precipitation in sol–gel-derived silica gel. *Appl. Catal. A Gen.* **2005**, *286*, 142–147. [CrossRef]

36. Zhang, M.; Ji, S.; Hu, L.; Yin, F.; Li, C.; Liu, H. Structural characterization of highly stable Ni/SBA-15 catalyst and its catalytic performance for methane reforming with CO_2. *Chin. J. Catal.* **2006**, *27*, 777–781. [CrossRef]

37. Liu, D.; Quek, X.Y.; Cheo, W.N.E.; Lau, R.; Borgna, A.; Yang, Y. MCM-41 supported nickel-based bimetallic catalysts with superior stability during carbon dioxide reforming of methane: Effect of strong metal–support interaction. *J. Catal.* **2009**, *266*, 380–390. [CrossRef]

38. Liu, Z.; Zhou, J.; Cao, K.; Yang, W.; Gao, H.; Wang, Y.; Li, H. Highly dispersed nickel loaded on mesoporous silica: One-spot synthesis strategy and high performance as catalysts for methane reforming with carbon dioxide. *Appl. Catal. B Environ.* **2012**, *125*, 324–330. [CrossRef]

39. Li, L.; He, S.; Song, Y.; Zhao, J.; Ji, W.; Au, C.-T. Fine-tunable Ni@ porous silica core–shell nanocatalysts: Synthesis, characterization, and catalytic properties in partial oxidation of methane to syngas. *J. Catal.* **2012**, *288*, 54–64. [CrossRef]

40. Carrero, A.; Calles, J.; Vizcaíno, A. Hydrogen production by ethanol steam reforming over Cu-Ni/SBA-15 supported catalysts prepared by direct synthesis and impregnation. *Appl. Catal. A Gen.* **2007**, *327*, 82–94. [CrossRef]

41. Klimova, T.; Calderón, M.; Ramírez, J. Ni and Mo interaction with Al-containing MCM-41 support and its effect on the catalytic behavior in DBT hydrodesulfurization. *Appl. Catal. A Gen.* **2003**, *240*, 29–40. [CrossRef]

42. Zhang, S.; Muratsugu, S.; Ishiguro, N.; Tada, M. Ceria-doped Ni/SBA-16 catalysts for dry reforming of methane. *ACS Catal.* **2013**, *3*, 1855–1864. [CrossRef]

43. Kantorovich, D.; Haviv, L.; Vradman, L.; Landau, M. Behaviour of NiO and NiO pohases at high loadings, in SBA-15 and SBA-16 mesoporous silica matrices. *Stud. Surf. Sci. Catal.* **2005**, *156*, 147–154. [CrossRef]

44. Luna, A.E.C.; Iriarte, M.E. Carbon dioxide reforming of methane over a metal modified Ni-Al_2O_3 catalyst. *Appl. Catal. A Gen.* **2008**, *343*, 10–15. [CrossRef]

45. Hu, Y.H.; Ruckenstein, E. The characterization of a highly effective NiO/MgO solid solution catalyst in the CO_2 reforming of CH_4. *Catal. Lett.* **1997**, *43*, 71–77. [CrossRef]

46. Li, B.; Xu, X.; Zhang, S. Synthesis gas production in the combined CO_2 reforming with partial oxidation of methane over Ce-promoted Ni/SiO_2 catalysts. *Int. J. Hydrogen Energy* **2013**, *38*, 890–900. [CrossRef]

47. Qian, L.; Ma, Z.; Ren, Y.; Shi, H.; Yue, B.; Feng, S.; Shen, J.; Xie, S. Investigation of La promotion mechanism on Ni/SBA-15 catalysts in CH_4 reforming with CO_2. *Fuel* **2014**, *122*, 47–53. [CrossRef]

48. Meshksar, M.; Daneshmand-Jahromi, S.; Rahimpour, M.R. Synthesis and characterization of cerium promoted Ni/SBA-16 oxygen carrier in cyclic chemical looping steam methane reforming. *J. Taiwan Inst. Chem. Eng.* **2017**, *76*, 73–82. [CrossRef]

49. Shi, C.; Zhang, P. Effect of a second metal (Y, K, Ca, Mn or Cu) addition on the carbon dioxide reforming of methane over nanostructured palladium catalysts. *Appl. Catal. B Environ.* **2012**, *115*, 190–200. [CrossRef]

50. Li, J.; Xia, C.; Au, C.; Liu, B. Y_2O_3-promoted NiO/SBA-15 catalysts highly active for CO_2/CH_4 reforming. *Int. J. Hydrogen Energy* **2014**, *39*, 10927–10940. [CrossRef]

51. Yogamalar, R.; Srinivasan, R.; Vinu, A.; Ariga, K.; Bose, A.C. X-ray peak broadening analysis in ZnO nanoparticles. *Solid State Commun.* **2009**, *149*, 1919–1923. [CrossRef]

52. Kaneko, K. Determination of pore size and pore size distribution. *J. Membr. Sci.* **1994**, *96*, 59–89. [CrossRef]

53. Cohan, L.H. Hysteresis and the Capillary Theory of Adsorption of Vapors1. *J. Am. Chem. Soc.* **1944**, *66*, 98–105. [CrossRef]

54. Tsoncheva, T.; Linden, M.; Areva, S.; Minchev, C. Copper oxide modified large pore ordered mesoporous silicas for ethyl acetate combustion. *Catal. Commun.* **2006**, *7*, 357–361. [CrossRef]

55. Rodrıguez-Castellón, E.; Dıaz, L.; Braos-Garcıa, P.; Mérida-Robles, J.; Maireles-Torres, P.; Jiménez-López, A.; Vaccari, A. Nickel-impregnated zirconium-doped mesoporous molecular sieves as catalysts for the hydrogenation and ring-opening of tetralin. *Appl. Catal. A Gen.* **2003**, *240*, 83–94. [CrossRef]

56. Yamazaki, T.; Kikuchi, N.; Katoh, M.; Hirose, T.; Saito, H.; Yoshikawa, T.; Wada, M. Behavior of steam reforming reaction for bio-ethanol over Pt/ZrO$_2$ catalysts. *Appl. Catal. B Environ.* **2010**, *99*, 81–88. [CrossRef]

57. Profeti, L.P.; Dias, J.A.; Assaf, J.M.; Assaf, E.M. Hydrogen production by steam reforming of ethanol over Ni-based catalysts promoted with noble metals. *J. Power Sources* **2009**, *190*, 525–533. [CrossRef]

58. De Diego, LF.; Ortiz, M.; García-Labiano, F.; Adánez, J.; Abad, A.; Gayán, P. Synthesis gas generation by chemical-looping reforming using a Nibased oxygen carrier. *Energy Procedia* **2009**, *1*, 3–10. [CrossRef]

59. Akbari-Emadabadi, S.; Rahimpour, M.R.; Hafizi, A.; Keshavarz, P. Promotion of Ca-Co bifunctional catalyst/sorbent with yttrium for hydrogen production in modified chemical looping steam methane reforming. *Catalysts* **2017**, *7*, 270. [CrossRef]

60. Antzara, A.; Heracleous, E.; Bukur, D.B.; Lemonidou, A.A. Thermodynamic analysis of hydrogen production via chemical looping steam methane reforming coupled with in situ CO$_2$ capture. *Int. J. Greenh. Gas Control* **2015**, *32*, 115–128. [CrossRef]

61. Adiya, Z.I.; Dupont, V.; Mahmud, T. Chemical equilibrium analysis of hydrogen production from shale gas using sorption enhanced chemical looping steam reforming. *Fuel Process. Technol.* **2017**, *159*, 128–144. [CrossRef]

62. Silva, J.M.; Soria, M.A.; Madeira, L.M. Thermodynamic analysis of Glycerol Steam Reforming for hydrogen production with in situ hydrogen and carbon dioxide separation. *J. Power Sources* **2015**, *273*, 423–430. [CrossRef]

63. He, H.; Dai, H.; Wong, K.W.; Au, C. RE$_{0.6}$Zr$_{0.4-x}$Y$_x$O$_2$ (RE = Ce, Pr; x = 0, 0.05) solid solutions: An investigation on defective structure, oxygen mobility, oxygen storage capacity, and redox properties. *Appl. Catal. A Gen.* **2003**, *251*, 61–74. [CrossRef]

64. Oemar, U.; Hidajat, K.; Kawi, S. Role of catalyst support over PdO-NiO catalysts on catalyst activity and stability for oxy-CO$_2$ reforming of methane. *Appl. Catal. A Gen.* **2011**, *402*, 176–187. [CrossRef]

65. Belhadi, A.; Trari, M.; Rabia, C.; Cherifi, O. Methane steam reforming on supported nickel based catalysts. effect of oxide ZrO$_2$, La$_2$O$_3$ and nickel composition. *Open J. Phys. Chem.* **2013**, *3*, 89–96. [CrossRef]

66. Wan, H.; Li, X.; Ji, S.; Huang, B.; Wang, K.; Li, C. Effect of Ni loading and Ce$_x$Zr$_{i-x}$O$_2$ promoter on Ni-based SBA-15 catalysts for steam reforming of methane. *J. Nat. Gas Chem.* **2007**, *16*, 139–147. [CrossRef]

67. Rakass, S.; Oudghiri-Hassani, H.; Rowntree, P.; Abatzoglou, N. Steam reforming of methane over unsupported nickel catalysts. *J. Power Sources* **2006**, *158*, 485–496. [CrossRef]

 *catalysts*

Article

Promoting the Synthesis of Ethanol and Butanol by Salicylic Acid

Jinxin Zou, Lei Wang and Peijun Ji *

Department of Chemical Engineering, Beijing University of Chemical Technology, Beijing 100029, China; Jinxin_zhou@outlook.com (J.Z.); wanglei410@outlook.com (L.W.)
* Correspondence: jipj@mail.buct.edu.cn; Tel.: +86-010-6442-3254

Received: 3 September 2017; Accepted: 22 September 2017; Published: 1 October 2017

Abstract: Multiwalled carbon nanotubes (MWCNTs) were functionalized with salicylic acid (SA). The copper-cobalt catalyst was impregnated on the SA functionalized MWCNTs (SA-MWCNTs). The catalyst copper-cobalt/SA-MWCNTs was used to catalyze the synthesis of alcohols from synthesis gas. Salicylic acid can promote the synthesis of ethanol and butanol from synthesis gas, thus reducing the synthesis of methanol. This work demonstrated that salicylic acid not only can be used to functionalize carbon nanotubes, but also can enhance the production of ethanol and butanol from synthesis gas. On the other hand, the copper-cobalt catalyst supported on MWCNTs of 30 nm in diameter can synthesize more ethanol and butanol than supported on MWCNTs of 15 and 50 nm in diameter, indicating that the diameter of MWCNTs also has an effect on the synthesis of alcohols.

Keywords: alcohols; salicylic acid; multiwalled carbon nanotubes; synthesis gas

1. Introduction

The conversion of synthesis gas producing higher alcohols as fuel benefits sustainable development [1]. Investigation on catalysts with a higher selectivity and yield has been continuously carried out [2,3]. Researchers are focusing on heterogeneous catalysts rather than on homogeneous ones, as homogeneous catalysts are difficult to recycle. Metals including Re, Ru, Rh, Co, Cu, and Mo have been extensively used for preparing heterogeneous catalysts for the conversion of synthesis gas [4–8]. Re, Ru, and Rh are noble metals and are effective for catalyzing the synthesis of alcohols [4–7]. However, these metals are expensive. In contrast, Co, Cu, and Mo are much cheaper than the noble metals [8,9]. The non-noble metal based catalysts are being paid much attention [10–13]. For preparing a heterogeneous catalyst, one or several metals are deposited or impregnated on a support [13]. Particle sizes and the distribution of the particles are affected by catalyst supports.

With a large surface area, good thermal conductivity, strong mechanical strength, and excellent electrical properties, carbon nanotubes (CNTs) have been investigated as supports for preparing Rh, Co, Cu, and Mo-based catalysts [5,14,15]. CO hydrogenation is facilitated by carbon nanotubes and the formation of ethanol is promoted on Rh/CNTs [16]. A relatively higher activity and space yield of higher alcohols can be achieved by the metals supported on carbon nanotubes. The high catalysis efficiency is ascribed to that the metals interacted with carbon nanotubes, on the other hand the metal particles were well distributed on the supports [14]. In addition, carbon nanotubes are capable of adsorbing hydrogen gas. This facilitates the interaction of the hydrogen with the metals.

In this work, multiwalled carbon nanotubes (MWCNTs) have been functionalized with salicylic acid (SA), then copper and cobalt have been deposited on the SA functionalized MWCNTs (SA-MWCNTs). The aim of this research is to investigate the effect of salicylic acid on the synthesis of higher alcohols from synthesis gas. The catalyst copper-cobalt/SA-MWCNTs have been used to catalyze the synthesis of higher alcohols, especially ethanol and butanol, from synthesis gas, as ethanol and butanol not only can be used as fuels but also can be used as feedstocks for producing chemicals [17,18]. The effect of the diameter of MWCNTs and salicylic acid on the conversion of syngas and the selectivity of higher alcohols, especially ethanol and butanol, have been investigated.

2. Results and Discussion

2.1. Characterization of the Copper-Cobalt/SA-MWCNTs Catalyst

For the characterization of the copper-cobalt/SA-MWCNTs catalyst, the MWCNTs with a diameter of about 30 nm were used. Figure 1a shows the transmission electron microscope (TEM) image of purified MWCNTs, which exhibited a smooth surface. In contrast, the catalyst copper-cobalt/SA-MWCNT exhibited nanosize particles on its surface (Figure 1b) due to the deposition of copper-cobalt particles. Figure 1b shows that the nanoparticles (copper-cobalt oxides) were well distributed on the surface of SA-MWCNT. The aromatic ring of salicylic acid can have a strong interaction with the wall of MWCNTs. Thus, salicylic acid can be used to functionalize MWCNTs though adsorption. The SA functionalized MWCNTs (SA-MWCNTs) possess functional hydroxyl and carboxyl groups. These groups can interact strongly with copper and cobalt. This facilitates the deposition of copper and cobalt on SA-MWCNTs.

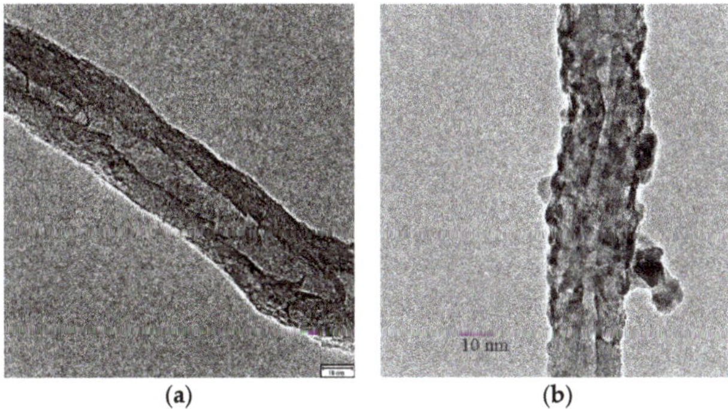

(a) (b)

Figure 1. Transmission electron microscope (TEM) images for purified multiwalled carbon nanotubes (MWCNT) (**a**) and the catalyst copper-cobalt/SA-MWCNT (**b**).

The X-ray diffraction (XRD) pattern of the catalyst is shown in Figure 2. Supplementary Materials Figure S1 shows the XRD patterns for SA-MWCNTs and purified MWCNTs. In Figure 2, peak 1 is for MWCNTs, and the peaks indicated by 2, 3, and 4 correspond to crystalline structures of the oxidized metals. Using the JCPDS chemical spectra data bank [19], the peaks were recognized representing CuO, $Cu_xCo_{3-x}O4$, and Cu_2O, respectively. The XRD pattern indicates the interaction between cobalt and copper species in the catalyst. This is possibly ascribed to that copper and cobalt ions can interact with salicylic acid of SA-MWCNTs, and further interacted with each other.

The spectra of X-ray photoelectron spectroscopy (XPS) for the catalyst are presented in Figure 3a,b. Supplementary Materials Figure S2 shows the XPS spectra for SA-MWCNTs and purified MWCNTs. The parent peak in Figure 3 was deconvoluted into three peaks. Figure 3a shows the XPS spectra for the copper states. The two peaks at 933.3 and 934.6 eV are ascribed to Cu(I) and Cu(II) oxides [20], respectively. The peak at 943.2 eV is ascribed to the satellite peak for the Cu(II) oxide. Figure 3a demonstrates two oxidation states of Cu(I) and Cu(II) for copper in the catalyst. Figure 3b shows the spectra for different cobalt states. The two peaks at 780 and 781.5 eV are ascribed to Co(III) oxide and Co(II) oxide [21], respectively, and the peak at 787.4 eV is due to the satellite peak for the Co(II) oxide. The results demonstrate the two oxidation states of Co(II) and Co(III) for cobalt in the catalyst. Figure 3 further confirms the interaction of copper and cobalt after bring supported on SA-MWCNTs.

Figure 4 shows the temperature programmed reduction (TPR) profile for copper-cobalt/SA-MWCNTs. Supplementary Materials Figure S3 shows the TPR profile for SA-MWCNTs. In Figure 4, there are two prominent peaks at 278 and 363 °C. They are acribed to the reduction of Cu^{2+} to Cu. The right shoulder (450–636 °C) is ascribed to the reduction of Co^{2+} to Co.

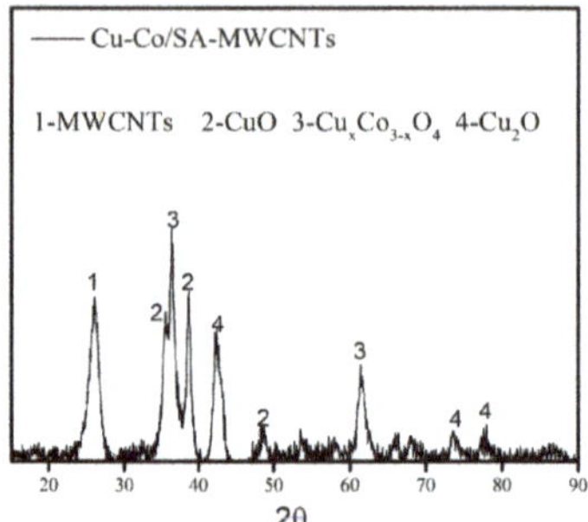

Figure 2. X-ray diffraction (XRD) pattern for the catalyst copper-cobalt/SA-MWCNTs.

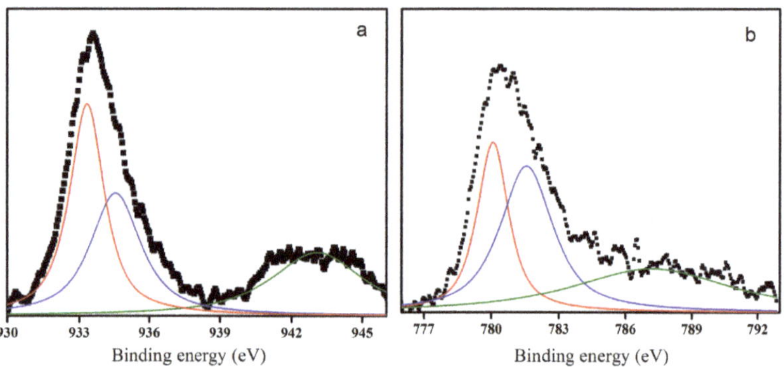

Figure 3. (**a**) XPS spectra of Cu 2p region of the catalyst; (**b**) XPS spectra of Co 2p region of the catalyst.

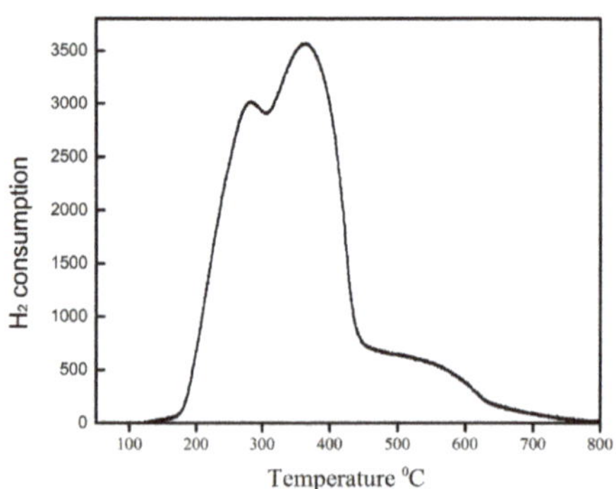

Figure 4. Temperature programmed reduction (TRP) profile for copper-cobalt/SA-MWCNTs.

2.2. Alcohol Synthesis from Syngas

Experimental results showed that both the purified MWCNTs and SA-MWCNTs cannot catalyze the reaction of synthesis of alcohols from syngas. That is no alcohols were produced when using the carbon nanotubes as catalysts. Prior to evaluating the catalyst copper-cobalt/SA-MWCNTs and studying the efffect of salicylic acid on the production of alcohols, the effect of purification of MWCNTs and diameter of MWCNTs were first investigated. MWCNTs were purified by refluxing in HNO_3. Copper-cobalt supported on the purified MWCNTs has exhibited a better conversion of syngas and selectivity of ethanol and butanol than the copper-cobalt supported on the non-purified MWCNTs (Figure 5). The result showed that for preparing copper-cobalt based catalyst, MWCNTs should be purified. The residues on non-purified carbon nanotubes are amorphous carbon [22]. Amorphous carbon on the surface of MWCNTs not only affects the interaction of SA with the wall of MWCNTs, but also affects the deposition of copper and cobalt. This is the reason that using non-purified MWCNTs has exhibited a lower alcohol production than using purified MWCNTs. The effect of the diameter of MWVNTs on the synthesis of alcohols was also investigated (Figure 6). Using the MWCNTs with a diameter of 30 nm, the conversion of syngas (Figure 6a) and selectivity for the alcohols of ethanol + butanol (Figure 6b) are larger than the MWCNTs with diameters of 15 nm and 50 nm, indicating that the MWCNTs of 30 nm in diameter are better as supports for preparing the copper-cobalt based catalyst. The diameter of carbon nanotubes can roughly reflect the surface area of carbon nanotubes. Possibly, the carbon nanotubes with a diameter of about 30 nm can provide an appropriate surface area. When investigating the effect of salicylic acid on the synthesis of alcohols, salicylic acid was added to the reaction solutions according to predetermined weight ratios. Figure 7 shows the effect of salicylic acid on the production of alcohols. With increasing the ratio of salicylic acid to MWCNTs, the conversion of syngas and the selectivity toward to ethanol and butanol are increased. The optimal ratio is 0.3, at which the syngas conversion and the selectivity toward ethanol and butanol have reached highest values.

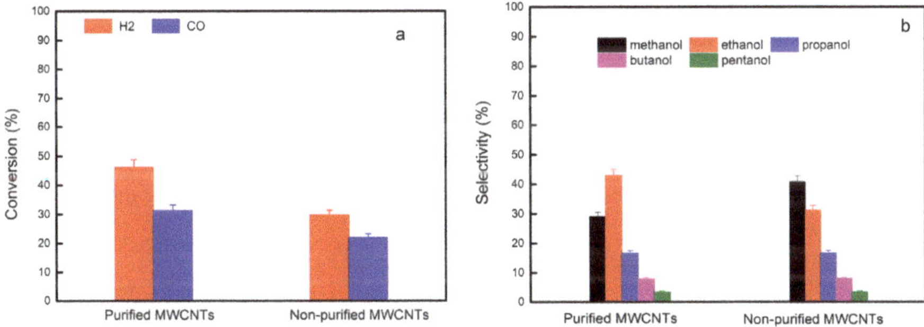

Figure 5. Effect of purification of MWCNTs on the synthesis of alcohols. Conversion of syngas (**a**) and Selectivity of alcohols (**b**).

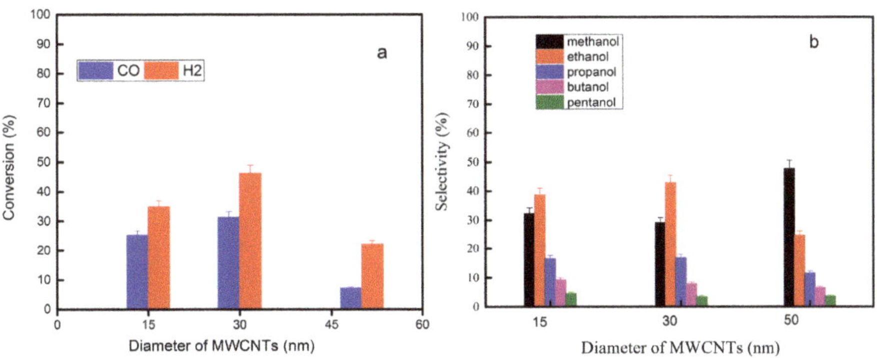

Figure 6. Effect of the diameter of MWCNTs on the syngas conversion (**a**) and alcohol selectivity (**b**).

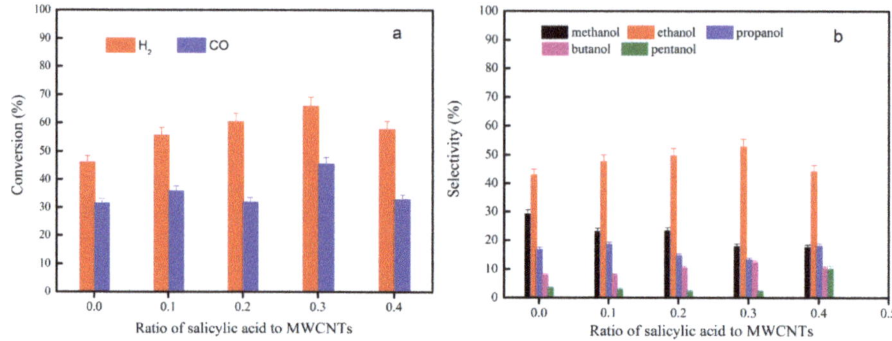

Figure 7. Effect of salicylic acid on the syngas conversion (**a**) and alcohol selectivity (**b**).

3. Experimental Section

3.1. Materials

Multiwalled carbon nanotubes (MWCNTs, diameters 15 ± 5 nm, 30 ± 10 nm, 50 ± 10 nm) were obtained from Nacen Nanotechnologies Inc. (Shenzhen, China). Salicylic acid (SA), ethanol, propanol, butanol, pentanol, $Cu(NO_3)_2$ $3H_2O$, and Co-$(NO_3)_2$ $6H_2O$ were obtained from Sinopharm Chemical Reagent Co. (Shanghai, China). All chemicals were of analytical reagent grade.

3.2. Catalyst Preparation

MWCNTs were refluxed in HNO_3 (2.5 M) at 70 °C for 10 h. Then, the MWCNTs suspensions were filtered by a 450 nm polycarbonate membrane. The purified MWCNTs were then mixed with the salicylic acid solutions with different concentrations. The mixtures were sonicated for 20 min. Then, salicylic acid functionalized MWCNTs (SA-MWCNTs) were filtered through a 450 nm polycarbonate membrane and rinsed with deionized water. The collected SA-MWCNTs were dried at 70 °C under vacuum.

The catalyst copper-cobalt/SA-MWCNTs was prepared as follows. 800 mg of SA-MWCNTs were dispersed in 200 mL of deionized water by sonication for 5 min. Then, the solution (115 mL) containing $Cu(NO_3)_2$ $3H_2O$ (10.2 mg/mL) and $Co(NO_3)_2$ $6H_2O$ (12.3 mg/mL), was added and sonicated for 3 min. After 50 min incubation, the mixture was dried at 50 °C under vacuum. The dried catalyst was calcined at 450 °C for 3 h by introducing nitrogen at a flow rate of 160 mL/min. The amounts of copper and cobalt deposited on MWCNTs-SA were determined using atomic absorption spectrometer (Model GGX-6). The concentrations of the metal ions in the stock solutions and in the residue solutions were measured. The difference between the concentrations was used to calculate the amount of copper and cobalt deposited on the support. The copper loading was finally determined to be 0.263 mg Cu/mg MWCNTs and that of cobalt was 0.271 mg Co/mg MWCNTs.

3.3. Catalyst Characterization

SA-MWCNTs and copper-cobalt/SA-MWCNTs were imaged by transmission electron microscopy (TEM) using a Hitachi H-800 system (Shanghai, China). Samples were first dispersed in ethanol by ultrasonication. Then the suspensions were dropped onto a carbon-coated copper grid. The X-ray diffraction (XRD) patterns of the catalysts were obtained with a diffractometer (Rigaku D/Max 2500 VBZ+/PC, Gu target at 35 kV, 30 mA) (Shanghai, China). The diffractograms were obtained by scanning at a rate of $2\theta = 1°/min$, the range was from $2\theta = 5°$ to $2\theta = 90°$. TPD/R/O 1100 (Shanghai, China) was used to record temperature programmed reduction (TPR) spectra for the catalysts, using a gas mixture of 10% hydrogen in argon. The flow rate was controlled at 30 mL/min. The samples were heated up to 800 °C at a heating rate of 5 °C/min. A thermal conductivity detector was used to monitor the sample pretreatment, adsorption, and desorption. A Thermo VG ESCALAB250 X-ray photoelectron spectrometer (Beijing, China) was used to obtain X-ray photoelectron spectroscopy (XPS) spectra for the catalysts. The measurements were carried out at a pressure of 2×10^{-9} Pa using Mg Kα X-ray as the excitation source.

3.4. Alcohol Synthesis from Syngas

The reactions were carried out in a fixed bed microreactor with a length of 500 mm and 9 mm in diameter. A temperature controller was used to control the temperature of the microreactor. Introducing the gases H_2, CO, and N_2 at certain rates to the reactor was controlled by mass flow controllers. In all of the reactions, nitrogen was used as an internal standard gas. The copper-cobalt/SA-MWCNTs catalyst (1.5 g) was placed in the reactor bed. The reactor was heated up to 300 °C at a rate of 3 °C/min. The catalyst copper-cobalt/SA-MWCNTs was reduced in situ at 300 °C for 15 h, and the flow rate of H_2 was controlled at 50 mL/min. Then, synthesis gas (H_2/CO ratio of 1.0) with a flow rate of 92 mL/min was introduced. The reactor pressure was controlled at 5 MPa. The composition of effluent gas stream was determined by an online GC-2014C Shimadzu Gas Chromatograph (Beijing, China), which was equipped with a TCD detector and a Porapak Q column. The produced alcohols were analyzed off-line using a FID detector and a PEG-20 M capillary column.

4. Conclusions

Salicylic acid has been used to functionalize multiwalled carbon nanotubes. SA-MWCNTs were used as supports for preparing copper-cobalt based catalyst. The catalyst copper-cobalt/SA-MWCNTs was used to catalyze the synthesis of alcohols from synthesis gas. Salicylic acid can promote the synthesis of ethanol and butanol from synthesis gas, reducing the synthesis of methanol. For preparing the copper-cobalt based catalyst, purified MWCNTs are better than non-purified MWCNTs, and the diameter of MWCNTs also has an effect on the alcohol production. The copper-cobalt catalyst supported on MWCNTs of 30 nm in diameter can synthesize more ethanol and butanol than supported on MWCNTs of 15 and 50 nm in diameter, indicating that the diameter of MWCNTs also has an effect on the synthesis of alcohols. This work demonstrated that salicylic acid not only can be used to functionalize carbon nanotubes, but also can enhance the production of ethanol and butanol from synthesis gas.

Supplementary Materials: The following are available online at www.mdpi.com/2073-4344/7/10/295/s1. Figure S1: XRD pattern for purified MWCNTs and SA-MWCNTs, Figure S2: XPS spectra for purified MWCNTs and SA-MWCNTs; Figure S3: TRP profile for SA-MWCNTs.

Acknowledgments: This work was supported by the National Science Foundation of China (21476023).

Author Contributions: Peijun Ji provided the idea and design for the study; Jinxin Zou and Lei Wang performed the experiments; Jinxin Zou drafted the manuscript; Peijun Ji revised it.

Conflicts of Interest: The authors declare no conflict of interest.

References

1. Xiong, H.; Jewell, L.L.; Coville, N.J. Shaped Carbons As Supports for the Catalytic Conversion of Syngas to Clean Fuels. *ACS Catal.* **2015**, *5*, 2640–2658. [CrossRef]
2. Yue, Y.; Ma, X.; Gong, J. An Alternative Synthetic Approach for Efficient Catalytic Conversion of Syngas to Ethanol. *Acc. Chem. Res.* **2014**, *47*, 1483–1492. [CrossRef] [PubMed]
3. Kim, T.W.; Kleitz, F.; Jun, J.W.; Chae, H.J.; Kim, C.U. Catalytic conversion of syngas to higher alcohols over mesoporous perovskite catalysts. *J. Ind. Eng. Chem.* **2017**, *51*, 196–205. [CrossRef]
4. Wang, J.; Liu, Z.; Zhang, R.; Wang, B. Ethanol Synthesis from Syngas on the Stepped Rh(211) Surface: Effect of Surface Structure and Composition. *J. Phys. Chem. C* **2014**, *118*, 22691–22701. [CrossRef]

5. Surisetty, V.R.; Dalai, A.K.; Kozinski, J. Effect of Rh Promoter on MWCNT Supported Alkali Modified MoS_2 Catalysts for Higher Alcohols Synthesis from CO Hydrogenation. *Appl. Catal. A* **2010**, *381*, 282–288. [CrossRef]

6. Lopez, L.; Velasco, J.; Montes, V.; Marinas, A.; Cabrera, S.; Boutonnet, M.; Järås, S. Synthesis of Ethanol from Syngas over Rh/MCM-41 Catalyst: Effect of Water on Product Selectivity. *Catalysts* **2015**, *5*, 1737–1755. [CrossRef]

7. Subramani, V.; Gangwal, S.K. A Review of Recent Literature to Search for an Efficient Catalytic Process for the Conversion of Syngas to Ethanol. *Energy Fuels* **2008**, *22*, 814–839. [CrossRef]

8. Zhao, L.; Li, W.; Zhou, J.; Mu, X.; Fang, K. One-step synthesis of Cu Co alloy/$Mn_2O_3Al_2O_3$ composites and their application in higher alcoholsynthesis from syngas. *Int. J. Hydrogen Energy* **2017**, *42*, 17414–17424. [CrossRef]

9. Gupta, M.; Smith, M.L.; Spivey, J.J. Heterogeneous Catalytic Conversion of Dry Syngas to Ethanol and Higher Alcohols on Cu-Based Catalysts. *ACS Catal.* **2011**, *1*, 641–656. [CrossRef]

10. Zuo, Z.J.; Wang, L.; Yu, L.M.; Han, P.D.; Huang, W. Experimental and Theoretical Studies of Ethanol Synthesis from Syngas over CuZnAl Catalysts without Other Promoters. *J. Phys. Chem. C* **2014**, *118*, 12890–12898. [CrossRef]

11. Yang, Q.; Cao, A.; Kang, N.; Ning, H.; Wang, J.; Liu, Z.T.; Liu, Y. Bimetallic Nano Cu–Co Based Catalyst for Direct Ethanol Synthesis from Syngas and Its Structure Variation with Reaction Time in Slurry Reactor. *Ind. Eng. Chem. Res.* **2017**, *56*, 2889–2898. [CrossRef]

12. Morrill, M.R.; Thao, N.T.; Shou, H.; Davis, R.J.; Barton, D.G.; Ferrari, D.; Agrawal, P.K.; Jones, K.W. Origins of Unusual Alcohol Selectivities over Mixed MgAl Oxide-Supported K/MoS_2 Catalysts for Higher Alcohol Synthesis from Syngas. *ACS Catal.* **2013**, *3*, 1665–1675. [CrossRef]

13. Hensley, J.E.; Lovestead, T.M.; Christensen, E.; Dutta, A.; Bruno, T.J.; McCormick, R. Compositional Analysis and Advanced Distillation Curve for Mixed Alcohols Produced via Syngas on a K-CoMoSx Catalyst. *Energy Fuels* **2013**, *27*, 3246–3260. [CrossRef]

14. Surisetty, V.R.; Dalai, A.K.; Kozinski, J. Synthesis of higher alcohols from synthesis gas over Co-promoted alkali-modified MoS_2 catalysts supported on MWCNTs. *Appl. Catal. A Gen.* **2010**, *385*, 153–162. [CrossRef]

15. Dong, X.; Liang, X.L.; Li, H.Y.; Lin, G.D.; Zhang, P.; Zhang, H.B. Preparation and characterization of carbon nanotube-promoted Co–Cu catalyst for higher alcohol synthesis from syngas. *Catal. Today* **2009**, *147*, 158–165. [CrossRef]

16. Sims, A.; Jeffers, M.; Talapatra, S.; Mondal, K.; Pokhrel, S.; Liang, L.; Zhang, X.; Elias, A.L.; Sumpter, B.G.; Meunier, V.; et al. Hydro-deoxygenation of CO on functionalized carbon nanotubes for liquid fuels production. *Carbon* **2017**, *121*, 274–284. [CrossRef]

17. Pomalaza, G.; Capron, M.; Ordomsky, V.; Dumeignil, F. Recent Breakthroughs in the Conversion of Ethanol to Butadiene. *Catalysts* **2016**, *6*, 203. [CrossRef]

18. Kim, M.; Park, J.; Kannapu, H.P.R.; Suh, Y.W. Cross-Aldol Condensation of Acetone and n-Butanol into Aliphatic Ketones over Supported Cu Catalysts on Ceria-Zirconia. *Catalysts* **2017**, *7*, 249. [CrossRef]

19. Eswaramoorthi, I.; Sundaramurthy, V.; Das, N.; Dalai, A.K.; Adjaye, J. Application of Multi-Walled Carbon Nanotubes as Efficient Support to Nimo Hydrotreating Catalyst. *Appl. Catal. A* **2008**, *339*, 187–195. [CrossRef]

20. Contarin, S.; Kevan, L. X-ray Photoelectron Spectroscopic Study of Copper-Exchanged X- and Y-Type Sodium Zeolites: Resolution of Two Cupric Ion Components and Dependence on Dehydration and X-Irradiation. *J. Phys. Chem.* **1986**, *90*, 1630–1632. [CrossRef]

21. Ernst, B.; Bensaddik, A.; Hilaire, L.; Chaumette, P.; Kiennemann, A. Study on a Cobalt Silica Catalyst during Reduction and Fischer-Tropsch Reaction: In Situ EXAFS Compared to XPS and XRD. *Catal. Today* **1998**, *39*, 329–341. [CrossRef]

22. Rinaldi, A.; Frank, B.; Su, D.S.; Hamid, S.B.A.; Schlögl, R. Facile Removal of Amorphous Carbon from Carbon Nanotubes by Sonication. *Chem. Mater.* **2011**, *23*, 926–928. [CrossRef]

 *catalysts*

Article

Conductive Cotton Filters for Affordable and Efficient Water Purification

Fang Li [1,2], Qin Xia [1], Qianxun Cheng [1], Mingzhi Huang [3] and Yanbiao Liu [1,2,*]

[1] School of Environmental Science and Engineering, Donghua University, Shanghai 201620, China;
    lifang@mail.dhu.edu.cn (F.L.); rachel.q.xia@gmail.com (Q.X.); chengqianxun@dhu.edu.cn (Q.C.)

[2] Textile Pollution Controlling Engineering Centre of Ministry of Environmental Protection,
    Shanghai 201620, China

[3] Department of Water Resources and Environment, Guangdong Provincial Key Laboratory of Urbanization
    and Geo-Simulation, Sun Yat-sen University, Guangzhou 510275, China; huangmzh6@mail.sysu.edu.cn

* Correspondence: yanbiaoliu@dhu.edu.cn; Tel.: +86-021-67792545

Received: 12 September 2017; Accepted: 26 September 2017; Published: 29 September 2017

Abstract: It is highly desirable to develop affordable, energy-saving, and highly-effective technologies to alleviate the current water crisis. In this work, we reported a low-cost electrochemical filtration device composing of a conductive cotton filter anode and a Ti foil cathode. The device was operated by gravity feed. The conductive cotton filter anodes were fabricated by a facile dying method to incorporate carbon nanotubes (CNTs) as fillers. The CNTs could serve as adsorbents for pollutants adsorption, as electrocatalysts for pollutants electrooxidation, and as conductive additives to render the cotton filters highly conductive. Cellulose-based cotton could serve as low-cost support to 'host' these CNTs. Upon application of external potential, the developed filtration device could not only achieve physically adsorption of organic compounds, but also chemically oxide these compounds on site. Three model organic compounds were employed to evaluate the oxidative capability of the device, i.e., ferrocyanide (a model single-electron-transfer electron donor), methyl orange (MO, a common recalcitrant azo-dye found in aqueous environments), and antibiotic tetracycline (TC, a common antibiotic released from the wastewater treatment plants). The devices exhibited a maximum electrooxidation flux of 0.37 mol/h/m^2 for 5.0 mmol/L ferrocyanide, of 0.26 mol/h/m^2 for 0.06 mmol/L MO, and of 0.9 mol/h/m^2 for 0.2 mmol/L TC under given experimental conditions. The effects of several key operational parameters (e.g., total cell potential, CNT amount, and compound concentration) on the device performance were also studied. This study could shed some light on the good design of effective and affordable water purification devices for point-of-use applications.

Keywords: conductive cotton filter; carbon nanotubes; low-cost; water purification; gravity feed

---

1. Introduction

One of the grand challenges of 21st century is to provide affordable and clean water to meet human needs. However, rapid industrial development, fast population growth, and global climate change have caused serious water pollution we are currently facing [1]. During the past few decades, numerous efforts have been devoted to developing feasible and sustainable technologies to alleviate the water crisis. To be mentioned are technologies such as membrane separations (e.g., reverse osmosis, RO) [2,3] and advanced oxidation processes (AOPs, e.g., UV-ozone and photocatalysis) [4–8]. Many of them only have limited success. For example, the efficacy of both RO and AOPs could be negatively affected by natural organic matters (NOM) due to membrane fouling and oxidant scavenging [9,10]. High energy consumption and high cost also restrict the wide application of other promising technologies such as electrodialysis. Therefore, it is highly desirable to develop efficient, affordable, scalable, and energy-saving water treatment devices and technologies.

Electrochemical oxidation has proven to be effective for the decomposition of organic pollutants and the key to this technology is a high-performance electrocatalyst [11–13]. In recent years, carbon-based materials have been widely applied as electrode materials for electrochemical processes. Among these newly developed electrode materials, carbon nanotubes (CNTs) are especially attractive due to their large specific surface area (50–500 m^2/g), excellent electrical conductivity (10^4–10^6 S/m), and desirable chemical resistivity and stability [14–17]. A highly porous CNT membrane or filter can be easily fabricated via vacuum filtration. These filters can be used for the removal of pollutants by physical adsorption [18]. A electrochemical CNT filter could bring this concept one step further by not only physically adsorb the pollutants but also electrochemically oxide the pollutants in situ [19]. However, most of the reported CNT filters are generally <10 cm in diameter and further upscaling of these CNT filters is challenging [20]. Another approach to utilize CNTs is to construct 3D macrostructures. For example, a CNT sponge synthesized by CVD method exhibits excellent rejection performance for several organic solvents and oils with different densities (e.g., ethanol, hexane, ethylene glycol, gasoline, and pump oil) [21]. The harsh synthesis conditions have, however, limited significantly the wide application of this promising design [22]. Also, the CNTs can be used as conductive fillers to boost the electrical conductivity of a designed composite. For example, Schoen et al. previously developed a composite filter material based on three one-dimensional (1D) materials, i.e., silver nanowires, CNTs, and cellulose cotton fibers. The fabricated composite filter could inactivate >98% of bacteria within only several seconds via electroporation [23]. To obtain a high-performance CNTs-based composite, an ideal support material is also of significance [24]. Among various support materials reported so far, the cotton-based support has attracted extensive interest from the community due to their desirable characteristics like mechanically and chemically robust, highly porous, readily available, and cheap [25]. Some reported cotton/CNTs composite materials have been used as electrode materials for supercapacitor [26] and sensor applications [27]. To the best of our knowledge, there are very limited reports on the simultaneous adsorption and electro-oxidation of organic pollutants using 3D conductive cotton filters in a continuous flow filtration design.

In this study, an efficient electrochemical water purification technology based on conductive cotton filters was developed. The composite filter can be fabricated via a facile and scalable dying method. In particular, the CNTs served as adsorbents and electrocatalysts for pollutant adsorption and electrooxidation. The device was operated by gravity feed (Figure 1a), so that the operation cost could be further decreased. The performance of the device was evaluated using three selected model compounds, e.g., ferrocyanide (a model single-electron-transfer electron donor), methyl orange (MO, a common recalcitrant azo-dye found in aqueous environments), and antibiotic tetracycline (TC, a common antibiotic released from the wastewater treatment plants). The effects of key operational parameters on the device performance were systematically studied. The details of this investigation presented below.

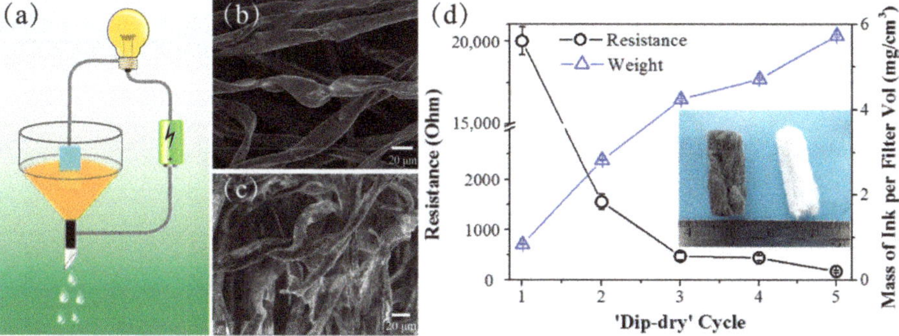

Figure 1. (**a**) Schematic illustration of the electrochemical cotton filtration device; FESEM images of (**b**) a pristine cotton and (**c**) a CNT-coated cotton filter; (**d**) change of filter resistance and weight of cotton filter as a function of 'dip–dry' cycles. The inset in (**d**) is digital pictures of a pristine cotton filter (white, right) and a CNT-coated cotton filter (black, left).

2. Results and Discussion

2.1. Fabrication of Conductive Cotton Filters

The conductive cotton filters were fabricated by a facile dying method. The CNT, SDBS, and cotton are key components for this design. The CNT could serve as high-performance sorbent and electrocatalyst for pollutants adsorption and electrooxidation. It could also facilitate the cotton as a conductive electrode material. Firstly, three different surfactants—polyvinyl pyrrolidone (PVP), sodium dodecyl sulphate (SDS), and SDBS—were employed to disperse CNTs in DI-H_2O. While only 10 mg mL^{-1} of SDBS could successfully disperse CNTs and a homogeneous solution can be obtained after 15 min probe sonication treatment. Visible agglomerates can be witnessed when using PVP and SDS of the same concentration. This finding suggests that SDBS is an ideal surfactant to facilitate the dispersion of CNT in aqueous solution via π–π interaction between the SDBS benzene rings and the aromatic structure of CNTs [28]. Thus, SDBS could provide good protection for CNTs to avoid agglomeration. Moreover, the SDBS could serve as a 'bridge' to connect CNTs with cotton, by bonding CNT with benzene ring moieties and bonding with the hydroxyl-groups of cotton cellulose fibers via van der Waals forces and/or hydrogen. The cotton could serve as macro-porous and low-cost support materials to 'host' these CNTs [29,30]. The CNTs might interwine each other to further enhance the stability of the as-fabricated 3D cotton filters. In a typical fabrication process, 100 mg cotton sample was immersed into a freshly-prepared CNT ink solution (composed of 1.5 mg/mL CNTs and 10 mg/mL SDBS), followed by a drying process at 120 °C for 30 min to remove water residue. The inset in Figure 1d compares the digital pictures of a pristine cotton (white, right) and a CNT-coated cotton (black, left). The black color provides supportive evidence for the successful CNT loading. Furthermore, field-emission scanning electron microscopy (FESEM) technique was employed to provide detailed morphological information. As displayed in Figure 1b,c, the pristine cotton sample showed a twisted and smooth fiber-like structure with an average fiber width of ~50 μm and an average pore size of ~110 μm. However, the surface became much rougher after CNT loading. A magnified image showed that the CNTs were distributed uniformly onto the cellulose fiber surface (Figure S1, Supplementary Materials). This data provides supportive evidence for the successful loading of CNT onto the cotton. Also, by varying the 'dip–dry' cycles and CNT content in the ink, the loading amounts of CNTs onto the cotton as well as the electrical conductivity of the as-fabricated filters can be controlled (Figure S2, Supplementary Materials). For example, as shown in Figure 1d, the accumulated ink mass adsorbed per volume of the cotton increased from 0.85 mg/cm^3 (one 'dip–dry' cycle, by dividing the filter weight gain with the filter volume) to 5.74 mg/cm^3 (five 'dip–dry' cycles) when using a 1.5 mg/mL

CNT ink. This indicates a quantitative sorption of conductive nanotubes onto the cotton surface. Meanwhile, as expected, the sheet resistance of the cotton filter samples decreased significantly from $20,000 \pm 850\ \Omega$ to $170 \pm 30\ \Omega$, mainly due to the continuous increase of highly conductive CNTs within the cotton filters. As a simple demonstration of the electrical conductivity, an LED lamp connected to a direct current (DC) power supply (4.0 V applied voltage) can be easily powered through the as-fabricated conductive cotton filters (Figure S3, Supplementary Materials). Given that the CNT loading amount became rather limited over five 'dip–dry' cycles, hence, this number was chosen for all subsequent experiments.

2.2. Electron Transfer

The electrochemical performance of a conductive cotton filter was firstly evaluated using ferrocyanide $(F_e(CN)_6^{4-})$ as model electron donor. The unique characteristics of single-electron transfer and negligible adsorption of ferrocyanide make it suitable candidate for electron transfer experiment [31,32]. The oxidation of ferrocyanide can be described by Equation (1).

$$Fe(CN)_6^{3-} + e^- \rightarrow Fe(CN)_6^{4-},\ E_0 = 0.139\ V\ vs.\ Ag/AgCl \tag{1}$$

Figure 2 shows the change of electrooxidation flux of ferrocyanide as a function of applied anode potential and ferrocyanide concentrations. At an anode potential range of 0.15–0.4 V (vs. Ag/AgCl), a linear relationship between electrooxidation flux of ferrocyanide and applied anode potential can be observed for all influent concentrations. Voltage-independent plateaus were achieved for all cases when anode potential above 0.4 V (vs. Ag/AgCl). This finding indicates the mass transport limitations. At an anode potential of 0.4 V, the electrooxidation rate of ferrocyanide was 0.037, 0.106, 0.145, and 0.373 mol/h/m^2 for 0.2, 0.5, 1.0, and 5.0 mmol/L ferrocyanide, respectively. The electrooxidation flux increased up to 10-fold by increasing the influent concentration and interval electrode convection. This value was lower than a reported graphene-CNT composite filter (e.g., 15-fold) [32]. The reason may be due to a much thinner thickness of the graphene-CNT filter (10 μm vs. 2.5 cm) and, hence, an increased transport resistance of ferrocyanide ions in the current design. As only one electron transfer was involved to oxide ferrocyanide to ferricyanide, the maximum electron transfers at an applied anode potential of 0.4 V (vs. Ag/AgCl) were calculated to be 1×10^{14}, 3×10^{14}, 1×10^{15} e/s for 0.2, 0.5, 1.0, and 5.0 mmol/L ferrocyanide, respectively.

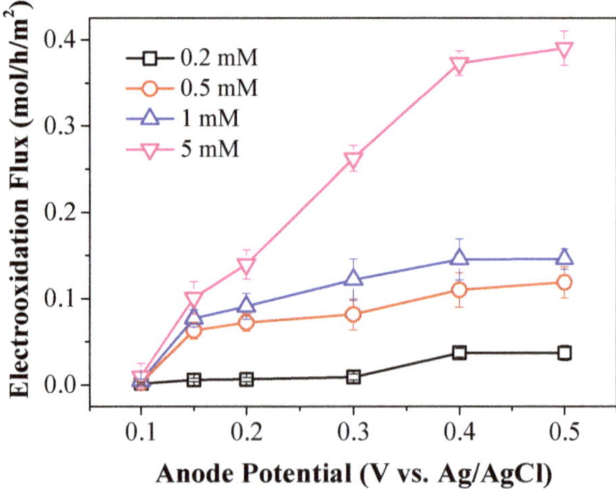

Figure 2. Concentration-dependent oxidation of ferrocyanide using an electrochemical cotton filter. Experimental conditions: $[CNT]_{ink}$ = 1.5 mg/mL, $[Na_2SO_4]$ = 10 mmol/L, and flow rate = 1.5 mL/min.

2.3. Performance of the Cotton Filter towards MO Removal

To further evaluate the feasibility of the as-fabricated conductive cotton filters for organic pollutants degradation, methyl orange (MO) was selected as a typical recalcitrant organic compound in aqueous environment. Firstly, the MO sorption process on a conductive cotton filter was examined by breakthrough curve analysis in the absence of electrochemistry. As displayed in Figure S4 (Supplementary Materials), the sorption behavior of cotton filters fabricated with different CNT concentrations varies significantly. The effluent MO concentration flow through the cotton filters fabricated by a 0.5 mg/mL and a 1.0 mg/mL CNT ink solution increased sharply at the initial 5–10 min and then maintained a stable concentration. For the cotton filter fabricated by a 1.5 mg/mL CNT ink, the effluent increased at a gentle slope in the first 18 min and then showed a steeper increase before breakthrough occurs. The cotton filter MO sorption capacity was 53.5 mg/g, 1.8 times higher than a pure CNT filter reported previously [33]. This increased sorption capacity observed in this study are likely due to the contribution of MO sorption by the cotton support. However, the limited area of the cotton filter results in an absolute sorption capacity that is relatively low and in turn MO breakthrough occurs within 1 h. Thus, further experiments were conducted to electrochemically degrade the adsorbed MO molecules and to regenerate adsorption sites.

The electrooxidative filtration of MO was evaluated as a function of total cell potential and CNT loading onto the cotton filters as displayed in Figure 3a. The MO oxidation flux increased with increasing total cell potential to a maximum of 0.26 ± 0.02 mol/h/m^2 at 3.5 V for the cotton filter fabricated using a 1.5 mg/mL CNT ink. Further increase of the total cell potential to 4.0 V did not improve the device performance due to other side reactions (e.g., water oxidation) occurred at this condition, resulting in the loss of electrochemical activity. Moreover, increased electrolytic gas bubble formation at a higher potential may block some active sites on the filter surface, or even degrade the filter integrity to some extent. It is of note that the optimal total cell potential of 3.5 V for MO electrooxidation was higher than that of a CNT electrochemical filter (e.g., 2.5 V). This difference may be due to an increased filter resistance (~175 Ω vs. ~50 Ω) and a reduced CNT content (~7 mg vs. 30 mg) of the cotton filters reported in this work. The increased resistance of the as-fabricated cotton filters may increase the resistance for the transport of electrons, so that an increased overpotential may be needed to overcome this barrier. Moreover, the CNT content in the ink is another important factor for the electro-oxidative process. As expected, the MO electrooxidation flux was increased with CNT amount for all cases. For example, at a given total cell potential of 3.5 V, the MO electrooxidation flux was 0.15, 0.23, and 0.26 mol/h/m^2 for CNT ink concentration of 0.5 mg/mL, 1.0 mg/mL, and 1.5 mg/mL, respectively. Since more CNT loading will lead to more active sites for the sorption and electrooxidation of MO molecules and increase the electrical conductivity of the as-fabricated filters (or decreased electron transport resistance). The 0.26 mol/h/m^2 electrooxidation flux of MO in a single pass through the cotton filter is of note since the device was running by gravity feed. These data suggest the potential of a cost-effective, energy-saving, and facile method for the efficient removal of organic compounds from water.

Open circuit potential measurements over a range of cell potentials for the device was conducted with a cotton filter anode, a Ti cathode, $[MO]_{in}$ = 0.06 mmol/L and $[Na_2SO_4]$ = 10 mmol/L. As shown in Figure 3b, a total cell potential of 3.0 V was required to achieve a high enough anode potential (>+0.8 V vs. Ag/AgCl) for MO oxidation as determined by the cyclic voltammogram measurement [33]. At a total cell potential of 4 V, the anode potential was as high as 1.5 V vs. Ag/AgCl which has exceeded the water oxidation potential of 1.23 V (vs. standard hydrogen electrode, SHE). This finding supports the change of MO electrooxidation flux with total cell potential as shown in Figure 3a.

The re-usability of the cotton filters is of great significance towards practical applications. Hence, additional experiments to evaluate the regeneration performance of the cotton filters were conducted. As shown in Figure 3c, the cotton filter exhibits an initial MO oxidation flux of 0.27 mol/h/m^2 and a MO removal efficiency of >98% in the first 30 min, which then slightly decreased in the following few hours. The decrement in MO oxidation can be due to the MO oxidation byproducts and/or precipitates accumulated onto the filter surface. This hypothesis was further confirmed by the SEM characterization results of the cotton filter after running. As shown in the inset, certain polydispersed nanoparticles and/or precipitates were observed on the cotton surface that might eventually foul the cotton filter. These materials could be organic polymers and/or inorganic sodium persulfate [34]. The build-up of polymer/precipitates may cause some adverse effects on the oxidative performance of the filter by blocking the active centers of the filter, significantly increasing resistance to water and electron transfer, thus reducing the reaction kinetics and electrochemistry. To eliminate the contribution of MO degradation by the possibly produced persulfate precipitates, another control experiment was conducted by mixing 0.1–5 mmol/L persulfate with 0.06 mmol/L MO. The results show that the MO concentration changed negligible (<0.4%) for all cases. This finding supports the conclusion that the MO removal was mainly due to the electrooxidation by the filter. A 100 mL of 1 mol/L HCl: ethanol mixture (50:50 vol %) washing of the cotton filter was found to be effective to restore the initial electrooxidation flux. An average electrooxidation flux of 0.2 mol/h/m^2 was achieved for the next two running cycles. Also, the lack of breakthrough during the 8 h continuous running at 3.5 V total cell potential suggests that the primary removal mechanism during electrochemical filtration is oxidative degradation rather than physical adsorption. These data suggest the potential of a cost-effective, energy-saving, and facile method for the efficient removal of organic compounds from water.

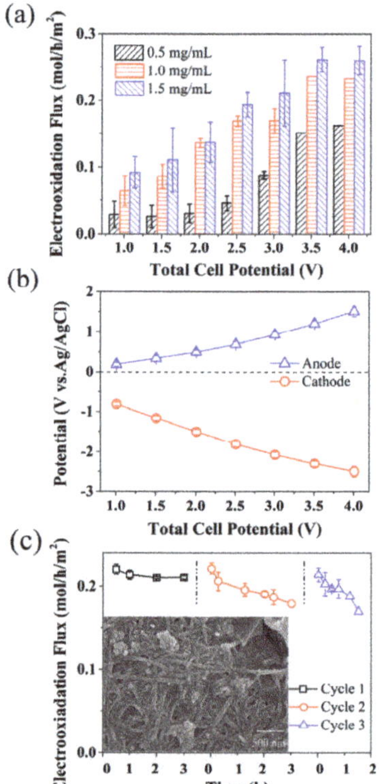

Figure 3. MO electrooxidation flux (**a**) and open-circuit potential (**b**) as a functional total cell potential. Experimental conditions: [MO]$_{in}$ = 0.06 mmol/L, [Na$_2$SO$_4$] = 10 mmol/L, [CNT]$_{ink}$ = 1.5 mg/mL, and flow rate = 1.5 mL/min; The regeneration performance of a cotton filter. The filter was washed with a 100 mL of 1 mol/L HCl: ethanol mixture (50:50 vol %) without electrochemistry before a next running cycle. The inset in (**c**) is SEM image of a cotton filter after 4 h of continuous operation.

2.4. Performance of the Cotton Filter towards TC Removal

The 0.26 mol/h/m^2 electrooxidation flux of MO in a single pass through the cotton filter is of note since the device was running under gravity feed. This also reveals the potential of a low-cost and energy-saving route for water purification. To further explore the potential for the treatment of other emerging organic contaminants, the as-fabricated conductive cotton filter was further challenged with a typical emerging contaminant, i.e., antibiotic tetracycline (TC, 0.2 mmol/L). TC is one of the commonly detected antibiotics in water and the wastewater treatment plant was considered as one major point sources for TC pollution. Control experiments in the absence of applied cell potential can only adsorb physically the TC molecules until all sorption sites were occupied after 70 min, i.e., adsorption saturation (Figure S5, Supplementary Materials). Figure 4a compares the TC electrooxidation flux as a function of total cell potential. The TC electrooxidation flux increased with increasing total cell potential from 1.0 V to 2.0 V with an electrooxidation flux of 0.9 ± 0.1 mol/h/m^2 for the cotton filter fabricated with a 1.5 mg/mL CNT ink. The TC electrooxidation flux changed negligibly with further increase in the total cell potential until 3.0 V. This finding can be explained by the open-circuit measurements as displayed in Figure 4b. At a total cell potential of 1.5 V, the anode potential was determined to be 0.7 ± 0.06 V (vs. Ag/AgCl), which is high enough to oxide TC molecules (e.g., dimethylamine group of TC at 0.5 V vs. Ag/AgCl) [35]. It is of note that the maximum TC electrooxidation kinetics was 2.0 V, which was quite different with that of MO. For example, an optimal total cell potential for MO electrooxidation was 3.5 V, while only 2.0 V was required for TC electrooxidation. This finding could be explained by their different molecular structures and physicochemical properties. TC is an amphoteric molecule with multiple functional groups/moieties (e.g., phenol, amino, alcohol, diketone). Moreover, compared with MO, the TC molecules tend to adsorb onto the sp^2-conjugated CNT sidewalls due to its relatively strong van der Waals, π–π, and cation–π interactions [35]. A recent report has demonstrated that TC has significant 3D molecular curvature and tend to adsorb onto the CNT surface until monolayer formation [35]. The LC-MS characterization results suggest that the characteristic TC peak observed in the influent solution was decreased by 63% and 96% at a total cell potential of 1.0 V and 2.0 V, respectively, indicates that the parent TC molecules has been degraded. TC was spiked into real surface water samples to further challenge the cotton filter. The results show that the TC electrooxidation flux decreased by 35% compared with that of model electrolyte solution. The lower conductivity (1408 μS/cm vs. 6289 μS/cm) and complex natural reservoir organic matrix (background chemical oxygen demand, COD = 47.1 mg/L) may account for the significant decrease in electrooxidation kinetics. The energy consumption of the developed electrochemical cotton filter technology for TC treatment is calculated at an applied total cell potential of 2.0 V, by assuming 31 electrons transferred per TC molecule, to be 1.2 kWh/kg COD (The COD used here is the theoretical COD). This value is comparative with state-of-the-art electrochemical oxidation processes with an energy consumption in the range of 5–100 kWh/kg COD [36–39]. Alternatively, the energy per volume treated is calculated to be only 0.05 kWh/m^3. The gravity feed could further save the pumping energy which was constantly required for conventional membrane separation processes, especially for some high pressure-driven membrane separation process like RO. Of course, the energy consumed for MO electrooxidation should be higher than that of TC, since a larger total cell potential will be required to achieve efficient MO degradation (e.g., 3.5 V vs. 2.0 V). The energy per MO volume treated is calculated to be 0.19 kWh/m^3, 3.7 times higher than that of TC. The experimental results have demonstrated that the cotton filter clogging caused by the accumulation of precipitates/polymers may greatly limit the practical applications of the developed device towards the treatment of real water samples. Hence, further studies will be necessary to enhance the electrooxidative capability and to address the filter regeneration issues. Additionally, the ubiquitous presence of dissolved natural organic matters may negatively affect the efficacy and efficiency of the device toward practical applications. Since there are only limited active sites available on the cotton filter surface, the pollutant concentration effect on the device performance also deserves future investigation. Especially for the treatment of trace organic contaminants, e.g., antibiotic tetracycline, their environmental concentration should be taken into consideration in future studies.

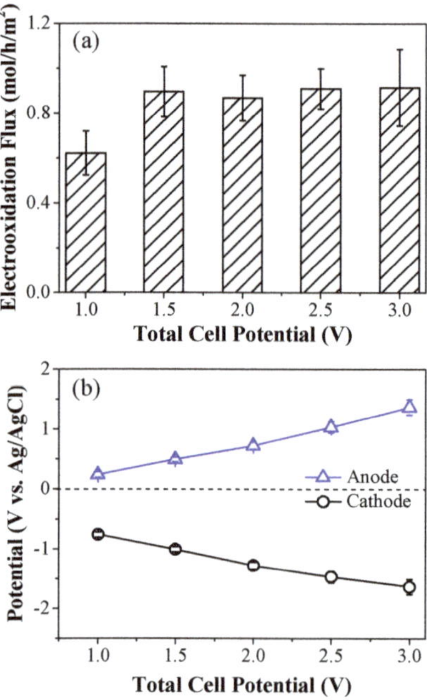

Figure 4. TC electrooxidation flux (**a**) and open-circuit potential (**b**) as a functional total cell potential. Experimental conditions: $[CNT]_{ink}$ = 1.5 mg/mL, $[TC]_{in}$ = 0.2 mmol/L, $[Na_2SO_4]$ = 10 mmol/L, and flow rate = 1.5 mL/min.

3. Materials and Methods

3.1. Materials

Multiwalled carbon nanotube networks (CNTs, <d> = 15 nm and <l> = 100 μm) were purchased from NanoTechLabs (Buckeye Composites, Yadkinville, NC, USA). Medical absorbent cottons were purchased from supermarket. Sodium sulfate (Na_2SO_4, ≥99.0%), methyl orange (MO, $C_{14}H_{14}N_3NaO_3S$, ACS reagent, dye content 85%), tetracycline (TC, $C_{22}H_{24}N_2O_8 \cdot xH_2O$, ≥98.0%), hydrochloric acid (HCl, ≥37%), ethanol (anhydrous, denatured), potassium hexacyanoferrate (II) trihydrate ($K_4Fe(CN)_6 \cdot 3H_2O$), and sodium dodecyl benzene sulfonate (SDBS) were purchased from Sigma-Aldrich (St. Louis, MO, USA). Deionized water used in the experiments was produced by using a Milli-Q ultrapure water system (Millipore, Billerica, MA, USA).

3.2. Fabrication of Conductive Cotton Filter

Firstly, the CNT ink was prepared by adding a certain amount of CNTs (0.5–1.5 mg/mL) and a 10 mg/mL SDBS to deionized water. The mixture was then bath sonicated (KQ3200E bath sonicator, Kunshan, China) for 10 min and probe sonicated (Branson SFX150 sonifier, St. Louis, MO, USA) for another 15 min to improve the CNT dispensability in water. A facile dying method was used to load CNTs onto the cellulose fiber-based cotton samples. In a typical fabrication process, 100 mg cotton sample was immersed into a freshly-prepared CNT ink for 5 min, following 30 min of drying in an oven (at 120 °C). Before drying, the cotton sample was pressed with finger to remove extra SDBS surfactant until no visible bubbles were observed and washed with DI water. Due to the strong absorption capability, the CNT ink was quickly coated onto the cotton. By varying the 'dip–dry' cycles and CNT content in the ink, different amounts of CNTs can be loaded onto the cellulose fibers of cotton. We assume the CNT ink coated uniformly onto the cotton surface. The mass of the loaded CNTs can be obtained from the mass difference before and after the dipping and drying of the cotton samples. The loading of CNTs onto the cotton could significantly boost the electrical conductivity of the as-fabricated filter. Finally, the conductive cotton filters (with 2.0 cm in the funnel and 0.5 cm connect to the power supply) were transferred into a plastic funnel (bottom diameter of 0.8 cm, funnel volume of 85 mL) for electrochemical filtration applications.

3.3. Electrochemical Filtration Device

The electrochemical filtration experiments were conducted using a conductive cotton filter (with a length of 2.5 cm) as anode, a Ti foil (2×5 cm) as cathode, and 10 mmol/L Na_2SO_4 as background electrolyte. The solution passed through the conductive cotton filter by only gravity feed. In a typical experiment of electrochemical filtration of methyl orange (MO), 0.06 mmol/L MO, and 10 mmol/L Na_2SO_4 were first flowed through the filter in the absence of applied voltage to achieve adsorption saturation of the filter, which could exclude the contribution of physical sorption to pollutants removal. Unless noted, the volume treated was 200 mL. Due to the limited space in the funnel, the solution was topped up every 15 min to maintain the 'driving force' (i.e., gravity). Then, an Agilent E3646A DC power supply was used to provide the voltage to induce the electro-oxidation of the pollutants. The effluent was collected at specific time intervals. Water flux was measured in a similar way by replacing the organic solution with DI-H_2O. When challenging with the organic solution, the flux will decrease to some extent due to the accumulation of precipitates and/or polymers on the filter surface. To evaluate the regeneration performance of the conductive cotton filters, a 100 mL of ethanol and 1 mol/L HCl mixture (50:50 vol %) was passed through the filter without electrochemistry after the MO electrochemical filtration. This could help to remove the organic residues from the filter surface. The oxidation flux was calculated by the Equation (2)

$$\text{Elextrooxidation Flux} = \frac{(C_{in} - C_{out})(mol/L) \times \text{flow rate}(L/h)}{\text{effective filter area } (m^2)} \tag{2}$$

where C_{in} is the initial influent compound concentration and C_{out} is the compound concentration after passing through the cotton filter. The effluent samples were collected after applying the external potential for 20 min. All these measurements were repeated at least three times for reproducibility.

3.4. Characterizations

The morphology of the as-fabricated conductive cotton filters was examined by a JEOL JSM-6700F filed-emission scanning electron microscopy (FESEM) (Carl Zeiss Supra55VP, Oberkochen, Germany). Micrographs were analyzed with *ImageJ* software (Bethesda, MD, USA) to get the average interfiber pore size of the as-fabricated filters, which was obtained from the average of at least 100 measurements from three FESEM images. The electrochemical characterizations of the samples were conducted on a CHI660E electrochemical workstation (Shanghai Chenhua Instrument Co. Ltd., Shanghai, China) using a three-electrode system. A conductive cotton filter, an AgCl/Ag electrode and a Ti foil served as working electrode, reference electrode and counter electrode, respectively. For the open-circuit voltage measurements, an Agilent E3646A DC power (Santa Clara, CA, USA) supply was used to provide an applied voltage of 0 to 4 V. The concentration of ferricyanide and methyl orange (MO) was determined by a Shimadzu UV-1800 UV–vis photometer (Kyoto, Japan) at their maximum absorbance values of 425 nm and 462 nm, respectively. The concentration of tetracycline (TC) was determined by using an Agilent 1290 UHPLC system (Waldbronm, Germany) coupled with 6540 quadrupole-time of flight (Q-TOF) mass detector equipped with a dual jet stream electrospray ionization source.

4. Conclusions

In conclusion, an affordable and effective electrochemical cotton-based filtration device for water treatment was developed. The CNT amount, total cell potential and surfactant were identified to be key parameters affecting the device performance. Moreover, the efficient electrooxidation of ferrocyanide, methyl orange, and antibiotic tetracycline suggest that the electrochemical cotton filters have good potential for water purification applications. Overall, the experiment results presented in this study quantitatively exemplified the advantages of a conductive cotton filter for water purification in a low-cost and energy-saving manner.

Supplementary Materials: The following are available online at www.mdpi.com/2073-4344/7/10/291/s1; Figure S1: FESEM images of conductive cotton filters fabricated using $[CNT]_{ink}$ = 1.5 mg/mL; Figure S2: Variation of filter resistance of as-fabricated cotton filters with CNT ink concentrations; Figure S3: Demonstration of a CNT cotton filter acts as a conducting path in the emission of an LED indicative lamp under applied voltage of 4 V; Figure S4: MO breakthrough curve under conditions of $[MO]_{in}$ = 0.06 mmol/L, $[Na_2SO_4]$ = 10 mmol/L, and flow rate = 1.5 mL/min; Figure S5. TC breakthrough curves under conditions of $[TC]_{in}$ = 0.2 mmol/L, $[Na_2SO_4]$ = 10 mmol/L, and flow rate = 1.5 mL/min.

Acknowledgments: This work was financially supported by the National Natural Science Foundation of China (No.51478099 and No. 51208206) and National Key Research and Development Program of China (No. 2016YFC0400501). Y.L. thanks Donghua University for the start-up grant (No. 113-07-005710).

Author Contributions: F.L. and Y.B.L. conceived and designed the experiments; Q.X. and Q.C. performed the experiments; M.Z.H. and Q.X. analyzed the data; Y.B.L. wrote the paper.

Conflicts of Interest: The authors declare no conflict of interest.

References

1. Zodrow, K.R.; Li, Q.; Buono, R.M.; Chen, W.; Daigger, G.; Dueñas-Osorio, L.; Elimelech, M.; Huang, X.; Jiang, G.; Kim, J.-H.; et al. Advanced materials, technologies, and complex systems analyses: Emerging opportunities to enhance urban water security. *Environ. Sci. Technol.* **2017**, *51*, 10274–10281. [CrossRef] [PubMed]
2. Zhang, R.; Liu, Y.; He, M.; Su, Y.; Zhao, X.; Elimelech, M.; Jiang, Z. Antifouling membranes for sustainable water purification: Strategies and mechanisms. *Chem. Soc. Rev.* **2016**, *45*, 5888–5924. [CrossRef] [PubMed]
3. Madaeni, S.S.; Ghaemi, N.; Rajabi, H. 1—*Advances in Polymeric Membranes for Water Treatment, in Advances in Membrane Technologies for Water Treatment*; Woodhead Publishing: Oxford, UK, 2015; pp. 3–41.
4. Liu, Y.; Zhang, Y.; Wang, L.; Yang, G.; Shen, F.; Deng, S.; Zhang, X.; He, Y.; Hu, Y.; Chen, X. Fast and large-scale anodizing synthesis of pine-cone TiO_2 for solar-driven photocatalysis. *Catalysts* **2017**, *7*, 229. [CrossRef]
5. O'Shea, K.E.; Dionysiou, D.D. Advanced oxidation processes for water treatment. *J. Phys. Chem. Lett.* **2012**, *3*, 2112–2113. [CrossRef]
6. Deng, Y.; Zhao, R. Advanced oxidation processes (AOPs) in wastewater treatment. *Curr. Pollution Rep.* **2015**, *1*, 167–176. [CrossRef]
7. Lazar, M.; Varghese, S.; Nair, S. Photocatalytic water treatment by titanium dioxide: Recent updates. *Catalysts* **2012**, *2*, 572–601. [CrossRef]
8. Santos, M.C.; Elabd, Y.A.; Jing, Y.; Chaplin, B.P.; Fang, L. Highly porous Ti_4O_7 reactive electrochemical water filtration membranes fabricated via electrospinning/electrospraying. *AIChE J.* **2016**, *62*, 508–524. [CrossRef]
9. Ikehata, K.; Jodeiri Naghashkar, N.; Gamal El-Din, M. Degradation of aqueous pharmaceuticals by ozonation and advanced oxidation processes: A review. *Ozone Sci. Eng.* **2006**, *28*, 353–414. [CrossRef]
10. Kim, K.-J.; Jang, A. Fouling characteristics of NOM during the ceramic membrane microfiltration process for water treatment. *Desalination Water Treat.* **2016**, *57*, 9034–9042. [CrossRef]
11. Chaplin, B.P. Critical review of electrochemical advanced oxidation processes for water treatment applications. *Environ. Sci. Process. Impacts* **2014**, *16*, 1182–1203. [CrossRef] [PubMed]
12. Anglada, Á.; Urtiaga, A.; Ortiz, I. Contributions of electrochemical oxidation to waste-water treatment: Fundamentals and review of applications. *J. Chem. Technol. Biotechnol.* **2009**, *84*, 1747–1755. [CrossRef]
13. Chen, G. Electrochemical technologies in wastewater treatment. *Sep. Purif. Technol.* **2004**, *38*, 11–41. [CrossRef]
14. Popov, V.N. Carbon nanotubes: Properties and application. *Mater. Sci. Eng. R* **2004**, *43*, 61–102. [CrossRef]

15. Wu, J.; Gerstandt, K.; Zhang, H.; Liu, J.; Hinds, B.J. Electrophoretically induced aqueous flow through single-walled carbon nanotube membranes. *Nat. Nanotechnol.* **2012**, *7*, 133–139. [CrossRef] [PubMed]

16. De Volder, M.F.L.; Tawfick, S.H.; Baughman, R.H.; Hart, A.J. Carbon nanotubes: Present and future commercial applications. *Science* **2013**, *339*, 535–539. [CrossRef] [PubMed]

17. Liu, Y.; Zheng, Y.; Du, B.; Nasaruddin, R.R.; Chen, T.; Xie, J. Golden carbon nanotube membrane for continuous flow catalysis. *Ind. Eng. Chem. Res.* **2017**, *56*, 2999–3007. [CrossRef]

18. Liu, H.; Zuo, K.; Vecitis, C.D. Titanium dioxide-coated carbon nanotube network filter for rapid and effective arsenic sorption. *Environ. Sci. Technol.* **2014**, *48*, 13871–13879. [CrossRef] [PubMed]

19. Gao, G.; Zhang, Q.; Hao, Z.; Vecitis, C.D. Carbon nanotube membrane stack for flow-through sequential regenerative electro-Fenton. *Environ. Sci. Technol.* **2015**, *49*, 2375–2383. [CrossRef] [PubMed]

20. Liu, Y.; Xie, J.; Ong, C.N.; Vecitis, C.D.; Zhou, Z. Electrochemical wastewater treatment with carbon nanotube filters coupled with in situ generated H_2O_2. *Environ. Sci. Water Res. Technol.* **2015**, *1*, 769–778. [CrossRef]

21. Gui, X.; Wei, J.; Wang, K.; Cao, A.; Zhu, H.; Jia, Y.; Shu, Q.; Wu, D. Carbon Nanotube Sponges. *Adv. Mater.* **2010**, *22*, 617–621. [CrossRef] [PubMed]

22. Li, X.; Xue, Y.; Zou, M.; Zhang, D.; Cao, A.; Duan, H. Direct oil recovery from saturated carbon nanotube sponges. *ACS Appl. Mater. Interfaces* **2016**, *8*, 12337–12343. [CrossRef] [PubMed]

23. Schoen, D.T.; Schoen, A.P.; Hu, L.; Kim, H.S.; Heilshorn, S.C.; Cui, Y. High speed water sterilization using one-dimensional nanostructures. *Nano Lett.* **2010**, *10*, 3628–3632. [CrossRef] [PubMed]

24. Arash, B.; Wang, Q.; Varadan, V.K. Mechanical properties of carbon nanotube/polymer composites. *Sci. Rep.* **2014**, *4*, 6479. [CrossRef] [PubMed]

25. Makowski, T.; Kowalczyk, D.; Fortuniak, W.; Jeziorska, D.; Brzezinski, S.; Tracz, A. Superhydrophobic properties of cotton woven fabrics with conducting 3D networks of multiwall carbon nanotubes, MWCNTs. *Cellulose* **2014**, *21*, 4659–4670. [CrossRef]

26. Pasta, M.; La Mantia, F.; Hu, L.; Deshazer, H.D.; Cui, Y. Aqueous supercapacitors on conductive cotton. *Nano Res.* **2010**, *3*, 452–458. [CrossRef]

27. Wang, Z.; Huang, Y.; Sun, J.; Huang, Y.; Hu, H.; Jiang, R.; Gai, W.; Li, G.; Zhi, C. Polyurethane/Cotton/Carbon Nanotubes Core-Spun Yarn as High Reliability Stretchable Strain Sensor for Human Motion Detection. *ACS Appl. Mater. Interfaces* **2016**, *8*, 24837–24843. [CrossRef] [PubMed]

28. Islam, M.F.; Rojas, E.; Bergey, D.M.; Johnson, A.T.; Yodh, A.G. High weight fraction surfactant solubilization of single-wall carbon nanotubes in water. *Nano Lett.* **2003**, *3*, 269–273. [CrossRef]

29. Qi, H.; Liu, J.; Mäder, E. Smart cellulose fibers coated with carbon nanotube networks. *Fibers* **2014**, *2*, 295. [CrossRef]

30. Rahman, M.J.; Mieno, T. Conductive cotton textile from safely functionalized carbon nanotubes. *J. Nanomater.* **2015**, *2015*. [CrossRef]

31. Liu, H.; Liu, J.; Liu, Y.; Bertoldi, K.; Vecitis, C.D. Quantitative 2D electrooxidative carbon nanotube filter model: Insight into reactive sites. *Carbon* **2014**, *80*, 651–664. [CrossRef]

32. Liu, Y.; Dustin Lee, J.H.; Xia, Q.; Ma, Y.; Yu, Y.; Lanry Yung, L.Y.; Xie, J.; Ong, C.N.; Vecitis, C.D.; Zhou, Z. A graphene-based electrochemical filter for water purification. *J. Mater. Chem. A* **2014**, *2*, 16554–16562. [CrossRef]

33. Vecitis, C.D.; Gao, G.; Liu, H. Electrochemical carbon nanotube filter for adsorption, desorption, and oxidation of aqueous dyes and anions. *J. Phys. Chem. C* **2011**, *115*, 3621–3629. [CrossRef]

34. Gao, G.; Pan, M.; Vecitis, C.D. Effect of the oxidation approach on carbon nanotube surface functional groups and electrooxidative filtration performance. *J. Mater. Chem. A* **2015**, *3*, 7575–7582. [CrossRef]

35. Liu, Y.; Liu, H.; Zhou, Z.; Wang, T.; Ong, C.N.; Vecitis, C.D. Degradation of the common aqueous antibiotic tetracycline using a carbon nanotube electrochemical filter. *Environ. Sci. Technol.* **2015**, *49*, 7974–7980. [CrossRef] [PubMed]

36. Avlonitis, S.A.; Kouroumbas, K.; Vlachakis, N. Energy consumption and membrane replacement cost for seawater RO desalination plants. *Desalination* **2003**, *157*, 151–158. [CrossRef]

37. Lee, Y.; Gerrity, D.; Lee, M.; Gamage, S.; Pisarenko, A.; Trenholm, R.A.; Canonica, S.; Snyder, S.A.; von Gunten, U. Organic contaminant abatement in reclaimed water by UV/H_2O_2 and a combined process consisting of O_3/H_2O_2 followed by UV/H_2O_2: Prediction of abatement efficiency, energy consumption, and byproduct formation. *Environ. Sci. Technol.* **2016**, *50*, 3809–3819. [CrossRef] [PubMed]

38. Zou, S.; Yuan, H.; Childress, A.; He, Z. Energy consumption by recirculation: A missing parameter when evaluating forward osmosis. *Environ. Sci. Technol.* **2016**, *50*, 6827–6829. [CrossRef] [PubMed]
39. Xiao, H.; Wu, M.; Zhao, G. Electrocatalytic oxidation of cellulose to gluconate on carbon aerogel supported gold nanoparticles anode in alkaline medium. *Catalysts* **2016**, *6*, 5. [CrossRef]

 *catalysts*

Article

A Demonstration of Pt L₃-Edge EXAFS Free from Au L₃-Edge Using Log–Spiral Bent Crystal Laue Analyzers

Yuki Wakisaka [1,*], Daiki Kido [1], Hiromitsu Uehara [1], Qiuyi Yuan [1], Satoru Takakusagi [1], Yohei Uemura [2], Toshihiko Yokoyama [2], Takahiro Wada [3], Motohiro Uo [3], Tomohiro Sakata [4], Oki Sekizawa [4,5], Tomoya Uruga [4,5], Yasuhiro Iwasawa [4] and Kiyotaka Asakura [1,*]

[1] Institute for Catalysis, Hokkaido University, Hokkaido 001-0021, Japan; d-kido@eis.hokudai.ac.jp (D.K.); uehara@oeic.hokudai.ac.jp (H.U.); qy.qiuyiyuan@gmail.com (Q.Y.); takakusa@cat.hokudai.ac.jp (S.T.)

[2] Institute for Molecular Science, Aichi 444-8585, Japan; y-uemura@ims.ac.jp (Y.U.); yokoyama@ims.ac.jp (T.Y.)

[3] Graduate School of Medical and Dental Sciences, Tokyo Medical and Dental University, Tokyo 113-8549, Japan; wada.abm@tmd.ac.jp (T.W.); uo.abm@tmd.ac.jp (M.U.)

[4] Innovation Research Center for Fuel Cells, The University of Electro-Communications, Tokyo 182-8585, Japan; s-tomohiro@uec.ac.jp (T.S.); sekizawa@uec.ac.jp (O.S.); urugat@spring8.or.jp (T.U.); iwasawa@pc.uec.ac.jp (Y.I.)

[5] Japan Synchrotron Radiation Research (JASRI), Hyogo 679-5148, Japan

[*] Correspondence: ywkskhkd@gmail.com (Y.W.); askr@cat.hokudai.ac.jp (K.A.); Tel.: +81-11-706-9113(K.A.)

Received: 30 March 2018; Accepted: 8 May 2018; Published: 13 May 2018

Abstract: Pt-Au nanostructures are important and well-studied fuel cell catalysts for their promising catalytic performance. However, a detailed quantitative local structure analysis, using extended X-ray absorption fine structure (EXAFS) spectroscopy, have been inhibited by interference between Pt and Au L₃-edges. In this paper, Pt L₃-edge XAFS analysis, free of Au L₃ edge, is demonstrated for a Pt-Au reference sample using a low-cost log–spiral bent crystal Laue analyzer (BCLA). This method facilitates the EXAFS structural analysis of Pt-Au catalysts, which are important to improve fuel cell catalysts.

Keywords: Pt-Au; XAFS; BCLA

1. Introduction

Platinum is one of the key elements for catalytic reactions in fuel cells. Although there are many studies in which authors suggest different methodologies to replace Pt with others low-cost metals, it is still difficult to substitute the catalytic performance of Pt. A practical approach, combining different metals with Pt have widely been adopted to reduce the amount of Pt and improve its activity and durability. Among those, Au is one of the interesting metals due to its superior oxygen-reduction-reaction activity and durability reported in Pt-Au nanostructures, where the Pt (shell)-Au (core) structures and the effect of Au decoration on the edges of Pt surfaces are used as a fuel cell catalyst [1–5]. Hence, it is essential to understand the local structures of both Pt and Au in an atomic scale to elucidate the mechanism of the catalytic reactions in Pt-Au nanostructures. Extended X-ray absorption fine structure (extended XAFS or EXAFS) spectroscopy is a suitable and widely-used method to investigate the local atomic structures of fuel-cell catalysts because of its atomic selectivity and applicability to nanoparticles under electrochemical environments [6]. However, in case of Pt-Au system, it is difficult to obtain a Pt L₃-edge EXAFS sufficient for its analysis due to the interference between Pt and Au, which are only separated by ~350 eV, so that Au L₃-edge appears at ~9.6 Å^{-1} in Pt L₃-edge EXAFS [7,8]. Although the problem can be solved by measuring K-edge XAFS, where Pt and Au K-edges are separated by

~2300 eV [9,10], the information in the long-range order is limited by the lifetime broadening, and Pt L$_3$-edge EXAFS measurement is preferable.

Glatzel et al. first demonstrated that EXAFS spectra sufficient for analysis under the existence of interfering absorption edges, called range-extended EXAFS, could be obtained by taking advantage of high-energy-resolution fluorescence detected XAFS (HERFD-XAFS) using crystal analyzers with an energy resolution of ~1 eV [11,12]. It was shown that HERFD-XAFS was not only useful for capturing the detailed structures of the X-ray absorption near edge structure spectra but also capable of obtaining the range extended EXAFS. Recently, this method was applied to the feasibility study of Pt L$_3$-edge EXAFS in the presence of Au [13]. In this paper, we demonstrated that range-extended Pt L$_3$-edge EXAFS can also be obtained under the existence of Au using a log–spiral bent crystal Laue analyzer (BCLA) [14]. Although the energy resolution of BCLA (>10 eV [15]) is generally less than the resolution of crystal analyzers used in HERFD-XAFS (~1 eV), the energy reolution of the BCLA is sufficiently small for discriminating Au fluorescence from Pt. On the other hand, adopting BCLA, one can expect a lower cost for experimental arrangement compared to HERFD-XAFS. Moreover, the emission energy scan of BCLA can be achieved by a vertical scan because it approximately corresponds to the change in the incident angle of the X-ray against the crystal face. These characteristics may facilitate the application of the BCLA to the range-extended EXAFS of Pt-Au catalysts.

2. Results

Figure 1 shows the emission X-ray intensity from a diluted Pt-Au reference sample measured through a BCLA moved to the vertical direction. The incident X-ray energy was 12.1 keV, which was corresponding to the energy above the Au L$_3$-edge. All four fluorescent peaks (Pt L$_{\alpha2}$, Pt L$_{\alpha1}$, Au L$_{\alpha2}$, Au L$_{\alpha1}$) were well resolved, and it is confirmed that there was a clear correspondence between the accepted X-ray fluorescent energy and the vertical position of the BCLA. Au fluorescent peaks were smaller compared to the expected molar ratio of the sample (Pt/Au = ~1/10); a solid-state detector (SSD) was used within the range of interest, only including the entire Pt L$_\alpha$ peaks. According to the full width half maximum (FWHM) of the Pt L$_{\alpha1}$ peak, the energy resolution is ~30 eV for this experimental arrangement.

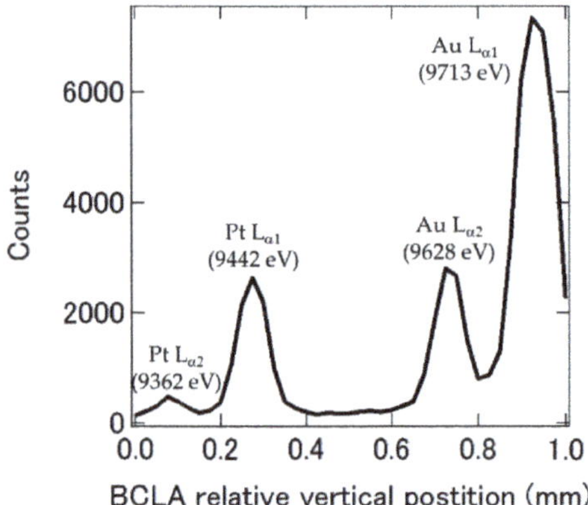

Figure 1. Emission X-ray intensity from a diluted Pt-Au sample measured through a BCLA moved to the vertical direction. Four peaks were assigned as Pt L$_{\alpha2}$, Pt L$_{\alpha1}$, Au L$_{\alpha2}$, Au L$_{\alpha1}$, from lower to higher positions of the BCLA.

Figure 2 shows the normalized XAFS spectra of a concentrated Pt-Au reference sample measured in a transmission mode and a fluorescence mode with the BCLA. The edge heights of the raw spectrum were 1 and 7.31.1 for Pt and Au, respectively. It was clearly observed by the transmission spectrum that the sample contained ~10 times more Au than Pt. In the fluorescence mode, apparently, no Au signal was observed due to the BCLA. However, the Pt fluorescence signal abruptly decreased at the Au L_3-edge; the incident X-rays were absorbed by the abundant Au atoms. Consequently, Pt atoms were less excited [11,16]. This effect can be avoided by a sufficient dilution of the sample.

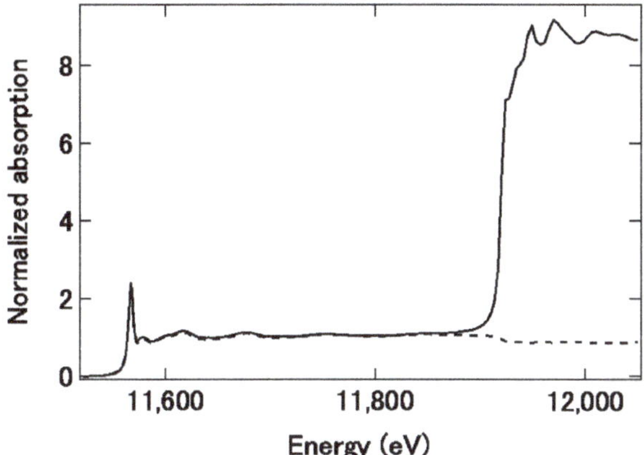

Figure 2. XAFS spectra of the concentrated Pt-Au reference sample measured in transmission (solid line) and fluorescence mode (dashed line) with the BCLA. The spectrum was normalized by the edge height of Pt = 1.

Figure 3a shows the XAFS spectrum of the diluted Pt-Au reference sample (see Materials and Methods) measured in the fluorescence mode with the BCLA. The edge heights of the raw data for this diluted sample measured in transmission mode was 0.007 for Pt and 0.05 for Au, respectively. As expected, no clear anomaly was observed near the region of Au L_3-edge, when the interference of Auand Pt L_3-edge EXAFS could be removed. Figure 3b shows the $k^3\chi$ plot with an accumulation time of less than 30 min. For comparison, the Pt L_3 transmission XAFS spectrum of a standard sample (PtCl$_4$) measured at Beam line (BL)14B2 in Super Photon ring-8 GeV (SPring-8) was overlaid as a red dashed line (see Figure 3b). There is a good agreement up to ~12 Å^{-1}. No clear edge was found even in this $k^3\chi$ plot. Here, we demonstrated that Pt L_3-edge EXAFS spectra, free of Au L_3-edge, could be obtained using a considerable amount of Au in BCLA.

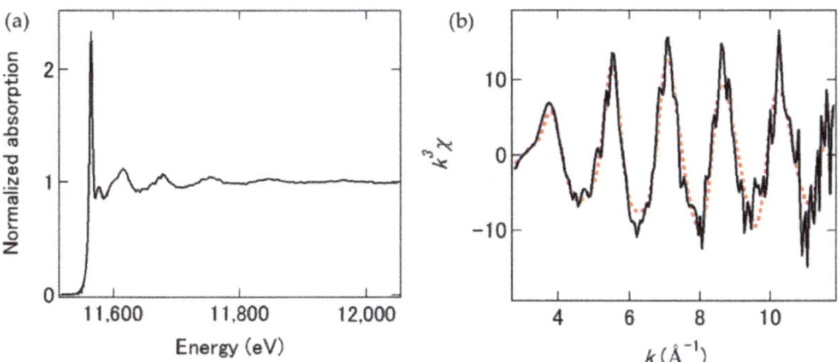

Figure 3. (**a**): Fluorescence XAFS spectrum of the diluted Pt-Au reference sample measured with the BCLA. The spectrum was normalized by the edge height of Pt = 1; (**b**): $k^3\chi$ EXAFS spectra of the diluted Pt-Au reference sample measured at BL36XU (black solid line) and of the standard sample PtCl$_4$ (red dashed line) measured at BL14B2.

3. Discussion

Previously, range-extended EXAFS was only achieved in HERFD-XAFS. In this work, we have successfully demonstrated the range-extended EXAFS analysis using BCLA is possible. BCLA has several advantages mentioned above compared to HERFD-XAFS method. In addition, the crystal alignment is quite simple. It is mainly achieved by a vertical scan followed by a precise two-dimensional linear scan of the BCLA [15,17]. The only constraint is the vertical size of the incident X-ray beam. In this study, it was ~50 μm, though this condition is not fixed, depending on the energy difference between the measuring (Pt) and interfering (Au) fluorescent X-rays (see Figure 1).

HERFD-XAFS is, in general, a powerful technique to detect the subtle spectral changes in X-ray absorption near to edge regions. By applying this technique to Pt catalysts in fuel cells, various adsorbates on Pt and its oxidation states have been discussed [18–20]. HERFD-XAFS can not be achieved by using the BCLA because of their moderate energy resolution larger than the core-hole lifetime broadening. However, this is preferable in case of direct comparison between the spectra measured in the transmission and the fluorescence mode using BCLA; which should have the same energy resolutions.

4. Materials and Methods

The concentrated Pt-Au reference sample was made by mixing $PtCl_4$ and AuCl powder. The mixture was then ground in a mortar and pestle together with BN (boron nitride) powder and pressed into a pellet with a size of 1 mm thick and 10 mm diameter. The Pt and Au concentration was Pt/Au ~ 1/10 and the Pt L_3-edge step ($\Delta\mu t$) was ~0.1. The diluted Pt-Au reference sample was made diluting the concentrated Pt-Au pellet by ~1/20 with additional BN powder.

The XAFS measurements were performed at BL36XU in SPring-8 (JASRI, Koto, Japan). The beam size of the incident X-rays was focused to ~50 μm (vertical) × ~500 μm (horizontal) by using 4 focusing mirrors equipped in the beamline. The photon flux was ~2×10^{13} photons/s, but it was reduced to ~2×10^{12} photons/s for the diluted Pt-Au sample using an Al attenuator. Ion chambers were used for the transmission measurement. A commercial BCLA (0095, FMB Oxford, UK) and a 25-element Ge SSD (Canberra, Coneticcut , USA) or a pixel-array detector (PILATUS 300K-W; Dectris, Baden-Daettwil, Switzerland) were used for detecting fluorescent X-rays. The sample and the BCLA/SSD (or PILATUS) were placed in the 45°/45° arrangement. The shaping time of the SSD was set to 0.5 μs, which resulted in an SSD energy resolution of 400 eV. The region of interest in the SSD was 9.09–9.76 keV. Thus, the Au fluorescence effect cannot fully suppress after the Au L_3-edge. The commercial BCLA was linearly scanned in two dimensions to find their optimum position so that the Pt $L_{\alpha1}$ fluorescent X-ray intensity became maximum in the multi-element SSD. Only the detector of elements in the multi-element SSD, which sufficiently suppressed Au L_α fluorescent X-rays, was used for the spectral analyses [21].

5. Conclusions

Pt L_3-edge XAFS analysis, free from Au L_3-edge, was demonstrated here for the first time using BCLA; a low-cost and high-sensitive crystal analyzers, which facilitates detail EXAFS analyses for Pt-Au fuel cell catalysts. Our results confirm the feasibility of the range-extended EXAFS using BCLA, we apply this technique to two interesting fuel cell models containing Pt and Au; Au-Pt-Co-N nanoparticles deposited on a highly oriented pyrolytic graphite [22] and monolayer Pt deposited on Au thin film with 60 nm thickness on a Si (100) substrate [23]. We will soon report these results.

Author Contributions: Y.W., H.U., and K.A. conceived, designed, and performed the experiments; D.K., Q.Y., Y.U., and T.W. performed the experiments; Y.W., S.T., M.U., T.Y., T.U., Y.I., and K.A. discussed the results; T.S., O.S., and T.U. supported the experiments; Y.W. analyzed the data; Y.W. and K.A. wrote the paper.

Acknowledgments: The authors would like to express their gratitude to the New Energy and Industrial Technology Development Organization (NEDO) Polymer Electrolyte Fuel Cell project for their financial support. We would like to thank Hiroyuki Asakura for discussing the range-extended EXAFS results of Pt-Au. T.W. and Y.U. were supported by the Cooperative Research Program of Institute for Catalysis, Hokkaido University. The XAFS measurements at SPring-8 were performed under project number 2016A7902 and 2016B7902. XAFS spectrum of standard sample $PtCl_4$ is utilized by SPring-8 BL14B2 XAFS database (2014S0000-000523). This work was supported by the Technical Division of Institute for Catalysis, Hokkaido University.

Conflicts of Interest: The authors declare no conflict of interest.

References

1. Kristian, N.; Wang, X. Pt shell–Au core/C electrocatalyst with a controlled shell thickness and improved Pt utilization for fuel cell reactions. *Electrochem. Commun.* **2008**, *10*, 12–15. [CrossRef]

2. Shuangyin, W.; Noel, K.; Sanping, J.; Xin, W. Controlled synthesis of dendritic Au@Pt core–shell nanomaterials for use as an effective fuel cell electrocatalyst. *Nanotechnology* **2009**, *20*, 025605. [CrossRef]

3. Xiu, C.; Shengnan, W.; Scott, J.; Zhibing, C.; Zhenghua, W.; Lun, W.; Li, Y. The deposition of Au-Pt core-shell nanoparticles on reduced graphene oxide and their catalytic activity. *Nanotechnology* **2013**, *24*, 295402. [CrossRef]

4. Dai, Y.; Chen, S. Oxygen Reduction Electrocatalyst of Pt on Au Nanoparticles through Spontaneous Deposition. *ACS Appl. Mater. Interfaces* **2015**, *7*, 823–829. [CrossRef] [PubMed]

5. Takahashi, S.; Chiba, H.; Kato, T.; Endo, S.; Hayashi, T.; Todoroki, N.; Wadayama, T. Oxygen reduction reaction activity and structural stability of Pt-Au nanoparticles prepared by arc-plasma deposition. *Phys. Chem. Chem. Phys.* **2015**, *17*, 18638–18644. [CrossRef] [PubMed]

6. Iwasawa, Y.; Asakura, K.; Tada, M. *XAFS Techniques for Catalysts, Nanomaterials, and Surfaces*; Springer: New York, NY, USA, 2016.

7. Nagamatsu, S.; Arai, T.; Yamamoto, M.; Ohkura, T.; Oyanagi, H.; Ishizaka, T.; Kawanami, H.; Uruga, T.; Tada, M.; Iwasawa, Y. Potential-Dependent Restructuring and Hysteresis in the Structural and Electronic Transformations of Pt/C, Au(Core)-Pt(Shell)/C, and Pd(Core)-Pt(Shell)/C Cathode Catalysts in Polymer Electrolyte Fuel Cells Characterized by in Situ X-ray Absorption Fine Structure. *J. Phys. Chem. C* **2013**, *117*, 13094–13107.

8. Yuan, Q.; Takakusagi, S.; Wakisaka, Y.; Uemura, Y.; Wada, T.; Ariga, H.; Asakura, K. Polarization-dependent Total Reflection Fluorescence X-ray Absorption Fine Structure (PTRF-XAFS) Studies on the Structure of a Pt Monolayer on Au(111) Prepared by the Surface-limited Redox Replacement Reaction. *Chem. Lett.* **2017**, *46*, 1250–1253. [CrossRef]

9. Kaito, T.; Mitsumoto, H.; Sugawara, S.; Shinohara, K.; Uchara, H.; Ariga, H.; Takakusagi, S.; Hatakeyama, Y.; Nishikawa, K.; Asakura, K. K-Edge X-ray Absorption Fine Structure Analysis of Pt/Au Core–Shell Electrocatalyst: Evidence for Short Pt–Pt Distance. *J. Phys. Chem. C* **2014**, *118*, 8481–8490. [CrossRef]

10. Kaito, T.; Mitsumoto, H.; Sugawara, S.; Shinohara, K.; Uehara, H.; Ariga, H.; Takakusagi, S.; Asakura, K. A new spectroelectrochemical cell for in situ measurement of Pt and Au K-edge X-ray absorption fine structure. *Rev. Sci. Instrum.* **2014**, *85*, 084104. [CrossRef] [PubMed]

11. Glatzel, P.; de Groot, F.M.F.; Manoilova, O.; Grandjean, D.; Weckhuysen, B.M.; Bergmann, U.; Barrea, R. Range-extended EXAFS at the L edge of rare earths using high-energy-resolution fluorescence detection: A study of La in LaOCl. *Phys. Rev. B* **2005**, *72*, 014117. [CrossRef]

12. Yano, J.; Pushkar, Y.; Glatzel, P.; Lewis, A.; Sauer, K.; Messinger, J.; Bergmann, U.; Yachandra, V. High-Resolution Mn EXAFS of the Oxygen-Evolving Complex in Photosystem II: Structural Implications for the Mn4Ca Cluster. *J. Am. Chem. Soc.* **2005**, *127*, 14974–14975. [CrossRef] [PubMed]

13. Asakura, H.; Kawamura, N.; Mizumaki, M.; Nitta, K.; Ishii, K.; Hosokawa, S.; Teramura, K.; Tanaka, T. A feasibility study of "range-extended" EXAFS measurement at the Pt L3-edge of Pt/Al_2O_3 in the presence of Au_2O_3. *J. Anal. At. Spectrom.* **2018**, *33*, 84–89. [CrossRef]

14. Zhong, Z.; Chapman, L.D.; Bunker, B.A.; Bunker, G.B.; Fischetti, R.; Segre, C.U. A bent Laue analyzer for fluorescence EXAFS detection. *J. Synchrotron Radiat.* **1999**, *6*, 212–214. [CrossRef] [PubMed]

15. Kujala, N.G.; Karanfil, C.; Barrea, R.A. High resolution short focal distance Bent Crystal Laue Analyzer for copper K edge X-ray absorption spectroscopy. *Rev. Sci. Instrum.* **2011**, *82*, 063106. [CrossRef] [PubMed]

16. Bianchini, M.; Glatzel, P. A tool to plan photon-in/photon-out experiments: Count rates, dips and self-absorption. *J. Synchrotron Radiat.* **2012**, *19*, 911–919. [CrossRef] [PubMed]

17. Karanfil, C.; Bunker, G.; Newville, M.; Segre, C.U.; Chapman, D. Quantitative performance measurements of bent crystal Laue analyzers for X-ray fluorescence spectroscopy. *J. Synchrotron Radiat.* **2012**, *19*, 375–380. [CrossRef] [PubMed]

18. Friebel, D.; Viswanathan, V.; Miller, D.J.; Anniyev, T.; Ogasawara, H.; Larsen, A.H.; O'Grady, C.P.; Nørskov, J.K.; Nilsson, A. Balance of Nanostructure and Bimetallic Interactions in Pt ModelFuel Cell Catalysts: In Situ XAS and DFT Study. *J. Am. Chem. Soc.* **2012**, *134*, 9664–9671. [CrossRef] [PubMed]

19. Merte, L.R.; Behafarid, F.; Miller, D.J.; Friebel, D.; Cho, S.; Mbuga, F.; Sokaras, D.; Alonso-Mori, D.; Weng, T.-C.; Nordlund, D.; et al. Electrochemical Oxidation of Size-Selected Pt Nanoparticles Studied Using in Situ High-Energy-Resolution X-ray Absorption Spectroscopy. *ACS Catal.* **2012**, *2*, 2371–2376. [CrossRef]

20. Cui, Y.-T.; Harada, Y.; Niwa, H.; Hatanaka, T.; Nakamura, N.; Ando, M.; Yoshida, T.; Ishii, K.; Matsumura, D.; Oji, H.; et al. Wetting Induced Oxidation of Pt-based Nano Catalysts Revealed by In Situ High Energy Resolution X-ray Absorption Spectroscopy. *Sci. Rep.* **2017**, *7*, 1482. [CrossRef] [PubMed]

21. Wakisaka, Y.; Iwasaki, Y.; Uehara, H.; Mukai, S.; Kido, D.; Takakusgi, S.; Uemura, Y.; Wada, T.; Yuan, Q.; Sekizawa, O.; et al. Approach to Highly Sensitive XAFS by Means of Bent Crystal Laue Analyzers. *J. Surf. Sci. Soc. Jpn.* **2017**, *38*, 378–383. [CrossRef]

22. Takahashi, S.; Takahashi, N.; Todoroki, N.; Wadayama, T. Dealloying of Nitrogen-Introduced Pt–Co Alloy Nanoparticles: Preferential Core–Shell Formation with Enhanced Activity for Oxygen Reduction Reaction. *ACS Omega* **2016**, *1*, 1247–1252. [CrossRef]

23. Liu, Y.; Hangarter, C.M.; Garcia, D.; Moffat, T.P. Self-terminating electrodeposition of ultrathin Pt films on Ni: An active, low-cost electrode for H_2 production. *Surf. Sci.* **2015**, *631*, 141–154. [CrossRef]

Review

Recent Advances in Graphene Based TiO$_2$ Nanocomposites (GTiO$_2$Ns) for Photocatalytic Degradation of Synthetic Dyes

Rita Giovannetti *, Elena Rommozzi, Marco Zannotti and Chiara Anna D'Amato

School of Science and Technology, Chemistry Division, University of Camerino, Via S. Agostino 1, 62032 Camerino, Italy; elena.rommozzi@unicam.it (E.R.); marco.zannotti@unicam.it (M.Z.); chiaraanna.damato@unicam.it (C.A.D.)
* Correspondence: rita.giovannetti@unicam.it; Tel.: +39-0737-402272

Academic Editors: Ewa Kowalska, Marcin Janczarek and Agata Markowska-Szczupak
Received: 24 August 2017; Accepted: 10 October 2017; Published: 16 October 2017

Abstract: Synthetic dyes are widely used in textile, paper, food, cosmetic, and pharmaceutical industries. During industrial processes, some of these dyes are released into the wastewater and their successive release into rivers and lakes produces serious environmental problems. TiO$_2$ is one of the most widely studied and used photocatalysts for environmental remediation. However, it is mainly active under UV-light irradiation due to its band gap of 3.2 eV, while it shows low efficiency under the visible light spectrum. Regarding the exploration of TiO$_2$ activation in the visible light region of the total solar spectrum, the incorporation of carbon nanomaterials, such as graphene, in order to form carbon-TiO$_2$ composites is a promising area. Graphene, in fact, has a large surface area which makes it a good adsorbent for organic pollutants removal through the combination of electrostatic attraction and π-π interaction. Furthermore, it has a high electron mobility and therefore it reduces the electron-hole pair recombination, improving the photocatalytic activity of the semiconductor. In recent years, there was an increasing interest in the preparation of graphene-based TiO$_2$ photocatalysts. The present short review describes the recent advances in TiO$_2$ photocatalyst coupling with graphene materials with the aim of extending the light absorption of TiO$_2$ from UV wavelengths into the visible region, focusing on recent progress in the design and applications in the photocatalytic degradation of synthetic dyes.

Keywords: titanium dioxide; graphene; photocatalysis; visible light; dyes

---

1. Introduction

One of the biggest problems that the world is facing today is environmental pollution, which increases every year and causes serious and irreparable damage to the Earth [1]. Therefore, environmental protection and a new approach in environmental remediation are important factors for a real improvement in quality of life and for sustainable development.

In recent years, a large number of research activities have been dedicated to environmental protection and remediation as a consequence of special attention from social, political, and legislative authorities, which has led to the delivery of very stringent regulations for the environment [2].

Due to industrialization and the lack of effective treatments of the effluents at the source, a severe deterioration of freshwater resources caused by the release of a wide range of hazardous substances into water bodies has occurred. Of these substances, synthetic dyes represent a large group and therefore deserve particular attention, due to the high quantity—more than 800,000 tons—that is produced annually worldwide [3]. About one third of these is released into receiving waters every year through industrial wastewater discharges [4], which may have a severe influence on both the environment and human health, affecting photosynthetic activity as well as dissolved oxygen concentration, and therefore causing serious ecologic problems [5].

Dyes present high toxicity, are carcinogenic and mutagenic in nature, and can also result in bioaccumulation in the food chain [6]. Because their origin is synthetic, these compounds are very stable in the presence of light and biodegradation [7]. Furthermore, very low concentrations of these are easily visible in contaminated waters; therefore, the total effect of the presence of dyes in the water ecosystem is both environmentally and aesthetically unacceptable.

Taking account of these considerations, there is an urgent need to develop new methods for the treatment of industrial wastewater containing synthetic dyes.

In recent years, photocatalysis studies have focused on the use of semiconductor materials as photocatalysts for environmental remediation. Semiconductor photocatalysis is a versatile, low-cost, and environmentally friendly treatment technology for many pollutants [8,9], and the use of heterogeneous photocatalysis is an emerging application in water decontamination.

Within a great number of oxide semiconductors, TiO_2 is the most widely used as promising photocatalyst, demonstrating a very important role in environmental remediation [10,11], solar energy [12], and other fields [13,14] due to its nontoxicity, low cost, corrosion resistance, abundant resources, and high photocatalytic efficiency [15–17].

A high number of studies concerning TiO_2 regard wastewater treatments [18–21] and, in many cases, total mineralization of pollutants without any waste disposal problem have been published [22]. On the other hand, the practical applications of this technique are limited to a narrow excitation wavelength because of a large band gap energy (3.2 eV), high recombination rate of the photo-produced electron–hole pair, and poor adsorption capacity [23–25]. In recent decades, various attempts have been applied to improve the catalytic efficiency of TiO_2 [18,26,27].

Due to its wide band-gap, TiO_2 is active only under UV-light irradiation; considering that the percentage of UV-light is less than 5% of the total incident solar spectrum on the Earth, in recent years research has focused on extending the light absorption of TiO_2 under visible light. The research goal of TiO_2 photocatalysis, as for other semiconductors, is represented by the combination of TiO_2 with other nanomaterials to achieve both visible light activation and adsorption capacity improvement, with the simultaneous limitation of the electron-hole recombination rate. For this reason, to improve the photocatalytic performances in the use of TiO_2, different approaches have been investigated, such as the use of co-catalysts, and loading with noble metal particles, dye, metallic or non-metallic doping [28–36]. Recently, enhancements of the photocatalytic activity of TiO_2 by visible light have been also demonstrated by the modification of TiO_2 with carbonaceous substances such as fullerenes, carbon nanotubes, and graphene to form carbon-TiO_2 composites [37–40].

In particular, graphene nanomaterials in combination with TiO_2 highlight new prospectives in the field of photocatalysis for their large specific surface area, flexible structure, extraordinary mobility of charge carriers at room temperature, high thermal and electrical conductivities, and high chemical stability [41–48], emerging as one of the most promising materials for enhancing the photocatalytic performance in the new generation of photocatalysts [49–56].

In this short review, the recent research advances in graphene-TiO_2 employed in photocatalysis are presented with the aim of extending the light absorption of TiO_2 from UV wavelengths into the visible region, as well as increasing the photocatalytic activity of the compound, focusing on the applications of synthetic dye degradation.

2. Graphene Materials: Concepts and Properties

Graphene (G) is the term used to indicate carbon atoms tightly packed in a plane monolayer into a two-dimensional (2D) "honeycomb lattice" that represents the building block for all graphitic materials such as fullerenes, nanotubes, and graphite [48,57].

The isolation of graphene from graphite in 2003, by Prof. Andre Geim and Prof. Kostya Novoselov, which was the first paper published in Science in 2004 [42] and was awarded the Nobel Prize in Physics in 2010, has fascinated and revolutionized the scientific community [42,58,59]. This is due to graphene's notable properties and to the high range of applications that these properties offer to technologies such as energy storage and photocatalysis [60–67].

The IUPAC commission proposed the term "graphene" to substitute the older term "graphite layers", which was inappropriate in the research of two-dimensional (2D) monolayers of carbon atoms because the three-dimensional (3D) stacking structure is identified as "graphite" [59].

After the first report on graphene, which was obtained manually by mechanical cleavage with a Scotch tape sample of graphite, various techniques developed to produce thin graphitic films have been reported [55,68,69]. These can be divided into two main groups: bottom-up and top-down methods [55,69], as schematized in Figure 1.

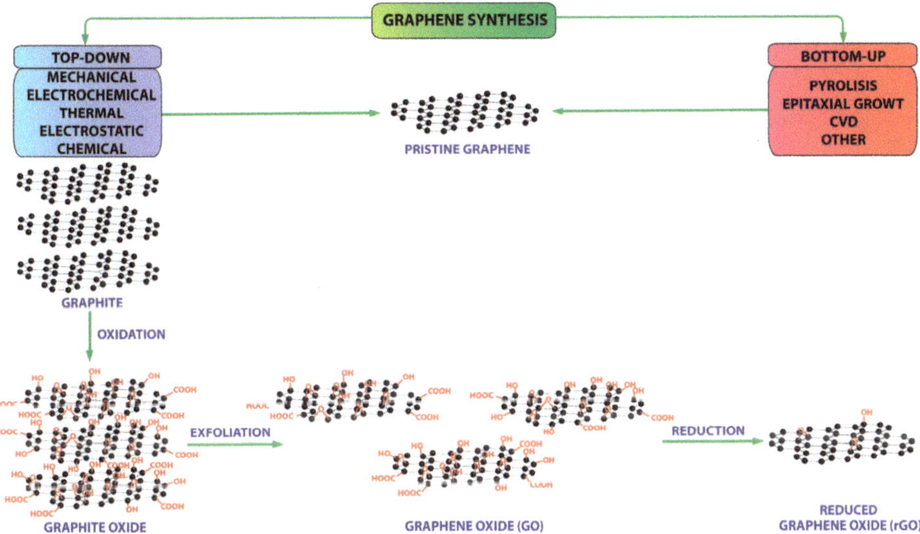

Figure 1. Schematic graphene synthesis: top-down and bottom-up methods.

In the bottom-up growth of graphene sheets (Table 1), the synthesis of graphene [70] can be obtained via epitaxial growth [71–73], chemical vapor deposition (CVD) [74–82], electrochemical reduction of CO and CO_2 [83,84], arc discharge [85,86], unzipping carbon nanotubes [70,87], organic synthesis [88], and pyrolysis [89–91].

The top-down approach (Table 2) offers considerable economic advantages over bottom-up methods [92,93], producing high quality graphene. In this case, the graphene is derived from the exfoliation of graphite, from mechanical [42,94–97], electrochemical expansion [98–100], thermal [101], electrostatic deposition [102], and chemical synthesis [103–105]. Some other techniques are unzipping nanotube [106–108] and microwave synthesis [109,110], or by chemical and/or thermal graphene oxide (GO) reduction [111]. This last method allows the production of low-cost and large-scale graphene, though with oxygen-containing groups and defects [112,113]. Much interest is directed at the study of proceedings to obtain graphene in the form of highly reduced graphene oxide (rGO) [114,115] or chemically modified graphene [116,117] from the oxidation and exfoliation of graphite and successive chemical reduction.

Table 1. Graphene synthesis with the bottom-up approach.

		Results	Highlight	Refs.
Epitaxial growth	-	General	Experimental and theoretical aspect	[71]
	-	Flower defects and effects of G grown on SiC substrate	Enhancements in the electronic properties of G-based devices	[72]
	-	G nanoribbons (GNR) on surface facets of SiC(0001)	Influence of the substrate step height on the energy barrier	[73]
Chemical vapor deposition (CVD)	-	Large-size, single-crystal, twisted bilayer G	Raman spectra measured as a function of the rotation angle	[74]
	-	G	Review: challenges and future perspective	[75]
	-	G on Ni and Cu substrates	Representative applications	[76]
	-	200 mm G on Ge(001)/Si(001) wafers	Absence of metallic contaminations	[77]
	-	Multilayer G films	Educational experiments: economical, safe, and simple technique in 30–45 min	[78]
	-	Large-scale G films on thin Ni layers	Two methods to shape and transfer films to specific substrates	[79]
	-	Monolayer G films on SiC(0001)	Ex-situ graphitization in argon atmosphere	[80]
	-	Centimeter-scale single- to few-layer G on Ni foils	Efficient roll-to-roll process	[81]
	-	N-doped graphene	N-type behavior useful to modulate G electrical properties	[82]
Electrochemical reduction of CO and CO_2	-	Several types of nanocarbons of controlled shape	Direct reaction of CO_2 with Mg metal	[83]
	-	G flakes	Room-temperature synthesis method on copper foil from different carbon sources using external charges	[84]
Arc discharge	-	Few-layered G	The reactivity of buffer gases (helium, oxygen-helium, and hydrogen-helium) is the key factor	[85]
	-	Few-layered G	Different mechanisms in the presence and absence of TiO_2 and ZnO catalysts	[86]
Unzipping carbon-nanotubes (CNT)	-	Chemical-free G	Radial and shear loading unzipping modes with cryomill method at 150 K	[70]
	-	G nanoribbons	Linear longitudinal opening of the Multi Wall CNT (MWCNT)	[87]
Organic synthesis	-	Two-dimensional G nanoribbons	Highly ordered monolayers (2D crystal) of larger G ribbons	[88]
Pyrolysis	-	General description	Pyrolysis of organic matter	[89]
	-	Metal-free catalyst of G sheets	Spray pyrolysis of iron carbonyl and pyridine	[90]
	-	G nanoporous with high specific surface area	Spray pyrolysis at different temperatures, graphene oxide-based precursor, nitrogen carrier gas	[91]

Table 2. Graphene synthesis with the top-down approach.

	Results	Highlight	Refs.
Mechanical	- Monocrystalline graphitic films	Exfoliation of small mesas of highly oriented pyrolytic graphite	[42]
	- G	Review: general description	[94]
	- G sheets	Dispersion and exfoliation of graphite in N-methyl-pyrrolidone	[95]
	- Stably single-layer G sheets in organic solvents	Exfoliation–reintercalation–expansion of graphite	[96]
	- G nanoplatelet	Thin films on a low-density polyethylene substrate	[97]
Electrochemical expansion	- Functional G sheets	Hyperexpanded graphite by electrolysis in a Li+ containing electrolyte and in situ electrochemical diazonium functionalization	[98]
	- G flakes	Cathodic graphite expansion in dimethylformamide (DMF) and functionalization by reducing aryl diazonium salts in organic solution	[99]
	- G	Exfoliation temperature increase from 25 to 95 °C reuslt in decrease of defects and increase of thermal stability (with H_2O_2 addition)	[100]
Chemical synthesis	- G	Review: general description	[103]
	- Carbon nanoscrolls	Low-temperature, catalyst-free graphite intercalation with alkali metals	[104]
	- G and chemically modified G	From colloidal suspensions	[105]
Unzipping nanotube	- Few-layer nanoribbons	Mechanical sonication and gas-phase oxidation in organic solvent of multiwalled carbon nanotubes	[106]
	- G nanoribbons	Lengthwise cutting of MWCNTs by a solution-based oxidative process	[107]
	- G nanoribbons	Plasma etching of CNT partly embedded in a polymer film	[108]
Microwave synthesis	- Flower-like G on hexagonal boron nitride crystals (h-BN)	Microwave G growth on polymethyl methacrylate (PMMA)-coated h-BN flakes	[109] [110]
Chemical and/or thermal graphene oxide (GO) reduction	- Polylactic acid (PLA)/rGO nanocomposites	Glucose and polhyvinilpyrrolidone (PVP) reduction of GO mixed with PLA	[111]
Reduction	- rGO - rGO	Two different reducing mixed reagents: HI/NH and NH/HI. Review: 50 types of reducing agents	[114] [115]
Other	- G - G networks	Review: different sizes and chemical compositions. Controlled segregation of G chemically modified on liquid interfaces	[116] [117]

The strong oxidation of graphite produces bulk solid graphite oxide that can be exfoliated in water or other suitable organic solvents to form GO [118–120]; several recent reviews report the synthesis of graphite oxide [121–123]. Graphite oxide can be prepared with Brodie [124], Staudenmaier [125], and Hummer's [126] methods. The synthetic technique and the extent of the reaction influence the degree of graphite oxidation; the most oxidized graphite oxide is produced with the Staudenmaier method, but this reaction may take several days. However, because both Staudenmaier and Brodie methods generate highly toxic ClO_2 gas that can decompose in air violently; thus, the most widely used method to prepare graphite oxide is Hummer's method.

With Hummer's method, the GO is prepared by exfoliating graphite oxide obtained from the oxidation of graphite powder with strong chemical oxidants, such as HNO_3, $KMnO_4$, and H_2SO_4 [112,113,126]. GO can be successively reduced to graphene by the partial restoration of the sp^2- hybridization by thermal [127], chemical [128], electrochemical [129], photothermal [130], photocatalytic [69,131], sonochemical [132,133], and microwave reduction methods [134]. The high number of oxygen-containing groups of the obtained GO product permits the interaction with the cations, providing important reactive molecules for the nucleation and growth of nanoparticles and therefore the formation of various graphene-based composites.

3. The Photocatalytic Process: Fundamentals of Graphene-TiO$_2$ Photocatalysts

Photocatalysis is based on the activation of a semiconductor (SC) by the sun or artificial light. When an SC material is irradiated with photons whose energy is higher or equal to its band gap energy, a promotion of an electron from the valence band (VB) to the conduction band (CB) occurs with the concomitant generation of a hole in the valence band VB [135].

In a water system, oxygen adsorbed on the surface of the semiconductor acts as an electron acceptor, while the adsorbed water molecules and hydroxyl anions act as electron donors, leading to the formation of a very powerful oxidizing $^{\bullet}$OH radical. The electrons react with oxygen molecules to form the superoxide anion, $O_2^{\bullet-}$. In the presence of a contaminant organic molecule (M), adsorbed on the catalyst surface, the $^{\bullet}$OH radical is the primary oxidant that reacts to produce adducts, followed by the fragmentation of the molecular structure into several intermediates species until the total mineralization of contaminant is completed with the formation of CO_2 and H_2O [136–138].

The successful of photocatalytic process is therefore strictly dependent on the competition between the reaction of the electron with water on the SC surface and the electron-hole recombination process that releases heat or radiation.

The overall process can be described by the following reactions:

- *electron $-$ hole pair generation :*

$$SC + hv \rightarrow SC\left(e_{CB}^- + h_{VB}^+\right)$$

- *hydroxil radicals formation :*

$$SC\left(h_{VB}^+\right) + H_2O_{ads} \rightarrow SC + {}^{\bullet}OH_{ads} + H_{ads}^+$$

- *superoxyde formation from electrons in CB and oxygen :*

$$SC(e_{CB}^-) + O_{2\,ads} \rightarrow SC + O_{2ads}^-$$

- *adsorption of M on catalyst :*

$$SC\left(h_{VB}^+\right) + M_{ads} \rightarrow SC + M_{ads}^+$$

- *the hydroxyl radical is the primary oxidant :*

$$OH_{ads} + M_{ads} \rightarrow intermediate\ products \rightarrow CO_2 + H_2O$$

As an example, Khan et al. [23] defined the different intermediate degradation products of Methyl Orange by using TiO$_2$ photocatalysts that are resumed in Figure 2. The reported compounds are derived from the reaction of $^{\bullet}$OH radicals with the final mineralization in CO_2 and water.

It is important to consider that, because the photocatalytic process is strictly dependent on the kinetics of charge carriers and redox reactions, the knowledge of the electronic processes that occur at the level of the SC surface are of great importance. In the SC, the oxidation fromphotogenerated holes and the reduction fromphotogenerated electrons may occur simultaneously and, to keep the photocatalyst electrically neutral, should occur at the same rate.

The photocatalytic activity is greatly limited by electron–hole recombination, and therefore several methods are used to increase the efficiency of charge carrier separation and therefore to improve the photocatalytic performance of the SC photocatalyst [139–141]. In this context, the combination of graphene with the SC photocatalyst represents an innovative strategy [55,69,142].

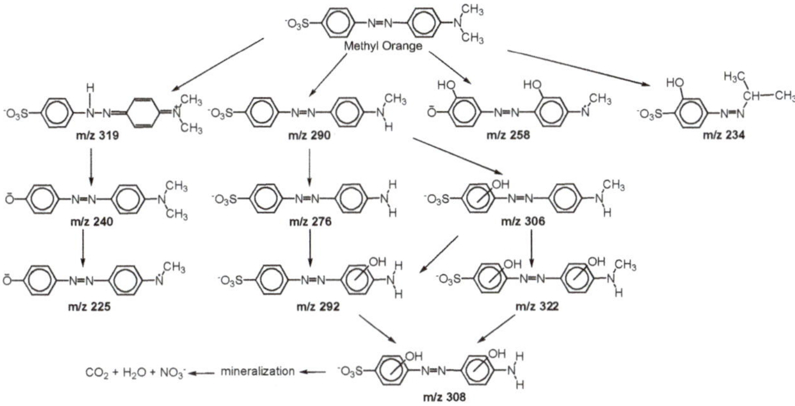

Figure 2. Proposed photodegradation mechanism of Methyl Orange [23].

Titanium dioxide (TiO2) is generally considered the best SC material that can be used as a photocatalyst [18,139,143], showing almost all of the required properties for an efficient photocatalytic process, except for activity under visible light irradiation.

From these considerations, TiO2/graphene nanocomposites (GTiO$_2$Ns) are widely used for photocatalytic applications, exploiting their potential in environmental applications.

The effective role of G in a photocatalytic event on GTiO$_2$Ns has not been completely investigated, or is not completely understood. In order to explain the possible mechanism, the light source (UV or VIS light) and the presence/absence of dye molecules, or rather the presence of a compound able to absorb visible light, are fundamental.

When the target molecule does not adsorb light and is efficiently adsorbed on the photocatalyst surface:

(1) UV light excitation of GTiO$_2$Ns photogenerates the electron–hole pairs and the electrons are then injected into graphene due to its more positive Fermi level [144]. This process is favored by the position of the work function of graphene that is −4.42 eV, with respect the conduction band of TiO$_2$ that is located at −4.20 eV [145]; from this consideration, the electron in the CB of TiO$_2$ is injected to G. Graphene scavengesphotogenerated electrons by dissolved oxygen; facilitates the hole-electron separation, reducing the recombination of e-(CB) and the holes (VB); and, due to its high carrier mobility, accelerates the electron transport, thus enhancing the photocatalytic performance [142,146].

(2) When the operational mechanism takes place via visible light, the electron transfer of thephotogenerated electron is promoted from the G photoexcited state and then delocalized to the TiO$_2$ surface. M.T Silva et al. indicated, by rGO photoluminescence study, that the photogenerated electrons under Vis or NIR laser can be transferred to the surface of TiO$_2$ with a consequent quenching of photoluminescence; also in this case, charge recombination is inhibited with a consequent increasing of photocatalytic activity under visible light [147]. It is important to know that the presence of G in GTiO$_2$Ns photocatalyst produces a red shift in the absorption, reducing its band gap and thus extending the photoresponse to a longer wavelength [148]. The explanation of visible light activation in the GTiO$_2$Ns composites is not clear, but it is possible to attribute this phenomena to the sensitization of TiO$_2$ due to the presence of graphene [147,149,150]. In this case, in the visible light excitation of GTiO$_2$Ns, graphene absorbs the light, and the photoexcited electrons in high energy G states are delocalized into the CB of the TiO$_2$ surface with the dissipation of excess energy due to electron vibrational interaction [151]; successively electrons

react with oxygen, resulting in the formation of superoxide radicals. In Figure 3, the activation mechanisms of GTiO$_2$Ns under UV and visible light are reported.

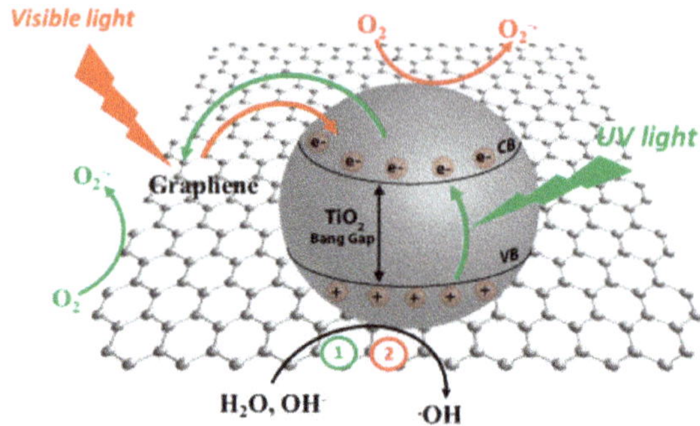

Figure 3. Mechanisms of UV and visible light activation of TiO$_2$ with G.

When a dye (D) as a target molecule is efficiently adsorbed onto the photocatalyst surface, the UV mechanism is the same as that reported before (1) while, under visible light, different activation pathways can be found. In this case, the dye acts both as a sensitizer for visible-activation and as a pollutant during the photocatalytic process, promoting the electron transfer from the excited dye molecules state to the TiO$_2$ CB while G acts as an electron scavenger, as in the UV mechanism [147].

The other possible pathway under visible light involves G which acts as visible light sensitizer promoting the electron transfer to the TiO$_2$ CB (as in 2) [152–154] (Figure 4). This process is less probable due to the competition between G and D as light sensitizers, but in any case is possible.

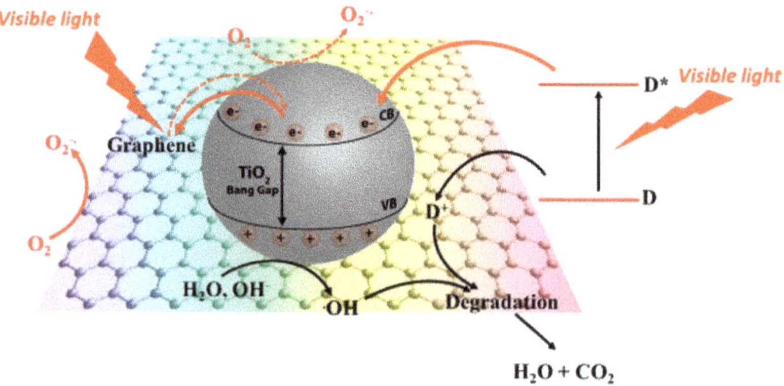

Figure 4. Mechanisms of UV and visible light activation of TiO$_2$ with G in the presence of dye (D).

4. Preparation Methods of GTiO$_2$Ns

According to the facts that graphene doping of TiO$_2$ generally contributes to the bathochromic shift of the absorption band, hinders the recombination of h$^+$/e$^-$ by transferring the photoexcited electron to the graphene surface, which also enhances the surface area of TiO$_2$ for better adsorptive properties [155], the investigation of GTiO$_2$Ns photoactivity under visible light seems to be crucial. The improvement of performance due to the presence of G in GTiO$_2$Ns photocatalyst is primarily attributed to the extension of the light absorption range, the increase of absorptivity, and the efficient charge separation and transportation. With GTiO$_2$Ns photocatalyst, the rate of h$^+$/e$^-$ recombination after light excitation decreases [156], while the charge transfer rate of electrons increases in addition to the surface-adsorption of chemical species thanks to π-π interactions [157].

However, the properties $GTiO_2Ns$ and different operational parameters, such as the characteristics of substrates, UV-vis or Vis light irradiation, etc., can affect the photocatalytic efficiency [157].

$GTiO_2Ns$ can be generally categorized into three kinds: TiO_2-mounted activated graphene [92], graphene-doped TiO_2, and graphene-coated TiO_2 [158]. Each of these types exhibits good photocatalytic activity. In order to improve the efficiency, the surface properties of graphene could be adjusted via chemical modification, which facilitates its use in composite materials [156,159,160].

$GTiO_2Ns$ can be generally realized in two different ways by "ex situ hybridization" or "in situ crystallization". In Figure 5, a schematic representation of ex situ hybridization and in situ crystallization in the synthesis of $GTiO_2NPs$ is reported.

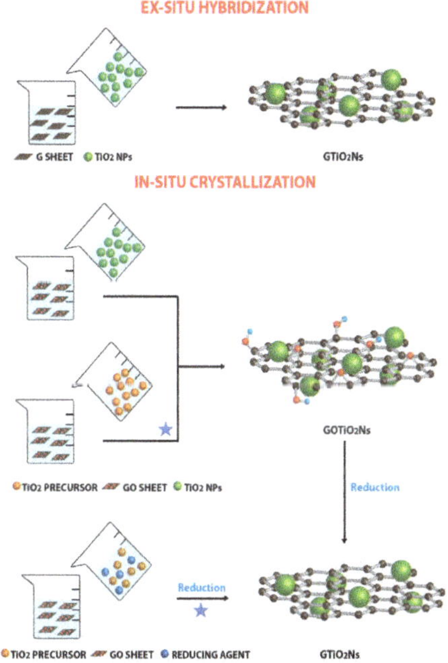

Figure 5. Schematic representation of ex situ hybridization and in situ crystallization in the synthesis of $GTiO_2Ns$.

Ex situ hybridization (Table 3) involves the mixing of graphene dispersions with pre-synthesized TiO_2 nanoparticles (TiO_2NPs) [147]. To enhance their solubility and improve the quality of $GTiO_2Ns$ composites, before mixing, TiO_2NPs and/or graphene sheets can be functionalized by covalent C–C coupling [59,161] or non-covalent π-π stacking reactions [162]. For example, Morawski et al. [163] prepared a visible light-active TiO_2-reduced GO photocatalyst by mechanically mixing TiO_2 with an appropriate mass ratio of reduced GO in 1-butyl alcohol and successive ultrasonication. Yong Liu et al. [164] prepared a graphene aqueous dispersion by sonicating a mixture of graphene raw materials with polyvinylpyrrolidone (PVP) as surfactant and a suspension of TiO_2 by dispersing TiO_2 powder in deionized water with ultrasonication. In this preparation, graphene-TiO_2 composite material was obtained by simple mechanical mixing and sonication using anhydrous ethanol to improve the wettability of the dispersion, while spray-coating was carried out both on polycarbonate and on rubber substrates to facilitate the testing of photocatalytic performance. Ramesh Raliya et al. [165] also prepared nanomaterial/composite solutions by mixing GO at different concentration ratios to TiO_2 and tested their effects on photocatalytic performance.

However, with ex situ hybridization, in some cases it is possible to obtain a low-density and non-uniform coverage of nanostructures by G sheets [113,166–169].

The most common strategy to synthetize $GTiO_2Ns$ nanocomposites is represented by *in situ crystallization* (Table 3). In this case, GO or rGO are usually used as starting materials for the presence of oxygen-containing functional groups on their surface. In fact, they act as a nucleation point for grooving and anchoring semiconductor nanocrystals. The successive reduction of GO generates $GTiO_2Ns$ with homogeneous distribution on 2D nanosheets, thereby promoting the direct interaction between semiconductor nanocrystals and G. For this reason, various methods, such as mixing, hydrothermal/solvothermal methods, sol-gel, electrochemical deposition, combustion, microwave, photo-assisted reduction, and self-assemble approaches, can be applied for the synthesis of $GTiO_2Ns$ [66,170,171].

In the *mixing method* (M), GO and semiconductors are mixed with stirring and ultrasonication in order to obtain the exfoliation of GO and, successively, GO/TiO_2 product with uniform distribution; in a second step the reduction of GO is performed [51,172–175].

A strong mixing and the best chemical interaction between GO and TiO_2 can be obtained by applying the sol-gel method (SG). As an example, Zhang et al. [176] used a sol-gel method to prepare a series of TiO_2 and graphene sheet composites by using tetrabutyl titanate and graphite oxide and the obtained precursors were calcinated at 450 °C for 2 h under air or nitrogen atmospheres. Zabihi et al. [177] prepared photocatalytic composite thin films of $GTiO_2Ns$ with a low-temperature synthetic process by using G dispersion and titanium isopropoxide bis (acetylacetonate) solution by using a sol-gel chemical method. Spin- and spray-coating was successively used for films deposition; the use of ultrasonic vibration demonstrated a positive effect from both sol-gel and deposition processes that produced rutile and anatase nanoparticles inclused in a matrix of few-layered graphene thin film.

Gopalakrishnan et al. [178] prepared TiO_2-reduced GO nanocomposites in situ from the incorporation of TiO_2 sol into GO sheets and the successive solvothermal method without the addiction of any chemicals. Shao et al. [179] reported a novel so-called gel-sol process for the synthesis of TiO_2 nanorods combining rGO composites by utilizing the triethanolamine as a shape controller and under specific conditions. The as-fabricated hybrid composites presented distinct advantages over the traditional methods in terms of the well-confined TiO_2 morphology and the formation of Ti-C bonds between TiO_2 nanoroads and rGO simultaneously. Long et al. [180] used the hydrothermal method to prepare efficient visible light active $GTiO_2Ns$ by using a GO suspension and TiO_2 sol with undergrown TiO_2 nanoparticles at 413 K. The study of obtained materials demonstrated high chemical interaction at the interface of GO sheets and the polymeric Ti-O-Ti structure that facilitated the retaining of more alkoxyl groups; the induced crystal disorders and oxygen vacancies contributed positively to the performance of the obtained photocatalyst.

In the *hydrothermal/solvothermal methods* (HD/SD), semiconductor nanoparticles or their precursors are loaded on GO sheets that are successively reduced to rGO involving reactions under controlled temperature and or pressure. For example, Hao Zhang et al. [181] obtained a $GTiO_2Ns$ (P25) nanocomposite photocatalyst with a high adsorptivity of dyes and a greater light absorption range by using a one-step hydrothermal reaction obtaining both GO reduction and P25 cover. Hu et al. [182] successfully synthesized different structures of TiO_2-graphene composites through the hydrothermal reaction by using GO and different titanium sources in hydrothermal conditions. The results showed the larger interfacial contact between TiO_2 and G, in addition to the greater surface area of the poriferous composite. Hamandi et al. [183] studied the influence of the photocatalytic performance of TiO_2 nanotubes (NT) resulting from the addition of GO. TiO_2 nanotubes were prepared using alkaline hydrothermal treatment of TiO_2 P25 followed by calcination at 400 °C under air, while GO-NT composites were obtained by the wet impregnation of the as-prepared TiO_2 nanotubes onto graphene oxide before reduction under H_2 at 200 °C. Sijia et al. [184] synthesized a novel composite photocatalyst, reduced graphene oxide (rGO)-modified superlong TiO_2 nanotubes with a length of about 500 nm, by an improved hydrothermal process and a heating reflux method.

A hybrid nanocomposite containing nanocrystalline TiO_2 and graphene-related materials (GO and rGO) was prepared from Kusiak et al. [185] by using the hydrothermal method under elevated pressure. In this preparation, the presence of graphitic carbon depends on temperature, the type of modification method, and the preparation conditions, obtaining multilayer GO flakes in comparison to rGO flakes. This multilayer character produced the most intensive time resolved microwave conductivity signals under both UV and visible illumination for materials modified with GO, indicating that more electrons are induced in the conduction band of hybrid nanocomposites. Li et al. [186] synthesized reduced graphene oxide-TiO_2 (rGO-TiO_2) composites with a sandwich-like structure by using a simple solvothermal method. Hemraj et al. [187] synthesized nanocrystalline anatase TiO_2-GO nanocomposites by two steps with the initial preparation of anatase TiO_2NPs by a simple sol-gel method and the successive decoration of GO on its surface by using a facile solvothermal method. Ju Hu et al. [188] synthesized TiO_2/reduced graphene oxide nanocomposites by using a one-step surfactant-assisted hydrothermal method. Compared with the control TiO_2/rGO nanocomposite, TiO_2/rGO-X (X = sodium dodecyl benzene sulfonate, Triton X-100, and acetyl trimethyl ammonium bromide) presented excellent photocatalytic activity. The study indicated that the surfactant-assisted hydrothermal method is an effective approach to improve the structure, morphology, and photocatalytic performance of TiO_2/rGO composites.

Semiconductor nanostructures can be *electrolytically deposited* on the surface of graphene-based substrates, at low temperature condition, from an aqueous solution containing the semiconductor. For example, Ming-Zheng Ge et al. [189] adopted both electrodeposition and carbonation methods to deposit rGO films on TiO_2 nanotubes obtained with two-step electrochemical anodization. Recently, Guimaraes de Oliveira et al. [190] used the electrochemical method to deposit GTiO$_2$Ns composite film on a Ti substrate by using GO and Ti(IV) aqueous solution, demonstrating a photocatalytic activity almost twice as high as that observed for the TiO_2 only film.

The *microwave-assisted strategy*, in this context, addresses the synthesis of semiconductor nanomaterials of a controlled size and shape. For example, Shanmugama et al. [191] synthesized GTiO$_2$Ns nanocomposites by a novel surfactant-free, environmentally friendly one-pot in situ microwave method. This method leads to the uniform distribution of TiO_2 nanoparticles on G sheets with the binding nature of TiO_2. Wang et al. [192] reduced GO with both direct and microwave-assisted reduction in the presence of Ti powders, showing that the microwave effect decreased the reduction time. In this case, Ti ions derived from the reaction of Ti powder with GO were hydrolytically transferred to TiO_2 with the formation of rGO-TiO_2, which was proven to be an active material in the removal of methylene blue. Yang et al. [193] prepared nanocomposites of TiO_2 and rGO, with a fast and simple microwave irradiation method by the reaction of graphene-oxide and commercial TiO_2 nanoparticles in water/ethanol solvent, examining different microwave powers with different time intervals. The effect of time and power on absorption wavelength and the creation of defects in the graphene layer was examined, obtaining an optimum irradiation time in which the nanocomposite presented the highest absorption wavelength (the smallest band gap), and the optimum value for microwave power in which the nanocomposite had the lowest number of defects. First, GO was prepared using a modified Hummer's method, and then TiO_2 and the mixture of water/ethanol and graphene-oxide (GO) was used for the hydrolysis of TiO_2, followed by its mounting on graphene-oxide. In this way, microwave irradiation reduced GO to graphene with the formation of TiO_2/rGO nanocomposites.

Other methods have been investigated, such as that proposed by Pu et al. [194] regarding the synthesis of TiO_2-rGO nanocomposites, conducted simultaneously with the *photoreduction* (PR) of GO nanosheets in a simple GO and TiO_2 ethanol system, demonstrating a facile and environmentally friendly strategy for the in situ preparation of the TiO_2-rGO hybrid "dyade". In Table 3, the principal synthesis methods of GTiO$_2$Ns are summarized.

Table 3. The synthesis methods of Graphene Based TiO_2 nanocomposites.

Ex-Situ Hybridization				
Photocatalyst	**Synthetic Route**	**Starting Graphite**	**Starting Semiconductor**	**Refs.**
rGO/TiO_2	Mechanical mixing and ultrasonication	rGO	TiO_2NPs in 1-butyl alcohol	[163]
G/TiO_2	Mechanical mixing and ultrasonication	G in PVP/water	TiO_2NPs anatase in water	[164]
GO/TiO_2	Mechanical mixing	GO	Titanium isopropoxide	[165]
In-Situ Crystallization				
rGO/TiO_2	M with GO and PR reduction	GO	TiO_2NPs	[174]
	M and GO and ST reduction	GO	TiO_2NPs	[175]
G/TiO_2	ST	GO	Tetrabutyl titanate	[145]
G/TiO_2	SG	rGO	Titanium isopropoxide	[146]
G/TiO_2	HD	GO	P25	[147]
rGO/TiO_2	SG	GO	Tetrabutyl titanate	[176]
G/TiO_2	SG	G dispersion	Titanium isopropoxide	[177]
rGO/TiO_2	Solvothermal SG	GO sheets	TiO_2 sol	[178]
rGO/TiO_2	SG	GO	Titanium isopropoxide	[179]
G/TiO_2	HD	GO suspension	TiO_2 sol	[180]
G/TiO_2	one-step HD	GO water/ethanol	TiO_2	[181]
G/TiO_2	HD	GO	Different Ti sources	[182]
G/TiO_2	Wet impregnation and thermal reduction (H_2)	GO	TiO_2 nanotubes from *HD* of TiO_2 P25	[183]
rGO superlong TiO_2	HD and heating reflux	GO	Super long TiO_2	[184]
GO and rGO/TiO_2 (nanocristals)	Elevated pressure HD	GO and rGO	TiO_2	[185]
rGO/TiO_2 sandwich-like structure	HD/ST	rGO	Butyl titanate	[186]
nanocrystalline anatase TiO_2-GO	SG and GO decoration by ST	GO	Anatase TiO_2NPs by *SG*	[187]
TiO_2/rGO-X nanocomposites	One-step surfactant (X)-assisted HD	GO	TiO_2 (P25)	[188]
rGO films on TiO_2 nanotubes	Two-step ED and carbonation techniques	rGO	Two-step anodized TiO_2 nanotubes from Ti foils	[189]
Ti plate deposited TiO_2 and GO film	ED	GO from nanographite	Ti plate and K_2TiF_6 aqueous solution	[190]
G/TiO_2	MW	GO	$TiCl_4$	[191]
rGO/TiO_2	MW	GO	Ti powder	[192]
rGO/TiO_2	MW	GO	TiO_2NPs	[193]
rGO/TiO_2	PR	GO	TiO_2NPs	[194]
rGO/TiO2	PR	GO	Colloidal TiO_2	[156]

5. Photocatalytic Degradation of Dyes with GTiO$_2$Ns

Dye molecules, which are solubilized during several application processes, present different colors derived from the selective absorption of light. Coloring properties of dyes depend on their chemical structure consisting of two fundamental components: the chromophore and the auxochrome. The chromophore is a covalently unsaturated group, which is responsible for the color production for the absorption in the UV or visible region; the auxochrome is supplementary to the chromophore, and influences the solubility in water and the affinity towards a particular support [195]. Concerning their uses, dyes can be anionic, cationic, and non-ionic; anionic dyes are divided into direct, acid, and reactive dyes, while cationic dyes are basic [196]. Water treatments by using $GTiO_2Ns$ in the photocatalytic degradations of dyes became very important due to the danger of dyes as environmental pollutants [142].

Methylene blue (MB) was first prepared by Caro in 1876 as an aniline-derived dye for textiles, and is largely used as a water-soluble cationic dye in the dyeing of paper, wood, plastic, silk, and cotton, as well as in scientific research and pharmaceutical industries [197,198]. Kyeong Min Cho et al. [8], to obtain MB photodegradation, used hydrothermal methods in the self-assembly of TiO_2 precursors, GO, and a surfactant to prepare rGO mesoporous TiO_2; this material showed high photocatalytic performances in the MB degradation test, good ability in the charge separation, and a large surface area compared to typical $GTiO_2Ns$.

Hanan H. Mohamed [199] recently described the simultaneous photoassisted oxidation of MB by using new synthesized hierarchical structures of anatase/rutile TiO_2 microsphere-rGO nanocomposites. Interestingly, rutile/anatase phase transformation was observed upon changing the GO content. The synthesized nanocomposites showed enhanced photocatalytic activity for the dye degradation compared to pristine TiO_2 nanoparticles that was attributed to the synergetic effect between anatase and rutile in the synthesized TiO_2 microsphere composites. Also, Zhang et al. [200] synthesized graphene-supported mesoporous titania nanosheets ($GTiO_2Ns$) by using liquid-phase exfoliated G as a template and sandwich-like G-silica as intermediates. $GTiO_2Ns$ showed a mesoporous structure formed by crystalline TiO_2 nanoparticles anchored on G nanosheets that showed high surface areas. These characteristics significantly enhanced the photocatalytic activity of TiO_2 in the MB degradation.

Najafi et al. [201] fabricated TiO_2-GO nanocomposites with different nanowire and nanoparticle morphologies of TiO_2 by using a one step hydrothermal method. The different morphologies of TiO_2 were grown on the surface of the GO sheets, previously synthesized by the modified Hummer's method. The obtained TiO_2-GO nanocomposites showed a tetragonal structure and covalent bonds between TiO_2 nanostructures and GO sheets. The best photodegradation rate of MB was found by using TiO_2 nanowire-GO nanocomposites. Divya et al. [202] used microwave irradiation of different GO weights and tetrabutyl titanate in isopropyl alcohol to obtain TiO_2 hybridation with the formation of $GTiO_2Ns$ nanocomposites. In this method, the irradiation with microwaves increase the temperature in a short time, favoring the GO reduction to G and the growth of TiO_2 nanoparticles on the G surface. The obtained $GTiO_2Ns$ nanocomposites presented efficient electron conductivity in G, therefore showing a reduced electron-hole recombination rate.

Another example is that of Sohail et al. [203], who prepared GO nanosheets by using a modified Hummer's technique with the successive conversion into particles by a spray-drying method. TiO_2 nanoparticles were grown on the surface of GO particles to obtain rGO-TiO_2 nanocomposites that showed better photocatalytic activities upon the degradation of MB compared to pure TiO_2, also showing in this case that rGO played an important role in the charge recombination to enhance the electron-hole separation.

Tseng et al. [204] studied a facile process to fabricate two-dimensional titania nanosheets (t-NS) by using GO as a support with in situ growth of anatase TiO_2 on GO suspended in butanol by combining the sol-gel and solvothermal processes without further calcination. Significant amounts of MB molecules can easily be adsorbed on the intrinsic graphitic structure. An optimum rate of hole titration by ethylenediaminetetraacetic acid was essential to efficiently reduce the recombination of charge carriers and consequently produce more active radicals to decompose MB. The strong interaction between graphitic and titania structures, rather than the crystallite size of anatase, dominated the photoreduction capability of t-NS.

Similarly, Minella et al. [205] obtained rGO-TiO_2 and rGO-SiO_2 hybrid materials by reducing different quantities of GO with hydrazine in the presence of TiO_2 and SiO_2 nanoparticles;

the photocatalytic activity of these nanomaterials was evaluated in the degradation of MB under both UV-vis and only visible light irradiation. The results showed that MB was strongly adsorbed on these new materials with an increase of degradation rates with respect to pristine semiconductors. The visible light degradation of MB was attributed to the dye-sensitization mechanism, while the UV-vis degradation was considered to be the typical semiconductor photocatalytic mechanism.

"Biphasic TiO_2 nanoparticles" and their G nanocomposites were synthesized by Raja et al. [206] by the hydrothermal method. Introducing high surface area, G suppressed the electron-hole pair recombination rate in the nanocomposites. Further, the nanocomposites showed a red-shift of the absorption edge with a contraction of the band gap from 2.98 eV to 2.85 eV. The characterization of photocatalytic activities under natural sunlight and UV-filtered sunlight irradiation revealed that the $GTiO_2Ns$ composite exhibited about a 15- and 3.5-fold increase in the degradability of Congo red and MB dyes, respectively, in comparison to pristine TiO_2. The authors therefore developed a visible light active G composite catalyst that can degrade both cationic and anionic dyes, making it potentially useful in environmental remediation and water splitting applications under direct sunlight.

The ex situ hydrothermal method was proposed from Verma et al. [207] to synthesize rGO mixed TiO_2 nano-composites. The study demonstrated that the cooperation between the optimal phase ratio of TiO_2 (anatase/rutile) and rGO produced a positive system in the photocatalytic degradation test with MB other than antibacterial activity. Atchudan et al. [208] used a solvothermal method to prepare GO grafting TiO_2 nanocomposites and studied it as a photoactive material under UV-light irradiation in the degradation of MB and methyl orange. The obtained degradation efficiency, compared with only TiO_2, demonstrated the important function that GO played in the increase of performance due to the increase of light absorption and reducing charge recombination.

Darvishi et al. [209] instead used titanium butoxide (TBO) as a TiO_2 precursor, commercial TiO_2, and GO in a water/ethanol mixture to produce hydrolytically, and under microwave irradiation, $rGO-TiO_2$ nanocomposites; these showed high conductive and light absorption properties, demonstrating an enhancement of photocatalytic MB degradation. In these nanocomposites, the content of G improved the photocatalytic performances of the photocatalysts.

An example of an eco-friendly method was studied by Rezaei et al. [210], who synthesized TiO_2-graphene nanocomposites by the addition of the "blackberry juice" to GO as a reducing agent to produce rGO nano-sheets. The obtained TiO_2 (anatase)-rGO materials exhibited an excellent photocatalytic activity toward MB degradation due to the enlarged surface area and the collaborative effect of rGO. Shanmugam et al. [191] synthesized $GTiO_2Ns$ nanocomposites by a surfactant-free, environmentally friendly one-pot in situ microwave method. TiO_2 nanoparticles of 5–10 nm were distributed on the G sheets. The photocatalytic activity of pure TiO_2 and $GTiO_2Ns$ nanocomposites was studied under UV and visible light irradiation sources with MB. The studies revealed the highest degradation efficiency of 97% with UV light and 96% with visible light irradiation by using the $GTiO_2Ns$ photocatalyst, opening the possibility of using this material for industrial wastewater treatment.

The sol-gel technique was used by Rezaei [211], who prepared $GO-TiO_2$ nanocomposites by evaluating the photocatalytic activities of MB aqueous solution under sunlight irradiation. The results showed that the nanocomposite containing 9.0 wt % of GO had the highest photocatalytic performance in the MB degradation to either "single-phase anatase" or other composites containing different GO quantities. The improvement of the photocatalytic activity was attributed to the favorable effect of the distribution of TiO_2 nanoparticles of less than 20 nm on GO sheets.

UV-assisted photocatalytic reduction of GO by TiO_2 nanoparticles in ethanol was used by Charoensuk et al. [212] in the preparation of $rGO-TiO_2$ nanocomposites. The photocatalytic activity of prepared rGO/TiO_2 and GO/TiO_2 nanocomposites was evaluated by the kinetics of the photocatalytic degradation of MB under UV irradiation. Significant roles in the MB photodegradation were played by important factors such as the bandgap, electron-hole recombination, characteristics of surface (area and functional groups), and adsorption capacity of nanocomposites. The results revealed that rGO/TiO_2 and GO/TiO_2 nanocomposites exhibited efficient charge separation, showing about 500% improvement of photocatalytic activity in MB photodegradation compared to pristine TiO_2.

The solvothermal technique was used by Wang et al. [213], who prepared an efficient photocatalyst of nano TiO_2-functionalized GO nanocomposites. This nanomaterial showed an quadruple increase in MB photodegradation activity with respect to that obtained by using the P25-graphene composite photocatalyst. Liu et al. [214] instead used an interesting impregnation-hydrothermal method to prepare highly distributed TiO_2 nanoparticles with in situ growth on functional G (FG), which was before obtained by modifying GO by using triethanolamine. The obtained FGTiO_2Ns photocatalyst showed better photocatalytic activity under UV light irradiation with respect to pure TiO_2 and GO-TiO_2 prepared by other similar methods, also revealing considerable photocatalytic skill under visible light irradiation.

A fruitful method to produce expanded exfoliated GO (EGO) was used by Baldissarelli et al. [215]. The use of ozone exposure and thermal treatment transformated the graphite surface to a G-like surface containing oxygen and sp^3 carbon. The EGO, obtained through thermal treatment, produced, with the addition of TiO_2 nanoparticles, photocatalytic TiO_2-EGO nanocomposites that demonstrated higher activity with respect to that produced by the common Hummer's method. In particular, TiO_2-EGO showed enhanced MB photodegradation under UV light compared to TiO_2 P25.

A new nanomaterial composed of ultrasmall TiO_2 nanoparticles and rGO nanosheets was prepared by Gu et al. [216], who used glucosamine in alkaline conditions under hydrothermal treatment. In this synthesis, glucosamine regulated the growth and homogeneity of TiO_2 nanoparticles dispersed on the G structure, permitting the formation of the expected GTiO_2Ns products, demonstrating that TiO_2 nanoparticles about 13 nm in diameter were strongly anchored on G. Different GTiO_2Ns samples were calcined at several temperatures; in particular, the GTiO_2Ns sample treated at 700 °C showed good photocatalytic activity in MB photodegradation compared to those produced at other calcination temperatures.

Moreover, Fan et al. [217] used the hydrothermal method to synthesize TiO_2 nanospindles, which are featured by large exposed {0 0 1} facets, and then fabricated TiO_2 nanospindles/reduced GO nanocomposites (rGO-TiO_2). The photocatalytic activity in the photodegradation of MB showed an impressive photocatalytic enhancement of rGO-TiO_2 with respect to pure TiO_2 nanospindles. In addition, Sun et al. [218] used a one-step in situ hydrothermal method to prepare chemically bonded GTiO_2 nanorods hybrid composites with high dispersity by using GO and TiO_2 (P25) as the starting materials without using reducing agents. Ultraviolet-visible diffuse reflectance measurements of the composites showed an enhanced light absorption and a red shift of absorption edge. When GTiO_2Ns nanorods hybrid composites were used as photocatalysts, they showed an enhancement of photocatalytic performance in the photodegradation of MB under visible light irradiation compared to that of pristine TiO_2 nanorods.

Rong et al. [219] prepared different GTiO_2Ns photocatalysts with a simple hydrothermal method. First, GO (obtained from graphite oxidation) was dispersed in a water/ethanol solution with ultrasonic treatment, after which TiO_2 was added and a thermal treatment at 120° for 12 h was used. A series of GTiO_2Ns photocatalysts obtained with different oxidation times (in the GO production from graphite) and GO contents were obtained. When these materials were applied to the photodegradation of MB under visible light irradiation, they exhibited excellent photocatalytic activities.

Gupta et al. [220] proposed a strategy to fabricate G quantum dots (GQDs) infilled TiO_2 nanotube arrays hybrid structure (GQDs-TiO_2NTAs) for the application of MB degradation under UV light irradiation. In particular, anodic oxidation of a Ti sheet by using an impregnation method produced the fill of GQDs into TiO_2 NTAs. The application of this material in the adsorption and photodegradation of MB in an aqueous solution under UV light irradiation showed high photocatalytic efficiency, which was attributed to the favorable visible light absorption and to efficiency in the transfer of photogenerated electrons from the TiO_2NTAs to GQDs that produced large photoinduced charge separation. No less important was the strong adsorption capacity of the GQDs to MB molecules.

GP-wrapped TiO_2 ($GwTiO_2$) hybrid material was fabricated by Ni et al. [221] with the simultaneous reduction and wrapping of GO on the surface of high-reactive anatase TiO_2, based on the surface negatively charged property of GO. $GwTiO_2$ gave a strong red-shifted light absorption edge and a contracted bandgap compared to that of GP randomly supported TiO_2; this behavior was attributed to the positive chemical interaction between TiO_2 and G. The photocatalytic MB degradation under a xenon lamp and visible light irradiation confirmed the best quality of this nanomaterial. Furthermore, Yang et al. [222] prepared TiO_2/G porous composites with rGO by using a template of G microsphere colloidal crystals that showed high light absorbing properties, and good MB adsorption properties, other than a fast ability in the charge transportation and separation. These porous composites showed high degradation activities in MB degradation under visible light irradiation with a constant rate almost 6.5 times higher than that of P25 under the same conditions.

Also, Suave et al. [223] prepared photocatalytic TiO_2 nanocomposites with different loads of ozonated graphene (OGn) and evaluated the performances in photocatalytic MB degradation and other dyes using UVC irradiation. Although the photocatalytic activity decreased with each cycle, it is possible to reuse the nanocomposite that facilitates its practical application for water treatment.

In Table 4, the different preparation methods of $GTiO_2Ns$ applied to MB photodegradation are listed.

Table 4. Photocatalytic applications of Graphene Based TiO_2 nanocomposites in the degradation of Methylene Blue.

Photoactive Nanomaterials	Dye Conc. (mg/L)	Catalyst Quantity (g/L)	Light Source	Irradiation Time (min)	Degradation (%)	Refs.
G/TiO_2	9.60	0.33	Visible	180	90	[8]
P25-G	8.64	0.6	UV	60	85	[199]
P25-G	8.64	0.6	Visible	60	65	[199]
G/TiO_2	10	0.6	UV	120	100	[200]
G/TiO_2	10	0.2	UV	40	85	[202]
TiO_2@rGO	–	0.1	UV	120	92	[203]
GO/TiO_2	15	0.2	UV	350	92	[204]
TiO_2-G	0.13	0.5	UV-Vis	450	100	[205]
rGO/TiO_2	320	0.5	Visible	90	95	[207]
TiO_2/GO	–	0.2	UV	25	100	[208]
TiO_2/G	0.13	0.4	UV	60	96	[209]
TiO_2/G	3	0.48	Visible	90	100	[210]
G/TiO_2	3.2	0.2	UV	180	97	[191]
G/TiO_2	3.2	0.2	Visible	240	96	[191]
TiO_2/GO	3	0.48	Visible	60	94	[211]
TiO_2/GO	5	0.1	UV	40	93	[213]
TiO_2/GO	5	0.1	Visible	40	70	[213]
TiO_2-Graphite Oxide	10	0.5	UV	60	100	[215]
TiO_2/G	10	0.2	UV	20	97	[216]
TiO_2/rGO	10	0.17	UV	60	100	[217]
G/TiO_2	5	0.5	Visible	100	70	[218]
TiO_2/G	10	0.8	Visible	100	98.8	[219]
Graphene quantum dots/TiO_2	6.4	–	UV	180	100	[220]
TiO_2/GO	10	0.1	UV	180	100	[221]
TiO_2/G	10	0.01	Visible	150	100	[222]
TiO_2/G	10	0.5	UV	90	100	[223]
Graphene quantum dots/TiO_2	6.4	–	UV	180	100	[220]

Methyl orange (MO) is a well-known anionic azo dye largely used in the textile, printing, paper, food, pharmaceutical, and research fields [224]. As an example, MO photodegradation was studied by Lavanya et al. [225], who combined novel reduced graphene oxide (rGO) with anatase/rutile mixed phase TiO_2 nanofibers (MPTNFs), by using electrospinning and easy chemical methods, to enhance the photocatalytic performance. The photocatalytic activity in the photodegradation of MO showed a significant increase in the reaction rate for the synergistic effect of anatase/rutile mixed phase in one-dimensional nanostructures, and the electronic interaction between TiO_2 and rGO that provided the improvement of electron transfer reduced the charge recombination for the enhancement of catalytic efficiency.

Xu et al. [63] used a two-step method to prepare G-pasted TiO_2 spheres. In the first step, the total cover of the surface of TiO_2 spheres with small GO sheets was obtained, while in the second step, GO was photocatalytically reduced in situ. G-pasted TiO_2 spheres presented more interfaces between G and TiO_2 with favorable interaction under the light irradiation compared to that of typical $GTiO_2Ns$ composites. As a result, these materials demonstrated higher photocatalytic activity in MO degradation. Hou et al. [226] instead used hydrothermal synthesis to obtain rGO-TiO_2 composites, and part of these were successively calcined at 450 °C (CS-rGO-TiO_2). Photocatalytic activities in the decomposition of MO under UV, visible, and solar light by using both RGO-TiO_2 and calcinated composites were tested. The amount of G apparently influenced the photocatalytic activities of both rGO-TiO_2 composites and CS-rGO-TiO_2, which showed better photocatalytic activity compared to that of rGO-TiO_2. These results were due to the calcination that favored the oxidation of residual organics in the rGO-TiO_2 and a better crystallization of TiO_2 nanoparticles with smaller diameters.

Zhao et al. [227] synthetized three-dimensional nanocomposites with TiO_2 nanotubes (TNTs) and rGO nanosheets by using the hydrothermal method. The obtained rGO/TNTs nanocomposites presented sufficient active sites and gave the path of good electron-transport. These nanomaterials showed better photocatalytic activity in MO degradation under UV-light irradiation with respect to traditional TiO_2 nanotubes. Also, Lu et al. [228] combined TiO_2 and GO and used UV irradiation to synthesize rGO-TiO_2 composites. The influence of TiO_2 quantity and UV irradiation time on the rGO formation during the composite synthesis was investigated. The results demonstrated that a longer UV irradiation time corresponded to a higher reduction degree of GO, and therefore to a higher photodegradation efficiency of the composites in MO photodegradation. However, the results also showed that excessive UV irradiation times gave a negative effect on the photodegradation efficiency and therefore proposed a mechanism that correlated the reduction degree of GO with the photodegradation efficiency of composites.

Chemisorption and successive heating at different temperatures was applied by Xia et al. [229] to combine TiO_2 in different anatase and rutile phases with G sheets. The photoactivities of these nanomaterials were tested by using MO under UV and visible light irradiation, demonstrating an enhanced respect to only TiO_2. Notably, the different crystallite phases of TiO_2 reacted with different behaviors under UV and visible light irradiation due to the different charge transfer mechanisms.

Another example was described by Ge et al. [230], who combined electrodeposition and carbonation techniques to obtain the deposition of rGO films on TiO_2 nanotube arrays (TiO_2 NTAs); these materials, prepared in ethylene glycol by using two-step electrochemical anodization, showed an enhanced MO photocatalytic degradation with respect to both pristine TiO_2 NTAs and annealed TiO_2 NTAs under the same conditions. The increase of efficiency in photocatalytic degradation was due to the increase of separation efficiency of photoinduced electrons and holes, to red-shift in the UV absorption and to the profitable adsorbent skill of rGO towards the MO dye.

$GTiO_2Ns$ nanocomposites were instead synthetized by Han et al. [231] by using the hydrothermal method and a successive annealing process. In this preparation, the loading of flocculent-like TiO_2 nanostructures onto G sheets was obtained. The evaluation of photoelectrochemical activities and the photocatalytic degradation of MO under UV light irradiation showed that the flocculent-like $GTiO_2Ns$ composites presented the greatest photocatalytic activity compared to that of commercial anatase TiO_2, due to the increase of light absorption and fruitful charge separation of the nanocomposite structure.

In order to synthesize rGO wrapped with anatase mesoporous TiO_2 nanofibers, Lavanya et al. [232] used electrospinning and easy chemical methods. The interface between rGO and TiO_2 nanofibers in these composites favored an efficient photogenerated charge carrier separation with an enhanced efficiency of photocatalytic MO degradation of 96% instead of only 43% of TiO_2 nanofibers. Wang et al. [233] applied a low-temperature solvothermal method using graphite oxide and $TiCl_3$ as precursors to prepare $GTiO_2Ns$ nanocomposites. During this solvothermal transformation, the synthesis of TiO_2 nanoparticles and G formation occurred. Compared to only TiO_2 nanoparticles, also in this case, the results revealed a greatly enhanced visible light photocatalytic MO degradation.

Athanasekou et al. [234] prepared rGO-TiO_2 composite membranes for incorporation into a water purification device for the application in the hybrid photocatalytic/ultrafiltration process. For this, GO sheets were decorated with TiO_2 nanoparticles and deposited into the pores of ultrafiltration mono-channel monoliths using the dip-coating technique and water filtration conducted in the dark, as well as UV and visible light irradiation. The obtained results, compared with that of standard nanofiltration, showed that the synergetic effects of GO on MO adsorption and photocatalytic degradation demonstrated that this material can be positively applied in nanofiltration technology.

Liu [235] prepared rGO-TiO_2 with the hydrothermal method by using GO and $Ti(OH)_4$ in ethanol/water. The results showed that rGO-TiO_2 presented a stratified structure composed of dispersed anatase TiO_2 on the surface of rGO; the larger surface area of the composite was favorable for the MO absorption; in addition, the extension of the visible light region absorption range was observed. All these factors positively favored the photocatalytic performances of the composites in the photodegradation of MO under visible light irradiation with respect to that of pure TiO_2.

In Table 5, the different preparation methods of $GTiO_2Ns$ applied in MO photodegradation are summarized.

Table 5. Photocatalytic applications of Graphene Based TiO_2 nanocomposites in the degradation of Methyl Orange.

Photoactive Nanomaterials	Dye Conc. (mg/L)	Catalyst Quantity (g/L)	Light Source	Irradiation Time (min)	Degradation (%)	Refs.
rGO/TiO_2 (mix anatase/ rutile nanofibers)	10	0.4	UV	120	97	[225]
G-pasted TiO_2 spheres	12	0.5	UV	75	95	[63]
GO/TiO_2	13	1	UV	60	88	[226]
		1	Vis	60	80	
rGO/TiO_2		1	UV	60	70	
		1	Vis	60	99	
rGO/TiO_2 anotube	20	0.25	UV	210	100	[227]
rGO/TiO_2	10	0.5	UV	75	70	[228]
rGO/TiO_2 mix anatasio/rutilio	6.55	0.6	UV	100	100	[229]
			Vis	100	50	
G/TiO_2/Magnetite	9.6	0.16	UV	90	99	[230]
Flocculent likeTiO_2/G	20	0.8	UV	60	70	[231]
rGO/TiO_2 nanofibers	15	0.4	UV	120	100	[232]
G/TiO_2	10	0.6	UV	60	80	[233]
TiO_2/rGO	10	0.5	Vis	240	90	[235]

In addition, other authors explored photocatalytic materials in the photodegradation of *Rhodamine-B* (RB), which is a brilliant soluble basic dye, most used in the dyeing of various products of the textile industry such as cotton, silk, paper, leather, and others. Notably, RB imparts an intense color to polluted wastewaters derived from different industries and from scientific laboratories [236]. In particular, for RhB photodegradation, Chen Q. et al. [237] used electrospinning and the hydrothermal reaction in mixed ethanol/water solution to prepare nanocomposites composed from rGO and TiO_2 nanotubes. In these nanocomposites, TiO_2 nanotubes combined with rGO sheets between Ti-C and Ti-O-C bonds and the absorption edge shifted to higher wavelengths, improving the visible light absorption with respect to pure TiO_2 nanotubes. The measurements of photocatalytic activities regarding RhB photodegradation under xenon lamp irradiation showed that the highest photocatalytic activities were obtained when the rGO ratio increased to 10%.

Furthermore, Chen Y. et al. [238] used a hydrothermal method to prepare rGO and TiO_2 hybrid of 10–20 nm, by using TiO_2 P25 nanoparticles and "liquid-exfoliated" GO. Also in this case, rGO-TiO_2, compared to only TiO_2, showed an increased photocatalytic RhB degradation under Xe lamp irradiation. The rGO-TiO_2 characterization confirmed the enhancement of photocatalytic and could be attributed to two reasons. The first was that rGO extended the path and lifetime of photogenerated electrons of TiO_2, minimizing the recombination of electron–hole pairs. while the second was that rG expanded the light absorption spectrum versus the visible light range. These effects were explained by the change of the energy gap and the likelihood of the up-conversion photoluminescence mechanism.

Another example was reported by Wang et al. [239] about the immobilization of anatase TiO_2 nanosheets on the magnetically actuated artificial cilia film by using rGO as the contact medium. The artificial cilia film was optimal in the immobilization of more powder of photocatalysts and, when used under s rotating magnetic field, exhibited an improvement of RhB degradation efficiency due to high mass transfer and to the efficiencies of photoproducts desorption. Biris et al. [240] synthesized core-shell nanostructural materials with multi-component architectures from TiO_2 and layers of graphitic materials by application CVD for 5, 10, 30, and 45 min with methane as the carbon source. The reaction time linearly regulated the quantity of G shells covering the TiO_2 surfaces, obtaining nanostructured materials with excellent stability and photocatalytic activity in the UV RhB degradation.

Moreover, Kim et al. [241] used synthesized composites of flower-like TiO_2 spheres and rGO (FTS-G). First TiO_2 spheres, with high specific surface area and good pore structure, and rGO without the use of strong reducing agents, were separately prepared; next, FTS-G were synthesized by using a solvothermal method. The photocatalytic performance of FTS-G was evaluated for RhB photodegradation under solar light irradiation, establishing that the rGO quantity was very important for the effects on photocatalytic activity. This preparation method could also be used to prepare other photocatalysts composed of flower-like TiO_2 spheres with carbon materials. Liu et al. [242] prepared a photocatalyst based on TiO_2 nanotubes and rGO with the hydrothermal method that was successively tested in the photocatalytic degradation of RhB under UV-light irradiation. Also with this photocatalyst, thanks to the introduction of rGO on TiO_2 nanotubes, the adsorption capacity as well as the photocatalytic activity in RhB photodegradation increased compared to the same without rGO.

Sedghi et al. [243], by using Titanium(IV) chloride as a photocatalyst precursor, prepared different nanocomposites of porous and magnetic porous GO in the mix with TiO_2 (anatase) and TiO_2 (mix phase), respectively. Between these nanocomposites, the mix of TiO_2 with magnetic porous GO showed enhanced efficiency in the RhB degradation under visible light irradiation compared with that of the other photocatalysts, obtaining 100% degradation in less than 20 min. Similarly, Zhang et al. [244] used the one-pot hydrothermal process, by using tetrabutyl titanate and GO, to prepare a series of self-assembled composites of anatase TiO_2 nanocrystals and three-dimensional graphene aerogel. The obtained composites showed higher adsorption capacities and good visible light efficiency in the RhB photodegradation. The results of this study clarified that the chemical-physical properties of these composites could be attributed to their 3D nanoporous structure with high surface areas and to the contemporary action of G nanosheets and TiO_2 nanoparticles.

An ultrasonication-assisted reduction process was used from He et al. [245] to produce rGO-TiO_2 nanocomposites. In this process, simultaneous GO reduction and TiO_2 crystals formation were obtained; photocatalytic studies demonstrated that the quantities rGO in these nanocomposites influenced the RhB degradation efficiency under visible light irradiation. The obtained increase in the photocatalytic activity was attributed in this case to the harvesting under visible-light irradiation and to the efficiency in the separation of the photogenerated charge carriers, also obtaining an improvement in the photoelectric conversion efficiency.

Fang et al. [246] exfoliated GO sheets from graphite and wrapped on the surfaces of polymer microspheres of about 2.5 µm in diameter. Successively, the solvothermal method permitted the nucleation and growth of anatase TiO_2 nanoparticles on the GO surfaces and the contemporary GO reduction to rGO. The as-prepared GO-skined GO/polymer hybrid microspheres composites tested in the photocatalytic activity with RhB solutions gave 96% degradation in 30 min under visible light irradiation, demonstrating an improvement with respect to commercial TiO_2. The results were attributed to the favorable interactions between rGO of singular electrical conductivity and TiO_2 nanoparticles.

Liang et al. [247] used a one-step hydrothermal approach to synthesize chemically bonded TiO_2/rGO nanocomposites by using $Ti(SO_4)_2$ and GO as precursors and ethanol/water solvent as a reducing agent, obtaining well-dispersed TiO_2 nanocrystals on the surface of rGO sheets with optimal interfacial contact. The obtained nanocomposites demonstrated higher photocatalytic activity with respect to only TiO_2 nanoparticles and to a simple mixing of TiO_2 and rGO samples in the RhB degradation which was attributed, also in this case, to the interaction between TiO_2 and rGO, with excellent electron trapping and transportation properties.

Li et al. [248] obtained different carbon materials-TiO_2 hybrid nanostructures incorporated into TiO_2, activated carbon, G, carbon nanotubes, and fullerene by using a solvothermal method and thermal annealing. The relationships between the interactions of carbon materials with TiO_2 nanoparticles, morphologies, structures, and increasing photodegradation performances in RhB degradation were clarified. The results of these photocatalysts were due to higher adsorption properties, the favorable formation of chemical bonds of Ti-O-C, reduced band gap, lower particle size, and charge-carrier qualities. In Table 6, the different preparation methods of $GTiO_2Ns$ applied in RhB photodegradation are listed.

Table 6. Photocatalytic applications of Graphene Based TiO_2 nanocomposites in the degradation of Rhodamin B.

Photoactive Nanomaterials	Dye Conc. (mg/L)	Catalyst Quantity (g/L)	Light Source	Irradiation Time (min)	Degradation (%)	Refs.
TiO_2/rGO	30	0.2	Vis	40	100	[238]
rGO/TiO_2 nanosheets onmagnetically cilia film rGO/TiO_2-Au	20	0.2	UV	180	83 100	[239]
Core-shell TiO_2/G	4.79	0.6	UV	270	100	[240]
Flower-like TiO_2 sphere /rGO	15	0.4	Simulated solar	120	100	[241]
TiO_2 nanotubes/rGO	10	0.5	UV	20	100	[242]
TiO_2/magnetic porous GO	10	0. 1	Vis	20	100	[243]
3D TiO_2/G aerogel	20	0.2	Visible light	180	99	[244]
TiO_2/rGO	4.79	0.4	Visible light	180	100	[245]
TiO_2/rGO/polymer	8	0.2	Visible light	30	96	[246]
TiO_2/rGO	20	0.5	UV	30	100	[247]
$GTiO_2$	5	0.5	Visible light	60	80	[248]

Rhodamine 6G (Rh6G) is dark reddish-purple dye, mostly used in the textile industries, in biochemistry research laboratories, and in other applications. Rh6G is a non-volatile compound that presents high solubility in water causing skin, eye, and respiratory system irritations, and is also carcinogenic and poisonous to living organisms [249].

In order to obtain the photocatalytic degradation of Rh6G, Pu et al. [194] demonstrated a facile and environmentally friendly strategy for the simultaneous in situ preparation of the rGO-TiO_2 "dyade hybrid" obtained by the photoreduction of GO nanosheets in ethanolic solutions of GO and TiO_2 (P25) mix. Successively, the photodegradation efficiency of the resultant composite by utilizing

Rh6G as the target pollutant was systematically investigated. The obtained rGO-TiO$_2$ presented a significant enhancement in photo energy adsorption, leading to the effective photocatalytic degradation reactions which exhibited more than triple the higher photodegradation rate compared to commercial TiO$_2$ (P25) nanoparticles.

Reactive Black 5 (RB5), which has two azo groups, belongs to the class of azo reactive dyes, which are abundantly used in textile industries for dyeing.

The photocatalytic degradation of RB5 was studied by Day et al. [250], who prepared different types of GTiO$_2$Ns nanocomposites with G and P25 or Titanium (IV) n-butoxide as precursors by using the hydrothermal method. To investigate the differences of the photocatalytic properties of composites with various TiO$_2$ shapes, nanotubes, nanosheets, and nanoparticles of TiO$_2$ were prepared and combined with G. The photocatalytic activities of these composites were tested in the degradation of RB5 under UV lamp irradiation, indicating that the differences in the TiO$_2$ morphology and the characteristics of heterojunction between G and TiO$_2$ played an important role in the photocatalytic abilities. The results demonstrated that these composites exhibited excellent adsorption capacity and higher photocatalytic activity with respect to P25, and that the photocatalytic activity of GTiO$_2$Ns nanosheets was higher than those of the other tested photocatalysts. Another example was proposed by Liang et al. [251], who obtained an increase of photocatalysis efficiency for RB5 by using TiO$_2$ nanotubes and G nanocomposites prepared with the hydrothermal method. In particular, Zhang et al. [252] deciphered the mechanism of RB5 photodegradation, which was used as a model for other azo dyes by using G-loaded TiO$_2$ as a photocatalyst. They demonstrated that electronic transfer groups affect the degradability of azo dyes, that $^\bullet$OH radicals are the major oxidized species, and that substituents with high electron-density in the dye structure were more favorably attacked by $^\bullet$OH radicals.

The photodegradation of the non-biodegradable azo dye *acid orange 7* (AO7) in aqueous solution was studied by Posa et al. [253] who, by using GTiO$_2$Ns nanocomposites, obtained an improvement of photocatalytic performance compared with that of pure TiO$_2$, examining also the reusability of the photocatalyst. They suggested a reaction mechanism on the basis of the obtained results. Furthermore, Gao et al. [254] synthesized different TiO$_2$ nanostructures: one-dimensional TiO$_2$ nanotubes and nanowires, three-dimensional spheres assembled by nanoparticles and by nanosheets. The results of photodegradation activity in the AO7 degradation indicated that the photodegradation efficiency of TiO$_2$ spheres assembled by nanosheets was the highest with respect to the other TiO$_2$ nanostructures, while its specific surface area was lower than that of TiO$_2$ nanotubes. The results were attributed to the highest light harvesting capacity under solar light derived from the multiple reflections of light, and from the hierarchical mesoporous structure. The last example regards the study of Muthirulan et al. [255], who fabricated GTiO$_2$Ns by using a simple one-step chemical process, mixing TiO$_2$ nanoparticles suspended in ethanol with G followed by ultrasonication and successive treatment in a rotary evaporator under vacuum. The dried GTiO$_2$Ns showed improvement in the photocatalytic ability in the degradation of AO7 under both UV and solar light irradiations. The obtained results of photodegradation indicated that GTiO$_2$Ns nanocomposites exhibited higher photocatalytic activity than that of TiO$_2$, which was attributed to the role of G in the suppression of charge recombination and in the promotion of the charge transfer, permitting also a possible reaction pathway for the degradation of AO7.

6. Conclusions

In this short review, we have described the characteristics of both TiO$_2$ semiconductors and graphene materials and reported several examples focusing on the recent progress in the design, synthesis, and applications of these GTiO$_2$Ns nanocomposites used as photocatalysts in the photocatalytic degradation of synthetic dyes.

We have described the important effects of G in GTiO$_2$Ns nanocomposites, as shown in Figure 6. In fact, in conjunction with TiO$_2$ nanomaterials, G can act as a co-adsorbent thanks to the increase of adsorption surface and to π-π interactions with dye and as a sensitizer due to the electron transfer from the G photoexcited state, with delocalization and energy excess dissipation, to the TiO_2 surface with the extension of the light absorption towards the visible region. Notably, both the functions of conductive materials and co-catalysts favor the stabilization of excited electrons, thus limiting the electron-hole recombination in the photocatalytic process. In addition to these properties, TiO_2 band-gap narrowing due to the interaction between TiO_2 and G was demonstrated.

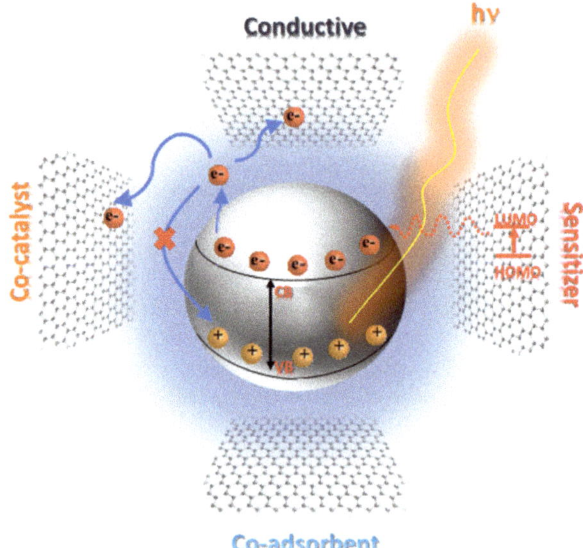

Figure 6. Schematic representation of the main features of graphene in the photocatalysis with TiO_2.

Thanks to these features, in order to obtain the best benefit of graphene in the enhancment of the photoactivity of TiO_2 nanomaterials, all of these factors that dominate the photocatalytic process must to be taken into consideration and optimized with the purpose of obtaining the best dye absorption and its complete mineralization.

Much research has been conducted regarding the photocatalytic degradation of synthetic dyes, and the high-quality results of these studies demonstrate that these materials could be promising for large-scale applications in wastewater depuration.

Acknowledgments: This work was partially supported from FAR Project NAMES: "Nanocomposite Materials for Energy and Environment Applications" of Camerino University.

Author Contributions: Rita Giovannetti organized and prepared the manuscript. Elena Rommozzi collected and analyzed the literature reports, Marco Zannotti, Chiara Anna D'Amato designed the figures and critically revised the manuscript.

Conflicts of Interest: The authors declare no conflict of interest.

References

1. Peirce, J.J.; Weiner, R.; Vesilind, P.A. *Environmental Pollution and Control*, 4th ed.; Elsevier: Oxford, UK, 1998; ISBN 978-0-7506-9899-3.
2. Andreozzi, R.; Caprio, V.; Insola, A.; Marotta, R. Advanced oxidation processes (AOP) for water purification and recovery. *Catal. Today* **1999**, *53*, 51–59. [CrossRef]
3. Chaukura, N.; Edna, C.; Murimba, E.C.; Gwenzi, W. Sorptive removal of methylene blue from simulated wastewater using biochars derived from pulp and paper sludge. *Environ. Technol. Innov.* **2017**, *8*, 132–140. [CrossRef]

4. Solís, M.; Solís, A.; Inés Pérez, H.; Manjarrez, N.; Flores, M. Microbial decolouration of azo dyes: A review. *Process. Biochem.* **2012**, *47*, 1723–1748. [CrossRef]

5. Zhang, Z.; Yang, R.Y.; Gao, Y.S.; Zhao, Y.F.; Wang, J.Y.; Huang, L.; Guo, J.; Zhou, T.T.; Lu, P.; Guo, Z.H.; et al. Novel $Na_2Mo_4O_{13}/\alpha$-MoO_3 hybrid material as highly efficient CWAO catalyst for dye degradation at ambient conditions. *Sci. Rep.* **2014**, *4*, 6797–6809. [CrossRef] [PubMed]

6. Santhosh, C.; Velmurugan, V.; Jacob, G.; Jeong, S.K.; Grace, A.N.; Bhatnagar, A. Role of nanomaterials in water treatment applications: A review. *Chem. Eng. J.* **2016**, *306*, 1116–1137. [CrossRef]

7. Padikkaparambil, S.; Narayanan, B.; Yaakob, Z.; Viswanathan, S.; Tasirin, S.M. Au/TiO_2 reusable photocatalysts for dye degradation. *Int. J. Photoenergy* **2013**, *2013*. [CrossRef]

8. Cho, K.M.; Kim, K.H.; Choi, H.O.; Jung, H.T. A highly photoactive, visible-light-driven graphene/2D mesoporous TiO_2 photocatalyst. *Green Chem.* **2015**, *17*, 3972–3978. [CrossRef]

9. Ibhadon, A.O.; Fitzpatrick, P. Heterogeneous photocatalysis: Recent advances and applications. *Catalysts* **2013**, *3*, 189–218. [CrossRef]

10. Wei, W.; Liu, D.; Wei, Z.; Zhu, Y. Short-range π-π stacking assembly on P25 TiO_2 nanoparticle for enhanced visible-light photocatalysis. *ACS Catal.* **2017**, *7*, 652–663. [CrossRef]

11. Jiang, W.J.; Liu, Y.F.; Wang, J.; Zhang, M.; Luo, W.J.; Zhu, Y.F. Separation-free polyaniline/TiO_2 3D hydrogel with high photocatalytic activity. *Adv. Mater. Interfaces* **2016**, *3*, 9.

12. Tatsuma, T.; Saitoh, S.; Ohko, Y.; Fujishima, A. TiO_2-WO_3 photoelectrochemical anticorrosion system with an energy storage ability. *Chem. Mater.* **2001**, *13*, 2838–2842. [CrossRef]

13. Sajan, C.P.; Wageh, S.; Al-Ghamdi, A.A.; Yu, J.G.; Cao, S.W. TiO_2 nanosheets with exposed {001} facets for photocatalytic applications. *Nano Res.* **2016**, *9*, 3–27. [CrossRef]

14. Chen, X.; Mao, S.S. Titanium dioxide nanomaterials: Synthesis, properties, modifications, and applications. *Chem. Rev.* **2007**, *107*, 2891–2959. [CrossRef] [PubMed]

15. Tong, H.; Ouyang, S.X.; Bi, Y.P.; Umezawa, N.; Oshikiri, M.; Ye, J.H. Nano-photocatalytic materials: Possibilities and challenges. *Adv. Mater.* **2012**, *24*, 229–251. [CrossRef] [PubMed]

16. Anpo, M.; Tackeuchi, M. The design and development of highly reactive titanium oxide photocatalysts operating under visible light irradiation. *J. Catal.* **2003**, *216*, 505–516. [CrossRef]

17. Al-Harbi, L.M.; El-Mossalamy, E.H.; Arafa, H.M.; Al-Owais, A.; Shah, M.A. TiO_2 Nanoparticles with Tetra-pad Shape Prepared by an Economical and Safe Route at very Low Temperature. *Mod. Appl. Sci.* **2011**, *5*, 130–135. [CrossRef]

18. Giovannetti, R.; D'Amato, C.A.; Zannotti, M.; Rommozzi, E.; Gunnella, R.; Minicucci, M.; Di Cicco, A. Visible light photoactivity of polypropylene coated nano-TiO_2 for dyes degradation in water. *Sci. Rep.* **2015**, *2*. [CrossRef] [PubMed]

19. Augugliaro, V.; Bellardita, M.; Loddo, V.; Palmisano, G.; Palmisano, L.; Yurdakal, S. Overview on oxidation mechanisms of organic compounds by TiO_2 in heterogeneous photocatalysis. *J. Photochem. Photobiol. C Photochem. Rev.* **2012**, *13*, 224–245. [CrossRef]

20. Sud, D.; Kaur, P. Heterogeneous photocatalytic degradation of selected organophosphate pesticides: A review. *Crit. Rev. Environ. Sci. Technol.* **2012**, *42*, 2365–2407. [CrossRef]

21. Wang, J.L.; Xu, L.J. Advanced oxidation processes for wastewater treatment: Formation of hydroxyl radical and application. *Crit. Rev. Environ. Sci. Technol.* **2012**, *42*, 251–325. [CrossRef]

22. Hashimoto, K.; Irie, H.; Fujishima, A. TiO_2 photocatalysis: A historical overview and future prospects. *Jpn. J. Appl. Phys.* **2005**, *12*, 8269–8285. [CrossRef]

23. Sheikh, M.U.D.; Naikoo, G.A.; Thomas, M.; Bano, M.; Khan, F. Solar-assisted photocatalytic reduction of methyl orange azo dye over porous TiO_2 nanostructures. *New J. Chem.* **2016**, *40*, 5483–5494. [CrossRef]

24. Zhang, N.; Zhang, Y.; Xu, Y.-J. Recent progress on graphene-based photocatalysts: Current status and future perspectives. *Nanoscale* **2012**, *4*, 5792–5813. [CrossRef] [PubMed]

25. Malato, S.; Fernández-Ibáñez, P.; Maldonado, M.I.; Blanco, J.; Gernjak, W. Decontamination and disinfection of water by solar photocatalysis: Recent overview and trends. *Catal. Today* **2009**, *147*, 1–59. [CrossRef]

26. Wang, W.K.; Chen, J.J.; Gao, M.; Huang, Y.X.; Zhang, X.; Yu, H.Q. Photocatalytic degradation of atrazine by boron-doped TiO_2 with a tunable rutile/anatase ratio. *Appl. Catal. B Environ.* **2016**, *195*, 69–76. [CrossRef]

27. Hao, R.R.; Wang, G.H.; Tang, H.; Sun, L.L.; Xu, C.; Han, D.Y. Template-free preparation of macro/mesoporous g-C_3N_4/TiO_2 heterojuction photocatalysts with enhanced visible light photocatalytic activity. *Appl. Catal. B Environ.* **2016**, *187*, 47–58. [CrossRef]

28. Gua, Y.; Xinga, M.; Zhang, J. Synthesis and photocatalytic activity of graphene based doped TiO_2 nanocomposites. *Appl. Surf. Sci.* **2014**, *319*, 8–15. [CrossRef]

29. Zaleska, A. Doped-TiO_2: A review. *Recent Pat. Eng.* **2008**, *2*, 157–164. [CrossRef]

30. Yang, H.G.; Sun, C.H.; Qiao, S.Z.; Zou, J.; Liu, G.; Smith, S.C.; Cheng, H.M.; Lu, G.Q. Anatase TiO_2 single crystals with a large percentage of reactive facets. *Nature* **2008**, *453*, 638–641. [CrossRef] [PubMed]

31. Adán, C.; Bahamonde, A.; Fernández-García, M.; Martínez-Arias, A. Structure and activity of nanosized iron-doped anatase TiO_2 catalysts for phenol photocatalytic degradation. *Appl. Catal. B Environ.* **2007**, *72*, 11–17. [CrossRef]

32. Cai, J.; Wu, X.; Li, S.; Zheng, F. Controllable location of Au nanoparticles as cocatalyst onto TiO_2@CeO_2 nanocomposite hollow spheres for enhancing photocatalytic activity. *Appl. Catal. B Environ.* **2017**, *201*, 12–21. [CrossRef]

33. Cheng, L.; Qiu, S.; Chen, J.; Shao, J.; Cao, S. A practical pathway for the preparation of Fe_2O_3 decorated TiO_2 photocatalyst with enhanced visible-light photoactivity. *Mater. Chem. Phys.* **2017**, *190*, 53–61. [CrossRef]

34. Xie, Y.; Meng, Y.; Wu, M. Visible-light-driven self-cleaning SERS substrate of silver nanoparticles and graphene oxide decorated nitrogen-doped Titania nanotube array. *Surf. Interface Anal.* **2016**, *48*, 334–340. [CrossRef]

35. Xing, Z.; Zong, X.; Zhu, Y.; Chen, Z.; Bai, Y.; Wang, L. A nanohybrid of CdTe@CdS nanocrystals and Titania nanosheets with p-n nanojunctions for improved visible light-driven hydrogen production. *Catal. Today* **2016**, *264*, 229–235. [CrossRef]

36. Hamzezadeh-Nakhjavani, S.; Tavakoli, O.; Akhlaghi, S.P.; Salehi, Z.; Esmailnejad-Ahranjani, P.; Arpanaei, A. Efficient photocatalytic degradation of organic pollutants by magnetically recoverable nitrogen-doped TiO_2 nanocomposite photocatalysts under visible light irradiation. *Environ. Sci. Pollut. Res.* **2015**, *22*, 18859–18873. [CrossRef] [PubMed]

37. Qi, K.; Selvaraj, R.; Al Fahdi, T.; Al-Kindy, S.; Kim, Y.; Wang, G.-C.; Tai, C.-W.; Sillanpää, M. Enhanced photocatalytic activity of anatase-TiO_2 nanoparticles by fullerene modification: A theoretical and experimental study. *Appl. Surf. Sci.* **2016**, *387*, 750–758. [CrossRef]

38. Qin, X.; He, F.; Chen, L.; Meng, Y.; Liu, J.; Zhao, N.; Huang, Y. Oxygen-vacancy modified TiO_2 nanoparticles as enhanced visible-light driven photocatalysts by wrapping and chemically bonding with graphite-like carbon. *RSC Adv.* **2016**, *6*, 10887–10894. [CrossRef]

39. Rahbar, M.; Behpour, M. Multi-walled carbon nanotubes/TiO_2 thin layer for photocatalytic degradation of organic pollutant under visible light irradiation. *J. Mater. Sci. Mater. Electron.* **2016**, *27*, 8348–8355. [CrossRef]

40. Lin, L.; Wang, H.; Xu, P. Immobilized TiO_2-reduced graphene oxide nanocomposites on optical fibers as high performance photocatalysts for degradation of pharmaceuticals. *Chem. Eng. J.* **2017**, *310*, 389–398. [CrossRef]

41. Stoller, M.D.; Park, S.; Zhu, Y.; An, J.; Ruoff, R.S. Graphene-based ultracapacitors. *Nano Lett.* **2008**, *8*, 3498–3502. [CrossRef] [PubMed]

42. Novoselov, K.S.; Geim, A.K.; Morozov, S.V.; Jiang, D.; Zhang, Y.; Dubonos, S.V.; Grigorieva, I.V.; Firsov, A.A. Electric field effect in atomically thin carbon films. *Science* **2004**, *306*, 666–669. [CrossRef] [PubMed]

43. Balandin, A.A.; Ghosh, S.; Bao, W.; Calizo, I.; Teweldebrhan, D.; Miao, F.; Lau, C.N. Superior thermal conductivity of single-layer graphene. *Nano Lett.* **2008**, *8*, 902–907. [CrossRef] [PubMed]

44. Cheng, Y.; Zhou, S.; Hu, P.; Zhao, G.; Li, Y.; Zhang, X.; Han, W. Enhanced mechanical, thermal, and electric properties of graphene aerogels via supercritical ethanol drying and high-temperature thermal reduction. *Sci. Rep.* **2017**, *7*. [CrossRef] [PubMed]

45. Sun, W.; Du, A.; Gao, G.; Shen, J.; Wu, G. Graphene-templated carbon aerogels combining with ultra-high electrical conductivity and ultra-low thermal conductivity. *Microporous Mesoporous Mater.* **2017**, *253*, 71–79. [CrossRef]

46. Tran, V.-T.; Saint-Martin, J.; Dollfus, P.; Volz, S. Optimizing the thermoelectric performance of graphene nano-ribbons without degrading the electronic properties. *Sci. Rep.* **2017**, *7*. [CrossRef] [PubMed]

47. Wei, A.; Li, Y.; Li, Y.; Ye, H. Thermal characteristics of graphene nanosheet with graphane domains of varying morphologies. *Comput. Mater. Sci.* **2017**, *138*, 192–198. [CrossRef]

48. Geim, A.K.; Novoselov, K.S. The rise of graphene. *Nat. Mater.* **2007**, *6*, 183–191. [CrossRef] [PubMed]

49. Leary, R.; Westwood, A. Carbonaceous nanomaterials for the enhancement of TiO_2 photocatalysis. *Carbon* **2011**, *49*, 741–772. [CrossRef]

50. Hamandia, M.; Berhault, G.; Guillardb, C.; Kochkara, H. Influence of reduced graphene oxide on the synergism between rutileand anatase TiO$_2$ particles in photocatalytic degradation of formic acid. *Mol. Catal.* **2017**, *432*, 125–130. [CrossRef]

51. Morales-Torres, S.; Pastrana-Martínez, L.M.; Figueiredo, J.L.; Faria, J.J.; Silva, A.M.T. Design of graphene-based TiO$_2$ photocatalysts—A review. *Environ. Sci. Pollut. Res.* **2012**, *19*, 3676–3687. [CrossRef] [PubMed]

52. Adamu, H.; Dubey, P.; Anderson, J.A. Probing the role of thermally reduced graphene oxide in enhancing performance of TiO$_2$ in photocatalytic phenol removal from aqueous environments. *Chem. Eng. J.* **2016**, *284*, 380–388. [CrossRef]

53. Giovannetti, R.; Rommozzi, E.; Zannotti, M.; D'Amato, C.A.; Ferraro, S.; Cespi, M.; Bonacucina, G.; Minicucci, M.; Di Cicco, A. Exfoliation of graphite into graphene in aqueous solution: An application as graphene/TiO$_2$ nanocomposite to improve visible light photocatalytic activity. *RSC Adv.* **2016**, *6*, 93048–93055. [CrossRef]

54. Truppi, A.; Petronella, F.; Placido, T.; Striccoli, M.; Agostiano, A.; Curri, M.L.; Comparelli, R. Visible-light-active TiO$_2$-based hybrid nanocatalysts for environmental applications. *Catalysts* **2017**, *7*, 100. [CrossRef]

55. Xiang, Q.; Yu, J.; Jaroniec, M. Graphene-based semiconductor photocatalysts. *Chem. Soc. Rev.* **2012**, *41*, 782–796. [CrossRef] [PubMed]

56. Zhang, Y.; Tang, Z.R.; Fu, X.Y.; Xu, J. TiO$_2$-graphene nanocomposites for gas-phase photocatalytic degradation of volatile aromatic pollutant: Is TiO$_2$-graphene truly different from other TiO$_2$-carbon composite materials? *ACS Nano* **2010**, *4*, 7303–7314. [CrossRef] [PubMed]

57. Surender Kumar, S. *Complex Magnetic Nanostructures: Synthesis, Assembly and Applications*; Springer: Cham, Switzerland, 2017.

58. Syväjärvi, M.; Tiwari, A. *Graphene Materials: Fundamentals and Emerging Applications*; Wiley: Hoboken, NJ, USA, 2015.

59. Singh, V.; Joung, D.; Zhai, L.; Das, S.; Khondaker, S.I.; Seal, S. Graphene based materials: Past, present and future. *Progress Mater. Sci.* **2011**, *56*, 1178–1271. [CrossRef]

60. Cong, H.-P.; Chena, J.-F.; Yu, S.-H. Graphene-based macroscopic assemblies and architectures: An emerging material system. *Chem. Soc. Rev.* **2014**, *43*, 7295–7325. [CrossRef] [PubMed]

61. Mahmood, N.; Zhang, C.; Yin, H.; Hou, Y. Graphene-based nanocomposites for energy storage and conversion in lithium batteries, supercapacitors and fuel cells. *J. Mater. Chem. A* **2014**, *2*, 15–32. [CrossRef]

62. Li, X.; Yu, J.; Wageh, S.; Al-Ghamdi, A.A.; Xie, J. Graphene in photocatalysis: A review. *Small* **2016**, *12*, 6640–6696. [CrossRef] [PubMed]

63. Xu, C.; Zhu, J.; Yuan, R.; Fu, X. More effective use of graphene in photocatalysis by conformal attachment of small sheets to TiO$_2$ spheres. *Carbon* **2016**, *96*, 394–402. [CrossRef]

64. Zhu, J.; Yang, D.; Yin, Z.; Yan, Q.; Zhang, H. Graphene and graphene-based materials for energy storage applications. *Small* **2014**, *10*, 3480–3498. [CrossRef] [PubMed]

65. Piao, Y. Preparation of porous graphene-based nanomaterials for electrochemical energy storage devices. In *Nano Devices and Circuit Techniques for Low-Energy Applications and Energy Harvesting*; Kyung, C.M., Ed.; KAIST Research Series; Springer: Dordrecht, The Netherland, 2016.

66. Li, N.; Liu, G.; Zhen, C.; Li, F.; Zhang, L.; Cheng, H.M. Battery performance and photocatalytic activity of mesoporous anatase TiO$_2$ nanospheres/graphene composites by template-free self-assembly. *Adv. Funct. Mater.* **2011**, *21*, 1717–1722. [CrossRef]

67. Maroni, F.; Raccichini, R.; Birrozzi, A.; Carbonari, G.; Tossici, R.; Croce, F.; Marassi, R.; Nobili, F. Graphene/silicon nanocomposite anode with enhanced electrochemical stability for lithium-ion battery applications. *J. Power Sources* **2014**, *269*, 873–882. [CrossRef]

68. Kaplas, T.; Kuzhir, P. Ultra-thin Graphitic Film: Synthesis and Physical Properties. *Nanoscale Res. Lett.* **2016**, *11*, 54. [CrossRef] [PubMed]

69. Allagui, A.; Abdelkareem, M.A.; Alawadhi, H.; Elwakil, A.S. Reduced graphene oxide thin film on conductive substrates by bipolar electrochemistry. *Sci. Rep.* **2016**, *6*. [CrossRef] [PubMed]

70. Bhuyan, M.S.A.; Uddin, M.N.; Islam, M.M.; Bipasha, F.A.; Hossain, S.S. Synthesis of graphene. *Int. Nano Lett.* **2016**, *6*, 65–83. [CrossRef]

71. Tetlow, H.; de Boer, J.P.; Ford, I.J.; Vvedenskyb, D.D.; Coraux, J.; Kantorovich, L. Growth of epitaxial graphene: Theory and experiment. *Phys. Rep.* **2014**, *542*, 195–295. [CrossRef]

72. Cui, Y.; Zhang, H.; Chen, W.; Yang, Z.; Cai, Q. Structural evolution of flower defects and effects on the electronic structures of epitaxial graphene. *J. Phys. Chem. C* **2017**, *121*, 15282–15287. [CrossRef]

73. Galves, L.A.; Wofford, J.M.; Soares, G.V.; Jahn, U.; Pfüller, C.; Riechert, H.; Lopes, J.M.J. The effect of the SiC(0001) surface morphology on the growth of epitaxial mono-layer graphene nanoribbons. *Carbon* **2017**, *115*, 162–168. [CrossRef]

74. Ramnani, P.; Neupane, M.R.; Ge, S.; Balandin, A.A.; Lake, R.K.; Mulchandani, A. Raman spectra of twisted CVD bilayer graphene. *Carbon* **2017**, *123*, 302–306. [CrossRef]

75. Chena, X.; Zhangb, L.; Chena, S. Large area CVD growth of graphene. *Synth. Met.* **2015**, *210*, 95–108. [CrossRef]

76. Zhang, Y.; Zhang, L.; Zhou, C. Review of chemical vapor deposition of graphene and related applications. *Acc. Chem. Res.* **2013**, *46*, 2329–2339. [CrossRef] [PubMed]

77. Lukosius, M.; Dabrowski, J.; Kitzmann, J.; Fursenko, O.; Akhtar, F.; Lisker, M.; Lippert, G.; Schulze, S.; Yamamoto, Y.; Schubert, M.A.; et al. Metal-free CVD graphene synthesis on 200 mm Ge/Si(001) substrates. *ACS Appl. Mater. Interfaces* **2016**, *8*, 33786–33793. [CrossRef] [PubMed]

78. Robert, M.; Jacobberger, R.M.; Machhi, R.; Wroblewski, J.; Ben Taylor, B.; Anne Lynn Gillian-Daniel, A.L.; Arnold, M.S. Simple graphene synthesis via chemical vapor deposition. *J. Chem. Educ.* **2015**, *92*, 1903–1907.

79. Kim, K.S.; Zhao, Y.; Jang, H.; Lee, S.Y.; Kim, J.M.; Kim, K.S.; Ahn, J.H.; Kim, P.; Choi, J.Y.; Hong, B.H. Large-scale pattern growth of graphene films for stretchable transparent electrodes. *Nature* **2009**, *457*, 706–710. [CrossRef] [PubMed]

80. Emtsev, K.V.; Bostwick, A.; Horn, K.; Jobst, J.; Kellogg, G.L.; Ley, L.; McChesney, J.L.; Ohta, T.; Reshanov, S.A.; Rohrl, J.; et al. Towards wafer-size graphene layers by atmospheric pressure graphitization of silicon carbide. *Nat. Mater.* **2009**, *8*, 203–207. [CrossRef] [PubMed]

81. Juang, Z.Y.; Wu, C.Y.; Lu, A.Y.; Su, C.Y.; Leou, K.C.; Chen, F.R.; Tsai, C.H. Graphene synthesis by chemical vapor deposition and transfer by a roll-to-roll process. *Carbon* **2010**, *48*, 3169–3174. [CrossRef]

82. Wei, D.C.; Liu, Y.Q.; Wang, Y.; Zhang, H.L.; Huang, L.P.; Yu, G. Synthesis of N-doped graphene by chemical vapor deposition and its electrical properties. *Nano Lett.* **2009**, *9*, 1752–1758. [CrossRef] [PubMed]

83. Zhang, H.; Zhang, X.; Sun, X.; Ma, Y. Shape-controlled synthesis of nanocarbons through direct conversion of carbon dioxide. *Sci. Rep.* **2013**, *3*. [CrossRef] [PubMed]

84. Hajian, M.; Zareie, M.; Hashemiana, D.; Bahrami, M. Synthesis of graphene through direct decomposition of CO_2. *RSC Adv.* **2016**, *6*, 73331–73335. [CrossRef]

85. Qin, B.; Zhang, T.; Chen, H.; Ma, Y. The growth mechanism of few-layer graphene in the arc discharge process. *Carbon* **2016**, *102*, 494–498. [CrossRef]

86. Poorali, M.S.; Bagheri-Mohagheghi, M.M. Synthesis and physical properties of multi-layered graphene sheets by Arc-discharge method with TiO_2 and ZnO catalytic. *J. Mater. Sci. Mater. Electron.* **2017**, *28*, 6186–6193. [CrossRef]

87. Xiao, B.; Li, X.; Li, X.; Wang, B.; Langford, C.; Li, R.; Sun, X. Graphene Nanoribbons Derived from the Unzipping of Carbon Nanotubes: Controlled Synthesis and Superior Lithium Storage. *J. Phys. Chem. C* **2014**, *118*, 881–890. [CrossRef]

88. Yang, X.; Dou, X.; Rouhanipour, A.; Zhi, L.; Räder, H.J.; Mülle, K. Two-dimensional graphene nanoribbons. *J. Am. Chem. Soc.* **2008**, *130*, 4216–4217. [CrossRef] [PubMed]

89. Kharisov, B.I.; Kharissova, O.V. Synthesis of graphene by pyrolysis of organic matter. In *Graphene Science Handbook: Fabrication Methods*; Aliofkhazraei, M., Ali, N., Milne, W.I., Ozkan, C.S., Mitura, S., Gervasoni, J.L., Eds.; CRC Press: Boca Raton, FL, USA, 2016; pp. 345–360.

90. Zou, B.; Wang, X.X.; Huang, X.X.; Wang, J.N. Continuous synthesis of graphene sheets by spray pyrolysis and their use as catalysts for fuel cells. *Chem. Commun.* **2015**, *51*, 741–744. [CrossRef] [PubMed]

91. Jabari-Seresht, R.; Jahanshahi, M.; Rashidi, A.; Ghoreyshi, A.A. Fabrication and evaluation of nonporous graphene by a unique spray pyrolysis method. *Chem. Eng. Technol.* **2013**, *36*, 1550–1558. [CrossRef]

92. Khan, M.; Tahir, M.N.; Adil, S.F.; Khan, H.U.; Siddiqui, M.R.H.; Al-warthan, A.A.; Tremel, W. Graphene based metal and metal oxide nanocomposites: Synthesis, properties and their applications. *J. Mater. Chem. A* **2015**, *3*, 18753–18808. [CrossRef]

93. Kim, H.; Abdala, A.A.; Macosko, C.W. Graphene/polymer nanocomposites. *Macromolecules* **2010**, *43*, 6515–6530. [CrossRef]

94. Yi, M.; Shen, Z. A review on mechanical exfoliation for the scalable production of graphene. *J. Mater. Chem. A* **2015**, *3*, 11700–11715. [CrossRef]

95. Hernandez, Y.; Nicolosi, V.; Lotya, M.; Blighe, F.M.; Sun, Z.Y.; De, S.; McGovern, I.T.; Holland, B.; Byrne, M.; Gunko, Y.K.; et al. High-yield production of graphene by liquid-phase exfoliation of graphite. *Nat. Nanotechnol.* **2008**, *3*, 563–568. [CrossRef] [PubMed]

96. Li, X.; Zhang, G.; Bai, X.; Sun, X.; Wang, X.; Wang, E.; Dai, H. Highly conducting graphene sheets and Langmuir-Blodgett films. *Nat. Nanotechnol.* **2008**, *3*, 538–542. [CrossRef] [PubMed]

97. Coscia, U.; Palomba, M.; Ambrosone, G.; Barucca, G.; Cabibbo, M.; Mengucci, P.; de Asmundis, R.; Carotenuto, G. A new micromechanical approach for the preparation of graphene nanoplatelets deposited on polyethylene. *Nanotechnology* **2017**, *28*. [CrossRef] [PubMed]

98. Zhong, Y.L.; Swager, T.M. Enhanced electrochemical expansion of graphite for in situ electrochemical functionalization. *J. Am. Chem. Soc.* **2012**, *134*, 17896–17899. [CrossRef] [PubMed]

99. Leroux, Y.R.; Jean-François Bergamini, J.-F.; Ababou, S.; Le Breton, J.-C.; Hapiot, P. Synthesis of functionalized few-layer graphene through fast electrochemical expansion of graphite. *J. Electroanal. Chem.* **2015**, *753*, 42–46. [CrossRef]

100. Hossaina, S.T.; Wanga, R. Electrochemical exfoliation of graphite: Effect of temperature and hydrogen peroxide addition. *Electrochim. Acta* **2016**, *216*, 253–260. [CrossRef]

101. McAllister, M.J.; Li, J.L.; Adamson, D.H.; Schniepp, H.C.; Abdala, A.A.; Liu, J.; Herrera-Alonso, M.; Milius, D.L.; Car, R.; Prudhomme, R.K.; et al. Single sheet functionalized graphene by oxidation and thermal expansion of graphite. *Chem. Mater.* **2007**, *19*, 4396–4404. [CrossRef]

102. Beidaghi, M.; Wang, Z.; Gu, L.; Wang, C. Electrostatic spray deposition of graphene nanoplatelets for high-power thin-film supercapacitor electrodes. *J. Solid State Electrochem.* **2012**, *16*, 3341–3348. [CrossRef]

103. Allen, M.J.; Tung, V.C.; Kaner, R.B. Honeycomb carbon: A review of graphene. *Chem. Rev.* **2010**, *110*, 132–145. [CrossRef] [PubMed]

104. Viculis, L.M.; Mack, J.J.; Kaner, R.B. A chemical route to carbon nanoscrolls. *Science* **2003**, *299*. [CrossRef] [PubMed]

105. Park, S.; Ruoff, R.S. Chemical methods for the production of graphenes. *Nat. Nanotechnol.* **2009**, *4*, 217–224. [CrossRef] [PubMed]

106. Jiao, L.Y.; Wang, X.R.; Diankov, G.; Wang, H.L.; Dai, H.J. Facile synthesis of high-quality graphene nanoribbons. *Nat. Nanotechnol.* **2010**, *5*, 321–325. [CrossRef] [PubMed]

107. Kosynkin, D.V.; Higginbotham, A.L.; Sinitskii, A.; Lomeda, J.R.; Dimiev, A.; Price, B.K.; Tour, J.M. Longitudinal unzipping of carbon nanotubes to form graphene nanoribbons. *Nature* **2009**, *458*, 872–876. [CrossRef] [PubMed]

108. Jiao, L.Y.; Zhang, L.; Wang, X.R.; Diankov, G.; Dai, H.J. Narrow graphene nanoribbons from carbon nanotubes. *Nature* **2009**, *458*, 877–880. [CrossRef] [PubMed]

109. Lin, T.; Liu, Z.; Zhou, M.; Bi, H.; Zhang, K.; Huang, F.; Wan, D.; Zhong, Y. Rapid microwave synthesis of graphene directly on h-BN with excellent heat dissipation performance. *ACS Appl. Mater. Interfaces* **2014**, *6*, 3088–3092. [CrossRef] [PubMed]

110. Xin, G.Q.; Hwang, W.; Kim, N.; Cho, S.M.; Chae, H. A graphene sheet exfoliated with microwave irradiation and interlinked by carbon nanotubes for high-performance transparent flexible electrodes. *Nanotechnology* **2010**, *21*. [CrossRef] [PubMed]

111. Shen, Y.; Jing, T.; Ren, W.; Zhang, J.; Jiang, Z.-G.; Yu, Z.-Z.; Dasari, A. Chemical and thermal reduction of graphene oxide and its electrically conductive polylactic acid nanocomposites. *Compos. Sci. Technol.* **2012**, *72*, 1430–1435. [CrossRef]

112. Cote, L.J.; Kim, F.; Huang, J. Langmuir-Blodgett assembly of graphite oxide single layers. *J. Am. Chem. Soc.* **2009**, *131*, 1043–1049. [CrossRef] [PubMed]

113. Loh, K.P.; Bao, Q.L.; Ang, P.K.; Yang, J.X. The chemistry of graphene. *J. Mater. Chem.* **2010**, *20*, 2277–2289. [CrossRef]

114. Moon, I.K.; Lee, J. Highly qualified reduced graphene oxides: The best chemical reduction. *Chem. Commun.* **2011**, *47*, 9681–9683. [CrossRef] [PubMed]

115. Chua, C.K.; Pumera, M. Chemical reduction of graphene oxide: A synthetic chemistry viewpoint. *Chem. Soc. Rev.* **2014**, *43*, 291–312. [CrossRef] [PubMed]

116. Chen, L.; Hernandez, Y.; Feng, X.; Mullen, K. From nanographene and graphene nanoribbons to graphene sheets: Chemical synthesis. *Angew. Chem. Int. Ed. Engl.* **2012**, *51*, 7640–7654. [CrossRef] [PubMed]

117. Barg, S.; Perez, F.M.; Ni, N.; Pereira, P.V.; Maher, R.C.; Garcia-Tunon, E.; Eslava, S.; Agnoli, S.; Mattevi, C.; Saiz, E. Mesoscale assembly of chemically modified graphene into complex cellular networks. *Nat. Commun.* **2014**, *5*, 4328. [CrossRef] [PubMed]

118. Ruoff, R. Graphene: Calling all chemists. *Nat. Nano* **2008**, *3*, 10–11. [CrossRef] [PubMed]

119. Paredes, J.I.; Villar-Rodil, S.; Martinez-Alonso, A.; Tascon, J.M.D. Graphene oxide dispersions in organic solvents. *Langmuir* **2008**, *24*, 10560–10564. [CrossRef] [PubMed]

120. Bianco, A.; Cheng, H.-M.; Enoki, T.; Gogotsi, Y.; Hurt, R.H.; Koratkar, N.; Kyotani, T.; Monthioux, M.; Park, C.R.; Tascon, J.M.D.; et al. All in the graphene family—A recommended nomenclature for two dimensional carbon materials. *Carbon* **2013**, *65*, 1–6. [CrossRef]

121. Poh, H.L.; Sanek, F.; Ambrosi, A.; Zhao, G.; Sofer, Z.; Pumera, M. Graphenes prepared by Staudenmaier, Hofmann and Hummer's methods with consequent thermal exfoliation exhibit very different electrochemical properties. *Nanoscale* **2012**, *4*, 3515–3522. [CrossRef] [PubMed]

122. Compton, O.C.; Nguyen, S.T. Graphene oxide, highly reduced graphene oxide, and graphene: Versatile building blocks for carbon-based materials. *Small* **2010**, *6*, 711–723. [CrossRef] [PubMed]

123. Dreyer, D.R.; Park, S.; Bielawski, C.W.; Ruoff, R.S. The chemistry of graphene oxide. *Chem. Soc. Rev.* **2010**, *39*, 228–240. [CrossRef] [PubMed]

124. Brodie, B.C. On the atomic weight of graphite. *Philos. Trans. R. Soc. Lond.* **1859**, *149*, 249–259. [CrossRef]

125. Staudenmaier, L. Verfahren zur darstellung der graphitsaure. *Ber. Dtsch. Chem. Ges.* **1898**, *31*, 1481–1487. [CrossRef]

126. Hummers, W.S., Jr.; Offeman, R.E. Preparation of graphitic oxide. *J. Am. Chem. Soc.* **1958**, *80*, 1339. [CrossRef]

127. Schniepp, H.C.; Li, J.L.; McAllister, M.J.; Sai, H.; Herrera-Alonso, M.; Adamson, D.H.; Prud'homme, R.K.; Car, R.; Saville, D.A.; Aksay, I.A. Functionalized single graphene sheets derived from splitting graphite oxide. *J. Phys. Chem. B* **2006**, *110*, 8535–8539. [CrossRef] [PubMed]

128. Gao, X.F.; Jang, J.; Nagase, S. Hydrazine and thermal reduction of graphene oxide: Reaction mechanisms, product structures, and reaction design. *J. Phys. Chem. C* **2010**, *114*, 832–842. [CrossRef]

129. Ramesha, G.K.; Sampath, S. Electrochemical reduction of oriented graphene oxide films: An in situ Raman spectroelectrochemical study. *J. Phys. Chem. C* **2009**, *113*, 7985–7989. [CrossRef]

130. Abdelsayed, V.; Moussa, S.; Hassan, H.M.; Aluri, H.S.; Collinson, M.M.; El-Shall, M.S. Photothermal deoxygenation of graphite oxide with laser excitation in solution and graphene-aided increase in water temperature. *J. Phys. Chem. Lett.* **2010**, *1*, 2804–2809. [CrossRef]

131. Ng, Y.H.; Lightcap, I.V.; Goodwin, K.; Matsumura, M.; Kamat, P.V. To what extent do graphene scaffolds improve the photovoltaic and photocatalytic response of TiO_2 nanostructured films? *J. Phys. Chem. Lett.* **2010**, *1*, 2222–2227. [CrossRef]

132. Golsheikh, A.M.; Lim, H.N.; Zakaria, R.; Huang, N.M. Sonochemical synthesis of reduced graphene oxide uniformly decorated with hierarchical ZnS nanospheres and its enhanced photocatalytic activities. *RSC Adv.* **2015**, *5*, 12726–12735. [CrossRef]

133. Vinodgopal, K.; Neppolian, B.; Lightcap, I.V.; Grieser, F.; Ashokkumar, M.; Kamat, P.V. Sonolytic design of graphene-Au nanocomposites. Simultaneous and sequential reduction of graphene oxide and Au(III). *J. Phys. Chem. Lett.* **2010**, *1*, 1987–1993. [CrossRef]

134. Voiry, D.; Yang, J.; Kupferberg, J.; Fullon, R.; Lee, C.; Jeong, H.Y.; Shin, H.S.; Chhowalla, M. High-quality graphene via microwave reduction of solution-exfoliated graphene oxide. *Science* **2016**, *353*, 1413–1416. [CrossRef] [PubMed]

135. Mills, A.; Le Hunte, S. An overview of semiconductor photocatalysis. *J. Photochem. Photobiol. A* **1997**, *108*, 1–35. [CrossRef]

136. Mahlambi, M.M.; Ngila, C.J.; Mamba, B.B. Recent developments in environmental photocatalytic degradation of organic pollutants: The case of Titanium dioxide nanoparticles—A review. *J. Nanomater.* **2015**, *2015*. [CrossRef]

137. Nosaka, Y.; Nosaka, A. Understanding hydroxyl radical ($^{\bullet}$OH) generation processes in photocatalysis. *ACS Energy Lett.* **2016**, *1*, 356–359. [CrossRef]

138. Zhang, Y.; Tan, Y.; Stormer, H.L.; Kim, P. Experimental observation of the quantum Hall effect and Berry's phase in graphene. *Nature* **2005**, *438*, 201–204. [CrossRef] [PubMed]
139. Teh, C.M.; Mohamed, A.R. Roles of titanium dioxide and ion-doped titanium dioxide on photocatalytic degradation of organic pollutants (phenolic compounds and dyes) in aqueous solutions: A review. *J. Alloys Compd.* **2011**, *509*, 1648–1660. [CrossRef]
140. Hager, S.; Bauer, R. Heterogeneous photocatalytic oxidation of organics for air purification by near UV irradiated titanium dioxide. *Chemosphere* **1999**, *38*, 1549–1559. [CrossRef]
141. Faraldos, M.; Bahamonde, A. Environmental applications of titania-graphene photocatalysts. *Catal. Today* **2017**, *285*, 13–28. [CrossRef]
142. Upadhyay, R.K.; Soin, N.; Ro, S.S. Role of graphene/metal oxide composites as photocatalysts, adsorbents and disinfectants in water treatment: A review. *RSC Adv.* **2014**, *4*, 3823–3851. [CrossRef]
143. Fujishima, A.; Rao, T.N.; Tryk, D.A. Titanium dioxide photocatalysis. *J. Photochem. Photobiol. C Photochem. Rev.* **2000**, *1*, 1–21. [CrossRef]
144. Hu, G.; Tang, B. Photocatalytic mechanism of graphene/titanate nanotubes photocatalyst under visible-light irradiation. *Mater. Chem. Phys.* **2013**, *138*, 608–614. [CrossRef]
145. Cruz, M.; Gomez, C.; Duran-Valle, C.J.; Pastrana-Martínez, L.M.; Faria, J.L.; Silva, A.M.T.; Faraldosa, M.; Bahamondea, A. Bare TiO_2 and graphene oxide TiO_2 photocatalysts on the degradation of selected pesticides and influence of the water matrix. *Appl. Surf. Sci.* **2017**, *416*, 1013–1021. [CrossRef]
146. Wang, C.; Cao, M.; Wang, P.; Ao, Y.; Hou, J.; Qian, J. Preparation of graphene-TiO_2 composites with enhanced photocatalytic activity for the removal of dye and Cr (VI). *Appl. Catal. A Gen.* **2014**, *473*, 83–89. [CrossRef]
147. Pastrana-Martínez, L.M.; Morales-Torres, S.; Likodimos, V.; José, L.; Figueiredo, J.L.; Faria, J.L.; Falaras, P.; Silva, A.M.T. Advanced nanostructured photocatalysts based on reduced graphene oxide-TiO_2 composites for degradation of diphenhydramine pharmaceutical and methyl orange dye. *Appl. Catal. B* **2012**, *123*, 241–256. [CrossRef]
148. Liu, S.; Sun, H.; Liu, S.; Wang, S. Graphene facilitated visible light photodegradation of methylene blue over titanium dioxide photocatalysts. *Chem. Eng. J.* **2013**, *214*, 298–303 [CrossRef]
149. Zhang, L.-W.; Fu, H.-B.; Zhu, Y.-F. Efficient TiO_2 photocatalysts from surface hybridization of TiO_2 particles with graphite-like carbon. *Adv. Funct. Mater.* **2008**, *18*, 2180–2189. [CrossRef]
150. Ren, L.; Qi, X.; Liu, Y.; Huang, Z.; Wei, X.; Li, J.; Yang, L.; Zhong, J. Upconversion-P25–graphene composite as an advanced sunlight driven photocatalytic hybrid material. *J. Mater. Chem.* **2012**, *22*, 11765–11771. [CrossRef]
151. Long, R.; English, N.J.; Prezhdo, O.V. Photo-induced charge separation across the graphene-TiO_2 interface is faster than energy losses: A time-domain ab initio analysis. *J. Am. Chem. Soc.* **2012**, *134*, 14238–14248. [CrossRef] [PubMed]
152. Rehman, S.; Ullah, R.; Butt, A.M.; Gohar, N.D. Strategies of making TiO_2 and ZnO visible light active. *J. Hazard. Mater.* **2009**, *170*, 560–569. [CrossRef] [PubMed]
153. Wu, H.; Wu, X.-L.; Wang, Z.-M.; Aoki, H.; Kutsuna, S.; Keiko Jimura, K.; Hayashi, S. Anchoring titanium dioxide on carbon spheres for high-performancevisible light photocatalysis. *Appl. Catal. B Environ.* **2017**, *207*, 255–266. [CrossRef]
154. Warkhade, S.K.; Gaikwad, G.S.; Zodape, S.P.; Pratap, U.; Maldhure, A.V.; Wankhade, A.V. Low temperature synthesis of pure anatase carbon doped titanium dioxide: An efficient visible light active photocatalyst. *Mater. Sci. Semicond. Process.* **2017**, *63*, 18–24. [CrossRef]
155. Jiang, W.J.; Zhu, Y.F.; Zhu, G.X.; Zhang, Z.J.; Chen, X.J.; Yao, W.Q. Three-dimensional photocatalysts with a network structure. *J. Mater. Chem. A* **2017**, *5*, 5661–5679. [CrossRef]
156. Williams, G.; Seger, B.; Kamat, P.V. TiO_2–graphene nanocomposites. UV-assisted photocatalytic reduction of graphene oxide. *ACS Nano* **2008**, *2*, 1487–1491. [CrossRef] [PubMed]
157. Pan, X.; Yang, M.-Q.; Tang, Z.R.; Xu, Y.-J. Noncovalently functionalized graphene-directed synthesis of ultralarge graphene-based TiO_2 nanosheet composites: Tunable morphology and photocatalytic applications. *J. Phys. Chem. C* **2014**, *118*, 27325–27335. [CrossRef]
158. Kamegawa, T.; Yamahana, D.; Yamashita, H. Graphene coating of TiO_2 nanoparticles loaded on mesoporous silica for enhancement of photocatalytic activity. *J. Phys. Chem. C* **2010**, *114*, 15049–15053. [CrossRef]

159. Bekyarova, E.; Itkis, M.E.; Ramesh, P.; Berger, C.; Sprinkle, M.; de Heer, W.A.; Haddon, R.C. Chemical modification of epitaxial graphene: Spontaneous grafting of aryl groups. *J. Am. Chem. Soc.* **2009**, *131*, 1336–1337. [CrossRef] [PubMed]

160. Zhang, Q.; Bao, N.; Wang, X.; Hu, X.; Miao, X.; Chaker, M.; Ma, D. Advanced fabrication of chemically bonded graphene/TiO_2 continuous fibers with enhanced broadband photocatalytic properties and involved mechanisms exploration. *Sci. Rep.* **2016**, *6*. [CrossRef] [PubMed]

161. Yao, Y.; Jie, G.; Bao, F.; Jiang, S.; Zhang, X.; Ma, R. Covalent functionalization of graphene with polythiophene through a Suzuki coupling reaction. *RSC Adv.* **2015**, *5*, 42754–42761. [CrossRef]

162. He, F.; Fan, J.; Ma, D.; Zhang, L.; Leung, C.; Chan, H.L. The attachment of Fe_3O_4 nanoparticles to graphene oxide by covalent bonding. *Carbon* **2010**, *48*, 3139–3144. [CrossRef]

163. Morawskia, A.W.; Kusiak-Nejmana, E.; Wanaga, A.; Kapica-Kozara, J.; Wróbela, R.J.; Ohtanib, B.; Aksienionekc, M.; Lipinska, L. Photocatalytic degradation of acetic acid in the presence of visiblelight-active TiO_2-reduced graphene oxide photocatalysts. *Catal. Today* **2017**, *280*, 108–113. [CrossRef]

164. Liu, Y.; Zhang, D.; Shang, Y.; Zang, W.; Li, M. Construction of multifunctional films based on graphene-TiO_2 composite materials for strain sensing and photodegradation. *RSC Adv.* **2015**, *5*, 104785–104791. [CrossRef]

165. Raliya, R.; Avery, C.; Chakrabarti, S.; Biswas, P. Photocatalytic degradation of methyl orange dye by pristine titanium dioxide, zinc oxide, and graphene oxide nanostructures and their composites under visible light irradiation. *Appl. Nanosci.* **2017**, *7*, 253–259. [CrossRef]

166. Kuila, T.; Bose, S.; Mishra, A.K.; Kharna, P.; Kim, N.H.; Lee, J.H. Recent advances in graphene based polymer composites. *Progress Mater. Sci.* **2012**, *57*, 1061–1105. [CrossRef]

167. Huang, X.; Qi, X.; Freddy Boeyab, F.; Zhang, H. Graphene-based composites. *Chem. Soc. Rev.* **2011**, *41*, 666–686. [CrossRef] [PubMed]

168. Bai, H.; Li, C.; Shi, G. Functional composite materials based on chemically converted graphene. *Adv. Mater.* **2011**, *23*, 1089–1115. [CrossRef] [PubMed]

169. Sreeprasad, T.S.; Berry, V. How do the electrical properties of graphene change with its functionalization? *Small* **2013**, *9*, 341–350. [CrossRef] [PubMed]

170. Pan, X.; Zhao, Y.; Liu, S.; Korzeniewski, C.L.; Wang, S.; Fan, Z. Comparing graphene-TiO_2 nanowire and graphene-TiO_2 nanoparticle composite photocatalysts. *ACS Appl. Mater. Interfaces* **2012**, *4*, 3944–3950. [CrossRef] [PubMed]

171. Lee, J.S.; You, K.H.; Park, C.B. Highly photoactive, low bandgap TiO_2 nanoparticles wrapped by graphene. *Adv. Mater.* **2012**, *24*, 1084–1088. [CrossRef] [PubMed]

172. Wang, J.; Salihi, E.C.; Šiller, L. Green reduction of graphene oxide using alanine. *Mater. Sci. Eng. C* **2017**, *72*, 1–6. [CrossRef] [PubMed]

173. Pei, S.; Cheng, H.-M. The reduction of graphene oxide. *Carbon* **2012**, *50*, 3210–3228. [CrossRef]

174. Bell, N.J.; Ng, Y.H.; Du, A.; Coster, H.; Smith, S.C.; Amal, R. Understanding the enhancement in photoelectrochemical properties of photocatalytically prepared TiO_2- reduced graphene oxide composite. *J. Phys. Chem. C* **2011**, *115*, 6004–6009. [CrossRef]

175. Zhang, Y.; Pan, C. TiO_2/graphene composite from thermal reaction of graphene oxide and its photocatalytic activity in visible light. *J. Mater. Sci.* **2011**, *46*, 2622–2626. [CrossRef]

176. Zhang, X.-Y.; Li, H.-P.; Cui, X.-L.; Linb, Y. Graphene/TiO_2 nanocomposites: Synthesis, characterization and application in hydrogen evolution from water photocatalytic splitting. *J. Mater. Chem.* **2010**, *20*, 2801–2806. [CrossRef]

177. Zabihi, F.; Ahmadian-Yazdi, M.-R.; Eslamian, M. Photocatalytic graphene-TiO_2 thin films fabricated by low-temperature ultrasonic vibration-assisted spin and spray coating in a sol-gel process. *Catalysts* **2017**, *7*, 136. [CrossRef]

178. Gopalakrishnan, A.; Binitha, N.N.; Yaakob, Z. Excellent photocatalytic activity of titania-graphene nanocomposites prepared by a facile route. *J. Sol-Gel Sci. Technol.* **2016**, *80*, 189–200. [CrossRef]

179. Shao, L.; Quan, S.; Liu, Y.; Guo, Z.; Wang, Z. A novel "gel-sol" strategy to synthesize TiO_2 nanorod combining reduced graphene oxide composites. *Mater. Lett.* **2013**, *107*, 307–310. [CrossRef]

180. Long, M.; Qin, Y.; Chen, C.; Guo, X.; Tan, B.; Cai, W. Origin of visible light photoactivity of reduced graphene oxide/TiO_2 by in situ hydrothermal growth of undergrown TiO_2 with graphene oxide. *J. Phys. Chem. C* **2013**, *117*, 16734–16741. [CrossRef]

181. Zhang, H.; Lv, X.; Li, Y.; Wang, Y.; Li, J. P25-graphene composite as a high performance photocatalyst. *ACS Nano* **2010**, *4*, 380–386. [CrossRef] [PubMed]

182. Hu, G.; Yang, J.; Zhao, D.; Chen, Y.; Cao, Y. Research on photocatalytic properties of TiO_2-graphene composites with different morphologies. *J. Mater. Eng. Perform.* **2017**, *26*, 3263–3270. [CrossRef]

183. Hamandi, M.; Berhault, G.; Guillard, C.; Kochkar, H. Reduced graphene oxide/TiO_2 nanotube composites for formic acid photodegradation. *Appl. Catal. B Environ.* **2017**, *209*, 203–213. [CrossRef]

184. Lv, S.; Wan, J.; Shen, Y.; Hu, Z. Preparation of superlong TiO_2 nanotubes and reduced graphene oxide composite photocatalysts with enhanced photocatalytic performance under visible light irradiation. *J. Mater. Sci. Mater. Electron.* **2017**, *28*, 14769–14776. [CrossRef]

185. Kusiak-Nejmana, E.; Wanaga, A.; Kowalczyka, L.; Kapica-Kozara, J.; Colbeau-Justinb, C.; Mendez Medrano, M.G.; Morawsk, A.W. Graphene oxide-TiO_2 and reduced graphene oxide-TiO_2 nanocomposites: Insight in charge-carrier lifetime measurements. *Catal. Today* **2017**, *287*, 189–195. [CrossRef]

186. Li, T.; Wang, T.; Qu, G.; Qu, G.; Liang, D.; Hu, S. Synthesis and photocatalytic performance of reduced graphene oxide-TiO_2 nanocomposites for orange II degradation under UV light irradiation. *Environ. Sci. Pollut. Res.* **2017**, *24*, 12416–12425. [CrossRef] [PubMed]

187. Yadav, H.M.; Kim, J.-S. Solvothermal synthesis of anatase TiO_2-graphene oxide nanocomposites and their photocatalytic performance. *J. Alloys Compd.* **2016**, *688*, 123–129. [CrossRef]

188. Hu, J.J.; Lia, H.; Muhammada, S.; Wua, Q.; Zhaoa, Y.; Jiaoa, Q. Surfactant-assisted hydrothermal synthesis of TiO_2/reduced graphene oxide nanocomposites and their photocatalytic performances. *J. Solid State Chem.* **2017**, *253*, 113–120. [CrossRef]

189. Ge, M.-Z.; Li, S.-H.; Huang, J.-Y.; Zhang, K.-Q.; Al-Deyabc, S.S.; Lai, Y.-K. TiO_2 nanotube arrays loaded with reduced graphene oxide films: Facile hybridization and promising photocatalytic application. *J. Mater. Chem. A* **2015**, *3*, 3491–3499. [CrossRef]

190. De Oliveira, A.G.; Nascimento, J.P.; de Fátima Gorgulho, H.; Martelli, P.B.; Furtado, C.A.; Figueiredo, J.L. Electrochemical synthesis of TiO_2/Graphene oxide composite films for photocatalytic applications. *J. Alloys Compd.* **2016**, *654*, 514–522. [CrossRef]

191. Shanmugam, M.; Alsalme, A.; Alghamdi, A.; Jayavel, R. In-situ microwave synthesis of graphene-TiO_2 nanocomposites with enhanced photocatalytic properties for the degradation of organic pollutants. *J. Photochem. Photobiol. B Biol.* **2016**, *163*, 216–223. [CrossRef] [PubMed]

192. Wang, H.; Gao, H.; Chen, M.; Xu, X.; Wang, X.; Pan, C.; Gao, J. Microwave-assisted synthesis of reduced graphene oxide/titania nanocomposites as an adsorbent for methylene blue adsorption. *J. Appl. Surf. Sci.* **2016**, *360*, 840–848. [CrossRef]

193. Yang, Y.; Liu, E.; Fan, J.; Hu, X.; Hou, W.; Wu, F.; Ma, Y. Green and facile microwave-assisted synthesis of TiO_2/graphene nanocomposite and their photocatalytic activity for methylene blue degradation. *Russ. J. Phys. Chem. A* **2014**, *88*, 478–483. [CrossRef]

194. Pu, S.; Zhu, R.; Ma, H.; Deng, D.; Pei, X.; Qi, F.; Chu, W. Facile in-situ design strategy to disperse TiO_2 nanoparticles on graphene for the enhanced photocatalytic degradation of rhodamine 6G. *Appl. Catal. B Environ.* **2017**, *218*, 208–219. [CrossRef]

195. Pokhrel, D.; Viraraghavan, T. Treatment of pulp and paper mill wastewater—A review. *Sci. Total Environ.* **2004**, *333*, 37–58. [CrossRef] [PubMed]

196. Liu, L.; Zhang, B.; Zhang, Y.; He, Y.; Huang, L.; Tan, S.; Cai, X. Simultaneous removal of cationic and anionic dyes from environmental water using montmorillonite-pillared graphene oxide. *J. Chem. Eng. Data* **2015**, *60*, 1270–1278. [CrossRef]

197. Bhullar, N.; Sharma, S.; Sud, D. Studies of adsorption for dye effluent and decolonization of methylene blue and methyl blue based on the surface of biomass adsorbent. *Pollut. Res.* **2012**, *31*, 681–686.

198. Hameed, B.H.; Din, A.T.M.; Ahmad, A.L. Adsorption of methylene blue onto bamboo-based activated carbon: Kinetics and equilibrium studies. *J. Hazard. Mater.* **2007**, *141*, 819–825. [CrossRef] [PubMed]

199. Mohamed, H.H. Biphasic TiO microspheres/reduced graphene oxide for effective simultaneous photocatalytic reduction and oxidation processes. *Appl. Catal. A Gen.* **2017**, *541*, 25–34. [CrossRef]

200. Zhang, N.; Li, B.; Li, S.; Yang, S. Graphene-supported mesoporous Titania nanosheets for efficient photodegradation. *J. Colloid Interface Sci.* **2017**, *505*, 711–718. [CrossRef] [PubMed]

201. Najafi, M.; Kermanpur, A.; Rahimipour, M.R.; Najafizadeh, A. Effect of TiO_2 morphology on structure of TiO_2-graphene oxide nanocomposite synthesized via a one-step hydrothermal method. *J. Alloys Compd.* **2017**, *722*, 272–277. [CrossRef]

202. Divya, K.S.; Madhu, A.K.; Umadevi, T.U.; Suprabha, T.; Radhakrishnan Nair, P.; Suresh, M. Improving the photocatalytic performance of TiO_2 via hybridizing with graphene. *J. Semiconduct.* **2017**, *38*. [CrossRef]

203. Sohail, M.; Xue, H.; Jiao, Q.; Li, H.; Khan, K.; Wang, S. Synthesis of well-dispersed TiO_2@reduced graphene oxide (rGO) nanocomposites and their photocatalytic properties. *Mater. Res. Bull.* **2017**, *90*, 125–130. [CrossRef]

204. Tseng, I.-H.; Sung, Y.-M.; Chang, P.-Y.; Lin, S.-W. Photocatalytic performance of Titania nanosheets templated by graphene oxide. *J. Photochem. Photobiol. A Chem.* **2017**, *339*, 1–11. [CrossRef]

205. Minella, M.; Sordello, F.; Minero, C. Photocatalytic process in TiO_2/graphene hybrid materials. Evidence of charge separation by electron transfer from reduced graphene oxide to TiO_2. *Catal. Today* **2017**, *281*, 29–37. [CrossRef]

206. Raja, V.; Shiamala, L. Biphasic TiO_2 nanoparticles decorated graphene nanosheets for visible light driven photocatalytic degradation of organic dyes. *Appl. Surf. Sci.* **2017**, *31*. [CrossRef]

207. Verma, R.; Samdarshi, S.K.; Sagar, K. Nanostructured bi-phasic TiO_2 nanoparticles grown on reduced graphene oxide with high visible light photocatalytic detoxification. *Mater. Chem. Phys.* **2017**, *186*, 202–211. [CrossRef]

208. Atchudan, R.; Jebakumar Immanuel Edison, T.N.; Perumal, S.; Karthikeyan, D.; Lee, Y.R. Effective photocatalytic degradation of anthropogenic dyes using graphene oxide grafting titanium dioxide nanoparticles under UV-light irradiation. *J. Photochem. Photobiol. A Chem.* **2017**, *333*, 92–104. [CrossRef]

209. Darvishi, M.; Seyed-Yazdi, J. Characterization and comparison of photocatalytic activities of prepared TiO_2/graphene nanocomposites using titanium butoxide and TiO_2 via microwave irradiation method. *Mater. Res. Express* **2016**, *3*. [CrossRef]

210. Rezaei, M.; Salem, S. Photocatalytic activity enhancement of anatase-graphene nanocomposite for methylene removal: Degradation and kinetics. *Spectrochim. Acta A Mol. Biomol. Spectrosc.* **2016**, *167*, 41–49. [CrossRef] [PubMed]

211. Rezaei, M.; Salem, S. Optimal TiO_2-graphene oxide nanocomposite for photocatalytic activity under sunlight condition: Synthesis, characterization, and kinetics. *Int. J. Chem. Kinet.* **2016**, *48*, 573–583. [CrossRef]

212. Charoensuk, J.; Charupongtawitch, R.; Usakulwattana, A. Kinetics of photocatalytic degradation of methylene blue by TiO_2-graphene nanocomposites. *J. Nanosci. Nanotechnol.* **2016**, *16*, 296–302.

213. Wang, R.; Yang, R.; Wang, B. Efficient degradation of methylene blue by the nano TiO_2-functionalized graphene oxide nanocomposite photocatalyst for wastewater treatment. *Water Air Soil Pollut.* **2016**, *227*. [CrossRef]

214. Liu, G.; Wang, R.; Liu, H.; Han, K.; Cui, H. Highly dispersive nano-TiO_2 in situ growing on functional graphene with high photocatalytic activity. *J. Nanopart. Res.* **2016**, *18*, 1–8. [CrossRef]

215. Baldissarelli, V.Z.; De Souza, T.; Andrade, L.; Oliveira, L.F.C.D.; José, H.J. Preparation and photocatalytic activity of TiO_2-exfoliated graphite oxide composite using an ecofriendly graphite oxidation method. *Appl. Surf. Sci.* **2015**, *359*, 868–874. [CrossRef]

216. Gu, L.; Zhang, H.; Jiao, Z.; Li, M.; Wu, M.; Lei, Y. Glucosamine-induced growth of highly distributed TiO_2 nanoparticles on graphene nanosheets as high-performance photocatalysts. *RSC Adv.* **2016**, *6*, 67039–67048. [CrossRef]

217. Fan, J.; Liu, E.; Hu, X.; Ma, Y.; Fan, X.; Li, Y.; Tang, C. Facile hydrothermal synthesis of TiO_2 nanospindles-reduced graphene oxide composite with a enhanced photocatalytic activity. *J. Alloys Compd.* **2015**, *623*, 298–303.

218. Sun, M.; Li, W.; Sun, S.; He, J.; Zhang, Q.; Shi, Y. One-step in situ synthesis of graphene-TiO_2 nanorod hybrid composites with enhanced photocatalytic activity. *Mater. Res. Bull.* **2015**, *61*, 280–286. [CrossRef]

219. Rong, X.; Qiu, F.; Zhang, C.; Fu, L.; Wang, Y.; Yang, D. Preparation, characterization and photocatalytic application of TiO_2-graphene photocatalyst under visible light irradiation. *Ceram. Int.* **2015**, *41*, 2502–2511. [CrossRef]

220. Gupta, B.K.; Kedawat, G.; Agrawal, Y.; Kumar, P.; Dwivedi, J. A novel strategy to enhance ultraviolet light driven photocatalysis from graphene quantum dots infilled TiO_2 nanotube arrays. *RSC Adv.* **2015**, *5*, 10623–10631. [CrossRef]

221. Sha, J.; Zhao, N.; Liu, E.; Shi, C.; He, C.; Li, J. In situ synthesis of ultrathin 2-D TiO_2 with high energy facets on graphene oxide for enhancing photocatalytic activity. *Carbon* **2014**, *68*, 352–359. [CrossRef]

222. Yang, Y.; Xu, L.; Wang, H.; Wang, W.; Zhang, L. TiO_2/graphene porous composite and its photocatalytic degradation of methylene blue. *Mater. Des.* **2016**, *108*, 632–639. [CrossRef]

223. Suave, J.; Amorim, S.M.; Moreira, R.F.P.M. TiO_2-graphene nanocomposite supported on floating autoclaved cellular concrete for photocatalytic removal of organic compounds. *J. Environ. Chem. Eng.* **2017**, *5*, 3215–3223. [CrossRef]

224. Ai, L.; Zhang, C.; Meng, L. Adsorption of methyl orange from aqueous solution on hydrothermal synthesized Mg–Al layered double hydroxide. *J. Chem. Eng. Data* **2011**, *56*, 4217–4225. [CrossRef]

225. Lavanya, T.; Dutta, M.; Ramaprabhu, S.; Satheesh, K. Superior photocatalytic performance of graphene wrapped anatase/rutile mixed phase TiO_2 nanofibers synthesized by a simple and facile route. *J. Environ. Chem. Eng.* **2017**, *5*, 494–503. [CrossRef]

226. Hou, X.; Sun, T.; Zhao, X. Calcination of reduced graphene oxide decorated TiO_2 composites for recovery and reuse in photocatalytic applications. *Ceram. Int.* **2017**, *43*, 1150–1159.

227. Zhao, F.; Dong, B.; Gao, R.; Su, G.; Liu, W.; Shi, L.; Xia, C. A three-dimensional graphene-TiO_2 nanotube nanocomposite with exceptional photocatalytic activity for dye degradation. *Appl. Surf. Sci.* **2015**, *351*, 303–308. [CrossRef]

228. Lu, Z.; Chen, G.; Hao, W.; Sun, G.; Li, Z. Mechanism of UV-assisted TiO_2/reduced graphene oxide composites with variable photodegradation of methyl orange. *RSC Adv.* **2015**, *5*, 72916–72922. [CrossRef]

229. Xia, H.Y.; He, G.Q.; Min, Y.L.; Liu, T. Role of the crystallite phase of TiO_2 in graphene/TiO_2 photocatalysis. *J. Mater. Sci. Mater. Electron.* **2015**, *26*, 3357–3363. [CrossRef]

230. Jiang, Y.; Wang, W.-N.; Biswas, P.; Fortner, J.D. Facile aerosol synthesis and characterization of ternary crumpled graphene-TiO_2-magnetite nanocomposites for advanced watertreatment. *ACS Appl. Mater. Interfaces* **2014**, *6*, 11766–11774. [CrossRef] [PubMed]

231. Han, W.; Ren, L.; Zhang, Z.; Qi, X.; Liu, Y.; Huang, Z.; Zhong, J. Graphene-supported flocculent-like TiO_2 nanostructures for enhanced photoelectrochemical activity and photodegradation performance. *J. Ceram. Int.* **2015**, *41*, 7471–7477. [CrossRef]

232. Lavanya, T.; Satheesh, K.; Dutta, M.; Victor Jaya, N. Superior photocatalytic performance of reduced graphene oxide wrapped electrospun anatase mesoporous TiO_2 nanofibers. *J. Alloys Compd.* **2014**, *615*, 643–650. [CrossRef]

233. Wang, Y.; Li, Z.; He, Y.; Li, F.; Liu, X. Low-temperature solvothermal synthesis of graphene-TiO_2 nanocomposite and its photocatalytic activity for dye degradation. *Mater. Lett.* **2014**, *134*, 115–118. [CrossRef]

234. Athanasekou, C.P.; Morales-Torres, S.; Likodimos, V.; Romanos, G.E.; Pastrana-Martinez, L.M.; Falaras, P.; Faria, J.L.; Figueiredo, J.L. Prototype composite membranes of partially reduced graphene oxide/TiO_2 for photocatalytic ultrafiltration water treatment under visible light. *Appl. Catal. B Environ.* **2014**, *158–159*, 361–372. [CrossRef]

235. Liu, Y. Hydrothermal synthesis of TiO_2-RGO composites and their improved photocatalytic activity in visible light. *RSC Adv.* **2014**, *4*, 36040–36045. [CrossRef]

236. Shakir, K.; Elkafrawy, A.F.; Ghoneimy, H.F.; Elrab Beheir, S.G.; Refaat, M. Removal of rhodamine B (a basic dye) and thoron (an acidic dye) from dilute aqueous solutions and wastewater simulants by ion flotation. *Water Res.* **2010**, *44*, 1449–1461. [CrossRef] [PubMed]

237. Chen, Q.; Zhou, M.; Zhang, Z.; Tang, T.; Wang, T. Preparation of TiO_2 nanotubes/reduced graphene oxide binary nanocomposites enhanced photocatalytic properties. *J. Mater. Sci. Mater. Electron.* **2017**, *28*, 9416–9422. [CrossRef]

238. Chen, Y.; Dong, X.; Cao, Y.; Xiang, J. Enhanced photocatalytic activities of low-bandgap TiO_2-reduced graphene oxide nanocomposites. *J. Nanopart. Res.* **2017**, *19*. [CrossRef]

239. Wang, W.; Huang, X.; Lai, M. RGO/TiO_2 nanosheets immobilized on magnetically actuated artificial cilia film: A new mode for efficient photocatalytic reaction. *RSC Adv.* **2017**, *7*, 10517–10523. [CrossRef]

240. Biris, A.R.; Toloman, D.; Popa, A.; Lazar, M.D.; Kannarpady, G.K.; Saini, V.; Watanabe, F.; Chhetri, B.P.; Ghosh, A. Synthesis of tunable core-shell nanostructures based on TiO_2-graphene architectures and their application in the photodegradation of rhodamine dyes. *Phys. E Low Dimens. Syst. Nanostruct.* **2016**, *81*, 326–333. [CrossRef]

241. Kim, T.-W.; Park, M.; Kim, H.Y.; Park, S.-J. Preparation of flower-like TiO_2 sphere/reduced graphene oxide composites for photocatalytic degradation of organic pollutants. *J. Solid State Chem.* **2016**, *239*, 91–98. [CrossRef]

242. Liu, H.; Wang, Y.; Shi, L.; Xu, R.; Huang, L.; Tan, S. Utilization of reduced graphene oxide for the enhancement of photocatalytic property of TiO_2 nanotube. *Desalination Water Treat.* **2016**, *57*, 13263–13272. [CrossRef]

243. Sedghi, R.; Heidari, F. A novel & effective visible light-driven TiO_2/magnetic porous graphene oxide nanocomposite for the degradation of dye pollutants. *RSC Adv.* **2016**, *6*, 49459–49468.

244. Zhang, J.-J.; Wu, Y.-H.; Mei, J.-Y.; Zheng, G.-P.; Yan, T.-T.; Zheng, X.-C.; Liu, P.; Guan, X.-X. Synergetic adsorption and photocatalytic degradation of pollutants over 3D TiO_2-graphene aerogel composites synthesized: Via a facile one-pot route. *Photochem. Photobiol. Sci.* **2016**, *15*, 1012–1019. [CrossRef] [PubMed]

245. He, R.; He, W. Ultrasonic assisted synthesis of TiO_2-reduced graphene oxide nanocomposites with superior photovoltaic and photocatalytic activities. *Ceram. Int.* **2016**, *42*, 5766–5771. [CrossRef]

246. Fang, R.; Liang, Y.; Ge, X.; Du, M.; Li, S.; Li, T. Preparation and photocatalytic degradation activity of TiO_2/rGO/polymer composites. *Colloid Polym. Sci.* **2015**, *293*, 1151–1157. [CrossRef]

247. Liang, D.; Cui, C.; Hub, H.; Wang, Y.; Xu, S.; Ying, B.; Li, P.; Lu, B. One-step hydrothermal synthesis of anatase TiO_2/reduced graphene oxide nanocomposites with enhanced photocatalytic activity. *J. Alloys Compd.* **2014**, *582*, 236–240. [CrossRef]

248. Li, Q.; Bian, J.; Zhang, L.; Zhang, R.; Wang, G. Synthesis of carbon materials-TiO_2 hybrid nanostructures and their visible-light photo-catalytic activity. *Chem Plus Chem* **2014**, *79*, 454–461.

249. Rajoriya, S.; Bargole, S.; Saharan, V.K. Degradation of a cationic dye (Rhodamine 6G) using hydrodynamic cavitation coupled with other oxidative agents: Reaction mechanism and pathway. *Ultrason. Sonochem.* **2017**, *34*, 183–194. [CrossRef] [PubMed]

250. Dai, B.; Tao, H.; Lin, Y.-J.; Chang, C.-T. Study of various nanostructures Titania with graphene composites: The preparation and photocatalytic activities. *Nano* **2016**, *11*. [CrossRef]

251. Liang, X.; Tao, H.; Zhang, Q. High performance photocatalytic degradation by graphene/titanium nanotubes under near visible light with low energy irradiation. *J. Nanosci. Nanotechnol.* **2015**, *15*, 4887–4894. [CrossRef] [PubMed]

252. Zhang, Q.; Liang, X.; Chen, B.-Y.; Chang, C. Deciphering effects of functional groups and electron density on azodyes degradation by graphene loaded TiO_2. *Appl. Surf. Sci.* **2015**, *357*, 1064–1071. [CrossRef]

253. Posa, V.R.; Annavaram, V.; Somala, A.R. Fabrication of graphene-TiO_2 nanocomposite with improved photocatalytic degradation for acid orange 7 dye under solar light irradiation. *Bull. Mater. Sci.* **2016**, *39*, 759–767. [CrossRef]

254. Gao, P.; Li, A.; Sun, D.D.; Ng, W.J. Effects of various TiO_2 nanostructures and graphene oxide on photocatalytic activity of TiO_2. *J. Hazard. Mater.* **2014**, *279*, 96–104. [CrossRef] [PubMed]

255. Muthirulan, P.; Devi, C.N.; Sundaram, M.M. TiO_2 wrapped graphene as a high performance photocatalyst for acid orange 7 dye degradation under solar/UV light irradiations. *Ceram. Int.* **2014**, *40*, 5945–5957. [CrossRef]

 *catalysts*

Article

Photocatalytic Reduction of CO_2 from Simulated Flue Gas with Colored Anatase

Yebin Guan [1,2], Ming Xia [1], Alessandro Marchetti [1], Xiaohong Wang [2], Weicheng Cao [1], Hanxi Guan [1] and Xueqian Kong [1,*]

[1]   Center for Chemistry of Novel & High-Performance Materials, and Department of Chemistry, Zhejiang University, Hangzhou 310027, China; guanyb@aqnu.edu.cn (Y.G.); xiaming301@126.com (M.X.); alexmar983@yahoo.it (A.M.); caowc@zju.edu.cn (W.C.); guanhx0123@zju.edu.cn (H.G.)

[2]   Anhui Key Laboratory of Functional Coordination Compounds, School of Chemistry and Chemical Engineering, Anqing Normal University, Anqing 246011, China; wangxiaohong@aqnu.edu.cn

*   Correspondence: kxq@zju.edu.cn; Tel.: +86-571-8795-3189

Received: 29 December 2017; Accepted: 9 February 2018; Published: 13 February 2018

Abstract: Photocatalytic reduction with sunlight is an economical and promising advanced approach for reducing the excessive emissions of CO_2 from the combustion of fossil fuels. Aimed at practical applications, a type of inexpensive colored anatase material was used to evaluate CO_2 photoreduction performance on a platform with a continuous flow of gas mixtures (10 vol % CO_2, 90% N_2), which resembles realistic flue gas conditions. The results showed an enhanced photocatalytic activity compared with standard P25 and significant improvement over pristine anatase. Based on a series of characterization techniques, we propose that the oxygen vacancies and surface hydroxyl groups on colored anatase can reduce the bandgap and assist the binding of CO_2 molecules. Our results showed that photoreduction of CO_2 is feasible under practical conditions, and the efficiency could be improved through modification of catalysts.

Keywords: photocatalytic reduction; TiO_2; anatase; CO_2; flue gas

1. Introduction

The excessive emissions of CO_2 generated by the combustion of fossil fuels have a prolonged consequence on the global climate [1]. For instance, as one of the major contributors to CO_2 emissions, thermal power stations emit 10–20 volume percent of CO_2 in their flue gas. Although major research efforts have been aimed at CO_2 capture and sequestration, direct conversion of CO_2 into useable chemicals is a desired, yet more challenging approach. Various chemical or electrochemical processes have been developed to reduce CO_2 to CO or hydrocarbon compounds [2–4]; however, due to the chemical inertness of CO_2, most strategies for CO_2 reduction can only be achieved at an unaffordable cost of energy or reagents. Alternatively, photocatalytic reduction of CO_2 with sunlight would be much more economical, and it is being actively researched [5]. However, current photocatalytic processes either use strong ultraviolet light, or additional sacrificial agents, or conditions far from the industrial processes, while the development of solar-active photocatalysts is still at its early stage.

Till now, a great quantity of works has been endeavored to explore photocatalysts for selective CO_2 transformations [6–9]. For instance, a novel molecular Ru(II)−Re(I) dinuclear complex photocatalytic system with high reduction ability of CO_2 and capture ability of CO_2 has been developed [6]. Rosenthal et al. developed an inexpensive bismuth-carbon monoxide evolving catalyst (Bi-CMEC), which can selectively catalyze the conversion of CO_2 to CO without the need for a costly supporting electrolyte [7]. $BiFeO_3$ and $BiFeO_3$/single-walled carbon nanotubes (SWCNTs) were prepared by the sol-gel method, and the modification of the catalysts by single-walled carbon nanotubes not only greatly improved the response to visible light, but also doubled the catalytic reduction yield of

CO_2 [8]. Among these, TiO_2-based photocatalysts have been most widely studied as photocatalyst with different crystalline forms or modifications [10–13], and they show considerable prospect for the selective transformation of CO_2 to CO, HCOOH, CH_3OH and CH_4 [14]. The P25 TiO_2 nanoparticles branded by Degussa, which have mixed phases of rutile and anatase, are the benchmark material with high photocatalytic performance. However, P25 is relatively expensive, which precludes its use on the scale required for fuel production. Alternatively, loading noble metals such as Pt, Au and Ag can enhance its activities of CO_2 photoreduction for commercial anatase or rutile materials, but the cost is also high. Therefore, the development of inexpensive photocatalysts that can promote the reduction of CO_2 to CO with high selectivity and efficiency is an important step for commercialization.

In this report, we evaluated the CO_2 photoreduction performance on a platform with a continuous flow of gas mixtures (10 vol % CO_2, 90% N_2 at 80 °C) that resembles realistic flue gas conditions. In such conditions, energy-intensive CO_2 separation processes [15] may be avoided. The conversion of CO_2 to CO can be achieved on titanium oxide catalysts with additional water under xenon light in the UV-visible or visible-only range. Carbon monoxide is a key chemical feedstock and a component of syngas, which can be further converted to liquid hydrocarbon fuels. In order to reduce the cost for large-scale application, a colored anatase (g-anatase) material has been prepared, which showed a comparable photocatalytic activity with standard P25 and significant improvement over pristine anatase. Based on characterizations by X-ray diffraction (XRD), transmission electron microscopy (TEM), X-ray photoelectron spectroscopy (XPS), electron paramagnetic resonance (EPR), Fourier transforming infrared spectroscopy (FT-IR) and solid-state nuclear magnetic resonance (SSNMR), we propose that oxygen vacancies and surface hydroxyl groups on colored anatase can reduce the bandgap and assist the binding of CO_2 molecules. Our results showed that photo-reduction of CO_2 is feasible under flue gas conditions, and the efficiency could be improved through modification of catalysts.

2. Results and Discussion

2.1. CO_2 Photoreduction Performance

A number of photocatalysts has been reported for the conversion of CO_2 into CO under continuous flow conditions (Table 1) [11–13,16,17]. However, for flue gas, CO_2 concentration is around 10% in volume. If we used pure CO_2 for reduction, this would demand either intensive energy or special adsorbents for CO_2 separation [18]. For practical and economic considerations, we performed photocatalysis using low concentration CO_2 just like flue gas (i.e., 90% N_2 and 10% CO_2). Other features of flue gas such as temperature and humidity have also been simulated [19]. The temperature of the reaction chamber was maintained at 80 °C, which also approximates the temperature of flue gas. A certain degree of humidity was maintained with added water, and the gas in the reaction chamber was internally cycled with a pump to assist the flow of flue gas. We used xenon light with either the UV or UV-visible range to excite anatase and colored-anatase samples. Under our experimental conditions, CO is the primary product that we detected through an on-line gas chromatography.

Table 1. A list of defective TiO_2 catalytic performance for CO_2 photoreduction.

Catalysts	Light Source	CO_2 Concentration	CO Production (μmol/h)	Reference
TiO_{2-x}/exposed {001}/{101} facets (through $NaBH_4$ reduction)	UV: 100-W mercury vapor lamp; visible: 450-W Xe lamp ($\lambda > 400$ nm)	4 mL/min CO_2 + H_2O continuous flow	UV:CO = 0.44, Vis:CO = 0.208	[13]
Brookite TiO_{2-x} (thermal treatment in He)	150-W solar simulator	2 mL/min, CO_2 + H_2O continuous flow	CO + CH_4 = 0.315	[11]
2.5 at% Co-TiO_2 (sol-gel method)	Visible light of 300-W Xe arc lamp with an L-42 glass filter	3 mL DI H_2O and 80 kPa CO_2 gas	CO = 0.194, CH_4 = 0.009	[16]
4% Cu@TiO_2 (annealing at 450 °C for 2 h in high vacuum)	Visible light of 500-W Xe lamp ($\lambda > 400$ nm)	CO_2 generated from $NaHCO_3$/H_2SO_4 reaction	CO = 0.162, CH_4 = 0.027	[12]
10% I-TiO_2 hydrolysis of titanium isopropoxide (TTIP) in iodic acid solution followed by hydrothermal treatment	Visible light of 150-W solar simulator	CO_2 + H_2O mix gas	CO = 0.48	[17]

The production of CO (μmol/g) during a 6-h photo-illumination period was calculated, as shown in Figure 1. It can be seen that the CO production on colored-anatase greatly increased; the maximum CO production rate was up to 4.99 μmol g^{-1} (i.e., a rate of 0.83 μmol g^{-1} h^{-1}) under UV-visible light irradiation (the UV-vis column in Figure 1), which is circa 10.7-times that of anatase and higher than P25 [20]. If the UV light (wavelength smaller than 400 nm) was cut off, the overall production of CO decreased (the UV cut400 column in Figure 1), and this suggests that UV-light plays an important role in the catalyzing process. However, even with visible light, the colored anatase can still produce about 1.2 μmol g^{-1} CO (i.e., a rate of 0.2 μmol g^{-1} h^{-1}), which is ~20% higher than P25. Although such a conversion rate of CO_2 into CO does not seem so impressive compared to the literature, it should be pointed out that our experiments were performed with more stringent conditions at a much lower CO_2 concentration. As a proof of concept experiment, our results have confirmed the feasibility of direct photocatalytic conversion of CO_2 with flue gas for the first time. Our work also suggests that improvement of the existing catalysts can greatly benefit the conversion efficiency. Therefore, in the following section, we carried out further study on the colored anatase, which is also introduced in this work.

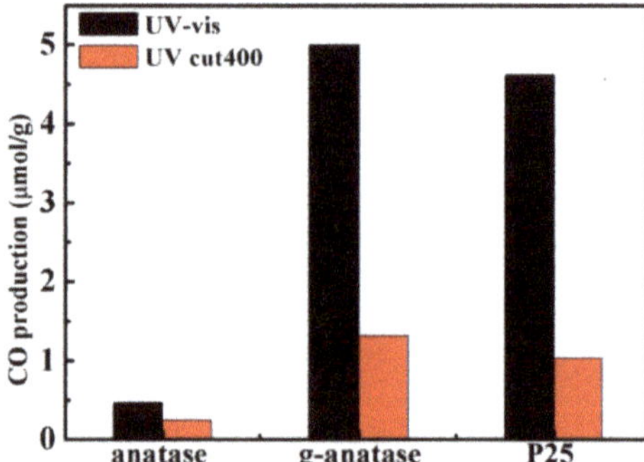

Figure 1. The histogram of CO maximum output with 6 h using anatase, colored anatase and P25 as the catalysts. g-anatase, gray anatase.

2.2. Characterization of Colored Anatase

It has been demonstrated that a higher density of oxygen vacancies in colored TiO_2 could contribute to enhanced photocatalytic activity [21–23]. For example, black TiO_2 nanoparticles with a defective core/disordered shell structure exhibited a bandgap of only 1.85 eV, and the bandgap narrowing is dictated by the synergistic presence of oxygen vacancies and surface disorder [22]. Figure 2a shows the commercial anatase (white color) and the treated anatase powder (gray color, named g-anatase). This distinct change of color indicates that the band structure of anatase nanocrystals has been modified by the $NaBH_4$ reduction treatment. The XRD patterns in Figure 2b showed that the gray TiO2 kept the anatase phase, suggesting this reduction treatment did not alter the crystalline core. The diffuse reflectance UV-Vis spectrum of colored anatase (Figure 2c) clearly shows a broad absorption from 400 nm to the near-infrared region. The relationship between absorbance $(Ah\nu)^{1/2}$ and photo energy ($h\nu$) is shown in Figure 2d, and the bandgap of g-anatase is calculated to be 2.04 eV (compared to 2.97 eV for pristine anatase), which allows the adsorption of a large part of visible light.

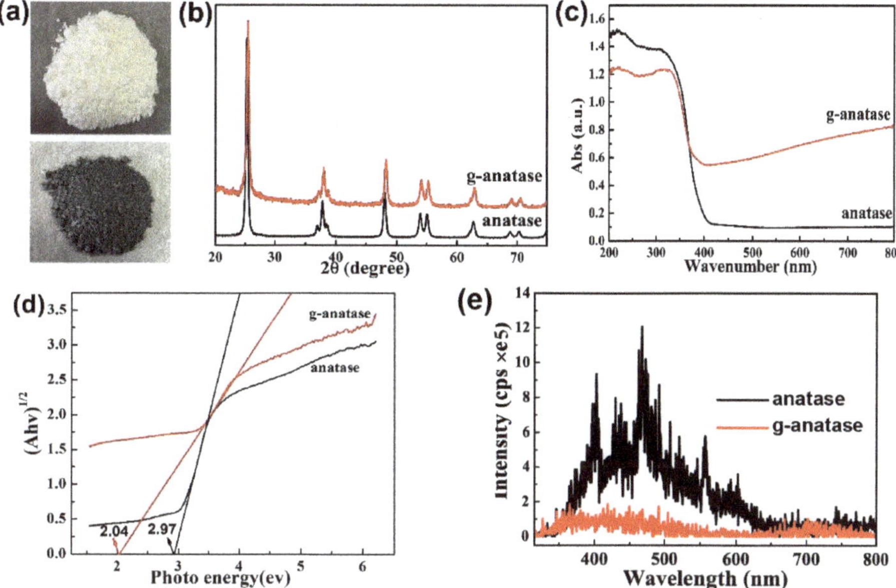

Figure 2. (**a**) Images of anatase (top) and colored anatase (bottom), (**b**) XRD patterns, (**c**) diffuse reflectance UV-Vis spectra, (**d**) relationship between (Ahν) $^{1/2}$ and hν and (**e**) photoluminescence (PL) spectra of anatase and colored anatase.

A photoluminescence (PL) experiment was used to track the recombination of charged carriers freed by photo excitation (325 nm) [24,25]. The PL intensity of colored anatase is much weaker compared to anatase, which indicates that the recombination rate of photo-generated electrons and holes had been reduced considerably (Figure 2e) [26]. Together, the UV-Vis and PL experiments explain the photoactivity of colored anatase in the visible and UV-visible range. In order to fully address the mechanism of its enhanced CO_2 reduction performance, we conducted further structural characterizations by TEM, EDX, XPS, IR, SSNMR, TG and CO_2 adsorption and desorption experiments.

TEM revealed that the average diameter of colored anatase nanocrystals is about 25 nm (Figure 3c), similar to pristine anatase (Figure 3a). Before the reduction treatment, TiO_2 nanocrystals exhibit a highly crystalline nature and well resolved lattice features throughout the whole particles (Figure 3b). However, after the reduction treatment, the anatase nanocrystals show a crystallized core with a disordered outer layer, which forms a shell of about 1.5 nm in thickness, as shown by the black arrows in Figure 3d, and this confirms that the obtained sample has a $TiO_2@TiO_{2-x}$ core/shell structure feature [27]. The corresponding EDX spectrum (Figure 4) showed that there are only Ti and O elements existing in the g-anatase (C and Cu are from the supporting super-thin carbon-coated copper grid). In addition, the atomic ratio of Ti to O is below the stoichiometric ratio of two, which further proved the formation of the amorphous shell.

The high-resolution spectra of Ti2p XPS of commercial anatase and colored anatase are shown in Figure 5a. The Ti2p 3/2 and Ti2p 1/2 peaks centered at the binding energies of 460.5 and 466.2 eV are typical characteristics of the Ti^{4+}–O bonds in anatase. Such peaks shift nevertheless to lower binding energies of 460.0 and 465.9 eV in the colored anatase, which indicates that Ti^{3+} species may have appeared in the colored anatase sample, as previously reported [28,29].

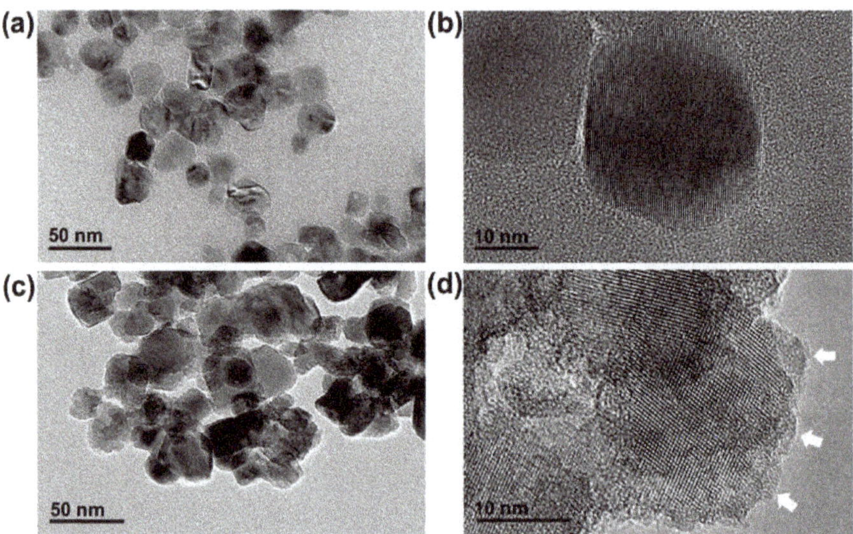

Figure 3. (**a**,**c**) TEM and (**b**,**d**) HRTEM images of anatase and colored anatase nanocrystals.

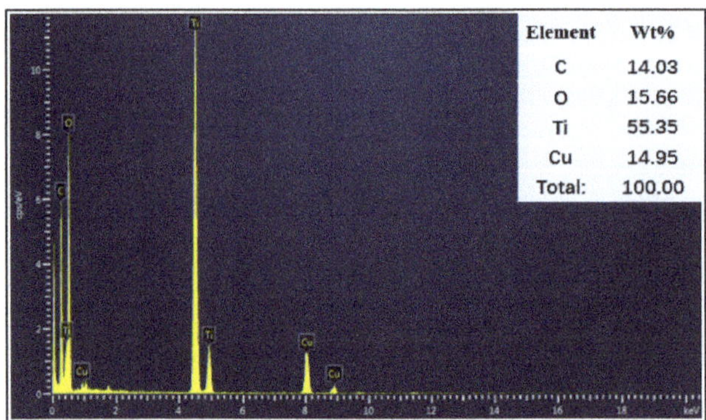

Figure 4. EDX of colored anatase nanocrystals.

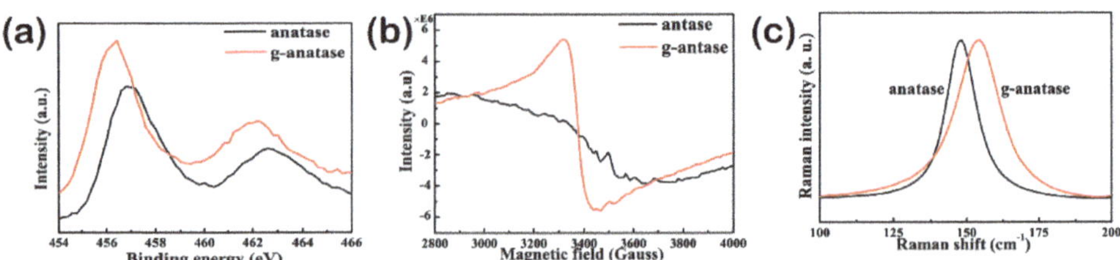

Figure 5. (**a**) High-resolution XPS spectra of Ti 2p, (**b**) electron paramagnetic resonance (EPR) and (**c**) Raman spectra of anatase and colored anatase.

The EPR technique was further employed to analyze the Ti^{3+}-related vacancy defects. From Figure 5b, there is an obvious signal at g = 2.058 for the g-anatase sample, which can be attributed to oxygen vacancy [30]; therefore, the defects probed from slow negative annihilation spectroscopy may be considered to be oxygen-related vacancy, which is consistent with the XPS results. Oxygen-related vacancy defects mainly exist in the bulk (or surface) lattice of the TiO_2 samples after reduction treatment. Raman spectroscopy (Figure 5c) reveals that the strongest vibration mode area (Eg) at 144 cm^{-1} had a significant peak broadening and blue shifting after reduction treatment, compared with commercial anatase. The peak broadening effect and shift have been observed in several recent studies on hydrogenated TiO_2 [22,30–32], which are attributed to the presence of lattice disorder resulting from phonon confinement or nonstoichiometry as the result of oxygen vacancy (VO) doping.

FT-IR spectra (Figure 6a,b) of both white and colored anatase were recorded to investigate light absorption in the mid-infrared light region, from 400 cm^{-1}–4000 cm^{-1}. This allowed gaining further insight into the molecular-level alterations induced by the hydrogenation process. Similar absorption features can be detected in the two samples, namely the OH of H_2O bands near 1635 cm^{-1} and 3400 cm^{-1} and the characteristic wide band in the region 510~650 cm^{-1}, which is due to the symmetric stretching vibrations of the Ti–O bonds of the TiO_4 tetrahedra. The differences between spectra (Figure 6b) reveal that hydrogenation leads to additional absorption peaks at 3648, 3672, and 3691 cm^{-1}, which can be associated with tetrahedral coordinated vacancies and designated as Ti^{4+}–OH [33]. Moreover, hydrogenation leads to another new absorbance at 3710 cm^{-1}, corresponding to terminal OH groups [34], which could be attributed to the hydrogen atoms embedded in the TiO_2 network.

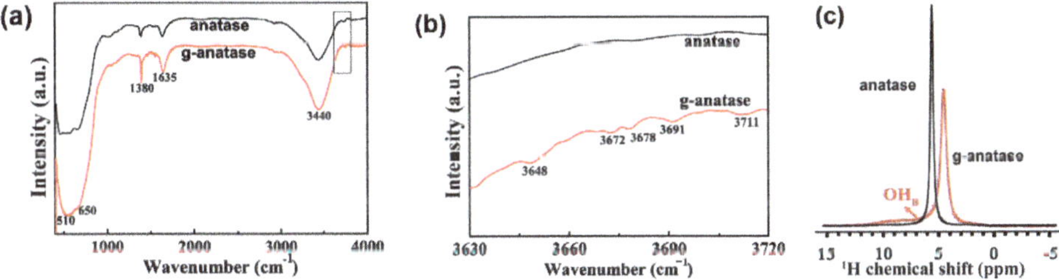

Figure 6. (a,b) FT-IR spectra and (c) ^{1}H solid-state NMR spectra of anatase and colored anatase.

Ultra-fast magic-angle spinning (MAS) solid-state NMR spectroscopy is a crucial tool for the effective characterization of the hydrogen species present on nanocrystals. We performed ^{1}H NMR measurements of anatase and colored anatase at an MAS rate of 60 kHz, as shown in Figure 6c. The dominant peaks at 5.5 and 4.8 ppm in anatase and colored anatase, respectively, can be assigned to the isolated water molecules trapped in anatase crystal defects. Such sharp ^{1}H peaks, which were also observed at the lower spinning speed of 12 kHz, are features of weak ^{1}H–^{1}H dipolar coupling. It is unlikely to be mobile water, however, as the sample has been treated in vacuum at elevated temperatures. Different from anatase, the colored anatase showed also a relatively broad signal at 6.7 ppm corresponding to bridging (i.e., non-terminal) surface hydroxyl groups (marked in Figure 6c as OH_B), which is consistent with the 3648 cm^{-1} FT-IR signal [35].

CO_2 adsorption and desorption curves of anatase and colored anatase are shown in Figure 7a. One can find that both CO_2 adsorbed and desorbed capacities detected at the low pressure range $(P/P_0 < 0.2)$ on g-anatase nanocrystals are higher than those on anatase nanocrystals; this feature is probably closely related to the more exposed alkaline sites, namely Ti^{3+} vacancies on g-anatase nanocrystals, which can absorb more CO_2 molecules [36]. The TGA curve reported in Figure 7b shows that the continuous weight losses of the anatase and g-anatase samples are 2.36 and 3.65% before 350 °C, which can be ascribed to the adsorbed water evaporation. Obviously, the greater weight loss of g-anatase indicates that more water molecules are adsorbed on the surface of the crystals, which is probably closely associated with the amorphous shell of colored anatase [37].

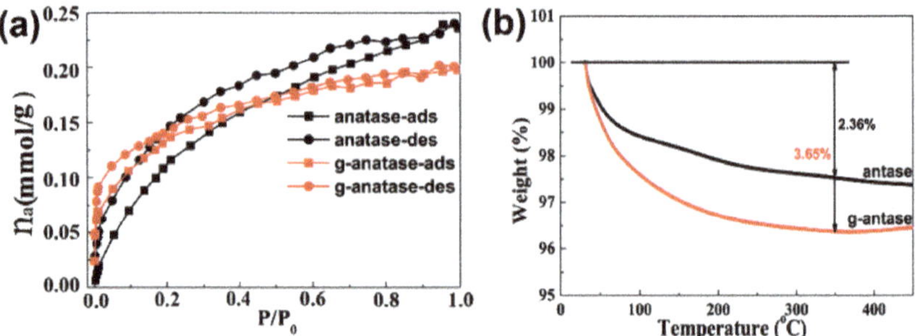

Figure 7. (a) CO_2 adsorption and desorption curves and (b) TGA curves of anatase and colored anatase samples.

3. Experimental Section

3.1. Surface Defective Anatase Preparation

Anatase (99.8%, 25 nm), $NaBH_4$ (98%) and ethanol (AR) were purchased from Aladdin Reagent Comp. (Shanghai, China) and used without purification. The surface defective anatase preparation process is reported elsewhere [20]. Typically, 2.40 g of anatase powder and 0.90 g of $NaBH_4$ were mixed thoroughly with 20 min if grinding. The mixture was then placed into a porcelain capsule inside a tubular furnace in Ar atmosphere and heated from ambient temperature with a rate of 10 °C/min until it reached 400 °C, then maintained at this level for 120 min. After self-cooling to ambient temperature, the obtained gray powder was washed seven times alternately with deionized water and ethanol to remove unreacted $NaBH_4$, then dried at 60 °C in air.

3.2. Sample Characterization

Diffraction patterns were collected using an X-ray diffractometer (XRD) (Bruker AXS D8 Focus, Karlsruhe, Germany) with Cu Ka radiation (λ = 1.54056 A). The UV-Vis absorption spectra were recorded on a Shimadzu UV 2600 UV/Vis spectrophotometer (SHIMADZU Corporation, Kyoto, Japan). Transmission electron microscope (TEM) images were taken using a JEOL 2100 operating at 200 kV (Japan Electron Optics Laboratory Co., Ltd., Kyoto, Japan). X-ray photoelectron spectra (XPS) analyses were recorded using an ESCALABMKII spectrometer with an AlKa (1486.6 eV) achromatic X-ray source (VG Scienta, East Sussex, UK). The EPR spectra were acquired using a Bruker EMX-8 spectrometer at 9.36 GHz at 298.5 K (Karlsruhe, Germany). Raman spectra were collected on a thermal dispersive spectrometer using laser light with an excitation wavelength of 532 nm at a laser power of 10 mW. ^{1}H SSNMR experiments were conducted at a probe temperature of 302 ± 1 K on an AVANCE 600-MHz wide bore spectrometer (Bruker, Karlsruhe, Germany) operating at 14.1 T (the Larmor frequency was 599.80 MHz for the proton) and equipped with a 1.3-mm triple-resonance magic angle spinning (MAS) probe. Free induction decays (FIDs) were not multiplied by an exponential anodization function before Fourier transformation. The spectrum of an empty NMR rotor was collected under the same conditions and subtracted to eliminate the contribution from background signals [35,38,39]. Chemical shifts were measured relative to secondary standard adamantane as an external reference [40,41].

3.3. The Gas-Solid Reaction System

The gas-solid reaction measurement is designed as shown in Scheme 1. A 300-W Xe lamp was used as the light source and eventually filtered to remove its UV part. As a preliminary step, we tested with GC/TCD-FID the eventual background level of carbon-containing products of a mixture of H_2O and He passing through the catalyst with the lamp on. This confirmed that CO_2 was indeed the only carbon source. The production of CO during a given period was calculated by measuring its concentrations at the reactor outlet.

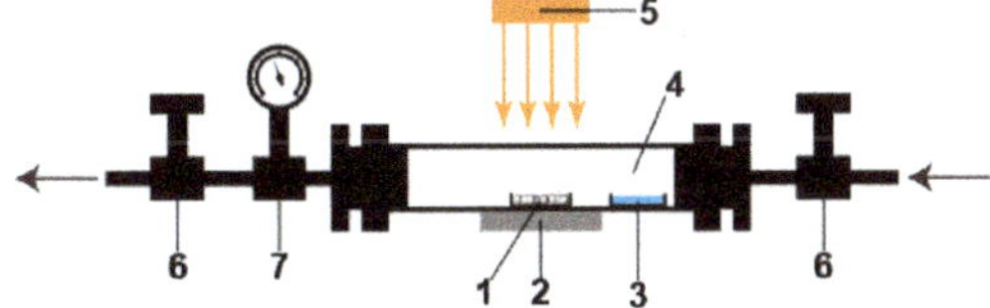

Scheme 1. Diagram of the photocatalytic equipment. (1) Quartz boat for the loading of the TiO_2 catalyst, (2) external heating sleeve, (3) Petri dish containing deionized water, (4) main device constituted by a high-permeability quartz tube and stainless seal flanges, (5) xenon light source, (6) needle valve for the control of the gas flow rate, (7) air gauge displaying the pressure variation in the reaction device. The arrows indicate the flow direction.

4. Conclusions

In summary, we firstly prepared the core/shell structure of the $TiO_2@TiO_{2-x}$ sample by mildly reducing commercial anatase. Such a colored anatase showed comparable photocatalytic activity towards CO_2 reduction under a simulated flue gas system. The enhanced activity is attributed to the formation of oxygen vacancies and Ti^{3+} on the amorphous surface shell, which promotes CO_2 activation under photo-illumination. Our work represents a further step to utilize post-combustion CO_2 in a sustainable manner, i.e., using sunlight and without further processing. Further refinement of reaction conditions and catalytic supports is still possible in order to improve the overall efficiency of photocatalysis

Acknowledgments: This work was financially supported by the National Natural Science Foundation of China (No. 41402037, 21573197) and the State Key Laboratory of Chemical Engineering No. SKL-ChE-16D03.

Author Contributions: X.K., Y.G. and M.X. conceived and designed the experiments; Y.G., M.X. and W.C. performed the experiments; M.X. and A.M. analyzed the data; H.G. and X.M. contributed reagents/materials/ analysis tools; Y.G. and X.K. wrote the paper.

Conflicts of Interest: The authors declare no conflict of interest.

References

1. Yu, J.G.; Low, J.X.; Xiao, W.; Zhou, P.; Jaroniec, M. Enhanced photocatalytic CO_2-reduction activity of anatase TiO_2 by coexposed {001} and {101} facets. *J. Am. Chem. Soc.* **2014**, *136*, 8839–8842. [CrossRef] [PubMed]
2. Zhu, Z.; Qin, J.; Jiang, M.; Ding, Z.; Hou, Y. Enhanced selective photocatalytic CO_2 reduction into CO over Ag/CdS nanocomposites under visible light. *Appl. Surf. Sci.* **2017**, *391*, 572–579. [CrossRef]
3. Tahir, M.; Tahir, B.; Amin, N.A.S.; Zakaria, Z.Y. Photo-induced reduction of CO_2 to CO with hydrogen over plasmonic Ag-NPs/TiO_2 NWs core/shell hetero-junction under UV and visible light. *J. CO2 Util.* **2017**, *18*, 250–260. [CrossRef]
4. Tahir, M.; Tahir, B.; Amin, N.A.S. Synergistic effect in plasmonic Au/Ag alloy NPs co-coated TiO_2 NWs toward visible-light enhanced CO_2 photoreduction to fuels. *Appl. Catal. B Environ.* **2017**, *204*, 548–560. [CrossRef]
5. House, R.L.; Iha, N.Y.M.; Coppo, R.L.; Alibabaei, L.; Sherman, B.D.; Kang, P.; Brennaman, M.K.; Hoertz, P.G.; Meyer, T.J. Artificial photosynthesis: Where are we now? Where can we go? *J. Photochem. Photobiol. C Photochem. Rev.* **2015**, *25*, 32–45. [CrossRef]
6. Nakajima, T.; Tamaki, Y.; Ueno, K.; Kato, E.; Nishikawa, T.; Ohkubo, K.; Yamazaki, Y.; Morimoto, T.; Ishitani, O. Photocatalytic reduction of low concentration of CO_2. *J. Am. Chem. Soc.* **2016**, *138*, 13818–13821. [CrossRef] [PubMed]

7. Medina-Ramos, J.; DiMeglio, J.L.; Rosenthal, J. Efficient reduction of CO_2 to CO with high current density using in situ or ex situ prepared Bi-Based materials. *J. Am. Chem. Soc.* **2014**, *136*, 8361–8367. [CrossRef] [PubMed]

8. Li, X.; He, S.Y.; Li, Z. Methanol synthesis in the catalytic reduction of CO_2 under the visible light by $BiFeO_3$ modified with carbon nanotubes. *J. Chin. Ceram. Soc.* **2009**, *37*, 1869–1872.

9. Ponzoni, C.; Rosa, R.; Cannio, M.; Buscaglia, V.; Finocchio, E.; Nanni, P.; Leonelli, C. Electrophoretic deposition of multiferroic $BiFeO_3$ submicrometric particles from stabilized suspensions. *J. Eur. Ceram. Soc.* **2012**, *33*, 1325–1333. [CrossRef]

10. Liu, G.D.; Xie, S.J.; Zhang, Q.H.; Tian, Z.F.; Wang, Y. Carbon dioxide-enhanced photosynthesis of methane and hydrogen from carbon dioxide and water over Pt-promoted polyaniline–TiO_2 nanocomposites. *Chem. Commun.* **2015**, *51*, 13654–13657. [CrossRef] [PubMed]

11. Liu, L.J.; Zhao, H.L.; Andino, J.M.; Li, Y. Photocatalytic CO_2 Reduction with H_2O on TiO_2 Nanocrystals: Comparison of Anatase, Rutile, and Brookite Polymorphs and Exploration of Surface Chemistry. *ACS Catal.* **2012**, *2*, 1817–1828. [CrossRef]

12. Zhao, J.; Li, Y.X.; Zhu, Y.; Wang, Y.; Wang, C. Enhanced CO_2 photoreduction activity of black TiO_2-coated Cu nanoparticles under visible light irradiation: Role of metallic Cu. *Appl. Catal. A* **2016**, *510*, 34–41. [CrossRef]

13. Liu, L.J.; Jiang, Y.Q.; Zhao, H.L.; Chen, J.T.; Cheng, J.L.; Yang, K.S.; Li, Y. Engineering co-exposed {001} and {101} facets in oxygen-deficient TiO_2 nanocrystals for enhanced CO_2 photoreduction under visible light. *ACS Catal.* **2016**, *6*, 1097–1108. [CrossRef]

14. Peng, C.; Reid, G.; Wang, H.F.; Hu, P. Perspective: Photocatalytic reduction of CO_2 to solar fuels over semiconductors. *J. Chem. Phys.* **2017**, *147*, 030901. [CrossRef] [PubMed]

15. IPCC. *Special Report on Carbon Dioxide Capture and Storage*; Prepared by Working Group III of the Intergovernmental Panel on Climate Change; Metz, B., Davidson, O., de Coninck, H.C., Loos, M., Meyer, L.A., Eds.; Cambridge University Press: Cambridge, UK; New York, NY, USA, 2005.

16. Wang, T.; Meng, X.; Liu, G.G.; Chang, K.; Li, P.; Kang, Q.; Liu, L.Q.; Li, M.; Ouyang, S.X.; Ye, J.H. In situ synthesis of ordered mesoporous Co-doped TiO_2 and its enhanced photocatalytic activity and selectivity for the reduction of CO_2. *J. Mater. Chem. A* **2015**, *3*, 9491–9501. [CrossRef]

17. Zhang, Q.Y.; Li, Y.; Ackerman, E.A.; Gajdardziska-Josifovska, M.; Li, H.L. Visible light responsive iodine-doped TiO_2 for photocatalytic reduction of CO_2 to fuels. *Appl. Catal. A* **2011**, *400*, 195–202. [CrossRef]

18. Sumida, K.; Rogow, D.L.; Mason, J.A.; McDonald, T.M.; Bloch, E.D.; Herm, Z.R.; Bae, T.H.; Long, J.R. Carbon Dioxide Capture in Metal–Organic Frameworks. *Chem. Rev.* **2012**, *112*, 724–781. [CrossRef] [PubMed]

19. McDonald, T.M.; Mson, J.A.; Kong, X.Q.; Bloch, E.D.; Gygi, D.; Dani, A.; Crocellà, V.; Giordanino, F.; Odoh, S.O.; Drisdell, W.S.; et al. Cooperative insertion of CO_2 in diamine-appended metal-organic frameworks. *Nature* **2014**, *519*, 303–308. [CrossRef] [PubMed]

20. Tan, H.Q.; Zhao, Z.; Niu, M.; Mao, C.Y.; Cao, D.P.; Cheng, D.J.; Feng, P.Y.; Sun, Z.C. A facile and versatile method for preparation of colored TiO_2 with enhanced solar-driven photocatalytic activity. *Nanoscale* **2014**, *6*, 10216–10223. [CrossRef] [PubMed]

21. Naldoni, A.; Allieta, M.; Santangelo, S.; Marelli, M.; Fabbri, F.; Cappelli, S.; Bianchi, C.L.; Psaro, R.; Santo, V.D. Effect of nature and location of defects on bandgap narrowing in black TiO_2 nanoparticles. *J. Am. Chem. Soc.* **2012**, *134*, 7600–7603. [CrossRef] [PubMed]

22. Janotti, A.; Varley, J.B.; Rinke, P.; Umezawa, N.; Kresse, G.; van de Walle, C.G. Hybrid functional studies of the oxygen vacancy in TiO_2. *Phys. Rev. B Condens. Matter Mater. Phys.* **2010**, *81*, 085212. [CrossRef]

23. Cronemeyer, D.C. Infrared absorption of reduced rutile TiO_2 single crystals. *Phys. Rev.* **1959**, *113*, 1222–1226. [CrossRef]

24. Yu, J.C.; Yu, G.J.; Ho, W.K.; Jiang, Z.T.; Zhang, L.Z. Effects of F-doping on the photocatalytic activity and microstructures of nanocrystalline TiO_2 powders. *Chem. Mater.* **2002**, *14*, 3808–3816. [CrossRef]

25. Xiang, Q.; Lv, K.; Yu, J. Pivotal role of fluorine in enhanced photocatalytic activity of anatase TiO_2 nanosheets with dominant (001) facets for the photocatalytic degradation of acetone in air. *Appl. Catal. B Environ.* **2010**, *96*, 557–564. [CrossRef]

26. Jiang, X.D.; Zhang, Y.P.; Jiang, J.; Rong, Y.S.; Wang, Y.C.; Wu, Y.C.; Pan, C.X. Characterization of oxygen vacancy associates within hydrogenated TiO_2: A positron annihilation study. *J. Phys. Chem. C* **2012**, *116*, 22619–22624. [CrossRef]

27. Chen, X.; Liu, L.; Peter, Y.Y.; Mao, S.S. Increasing solar absorption for photocatalysis with black hydrogenated titanium b nanocrystals. *Science* **2011**, *331*, 746–750. [CrossRef] [PubMed]

28. Zuo, F.; Bozhilov, K.; Dillon, R.J.; Wang, L.; Smith, P.; Zhao, X.; Bardeen, C.; Feng, P. Active facets on titanium(III)-doped TiO_2: An effective strategy to improve the visible-light photocatalytic activity. *Angew. Chem. Int. Ed.* **2012**, *51*, 6223–6226. [CrossRef] [PubMed]

29. Wang, Z.; Yang, C.; Lin, T.; Yin, H.; Chen, P.; Wan, D.; Xu, F.; Huang, F.; Lin, J.; Xie, X.; et al. H-doped black titania with very high solar absorption and excellent photocatalysis enhanced by localized surface plasmon resonance. *Adv. Funct. Mater.* **2013**, *23*, 5444–5450. [CrossRef]

30. Zou, X.X.; Liu, J.K.; Su, J.; Zuo, F.; Chen, J.S.; Feng, P.Y. Facile synthesis of thermal-and photostable titania with paramagnetic oxygen vacancies for visible-light photocatalysis. *Chem. Eur. J.* **2013**, *19*, 2866–2873. [CrossRef] [PubMed]

31. Venkatasubbu, G.D.; Ramakrishnan, V.; Sasirekha, V.; Ramasamy, S.; Kumar, J. Influence of particle size on the phonon confinement of TiO_2 nanoparticles. *J. Exp. Nanosci.* **2014**, *9*, 661–668. [CrossRef]

32. Zhang, W.; He, Y.; Zhang, M.; Yin, Z.; Chen, Q. Raman scattering study on anatase TiO_2 nanocrystals. *J. Phys. D Appl. Phys.* **2000**, *33*, 912–916. [CrossRef]

33. Kumar, P.M.; Badrinarayanan, S.; Sastry, M. Nanocrystalline TiO_2 studied by optical, FTIR and X-ray photoelectron spectroscopy: Correlation to presence of surface states. *Thin Solid Films* **2000**, *358*, 122–130. [CrossRef]

34. Szczepankiewicz, S.H.; Colussi, A.J.; Hoffmann, M.R. Infrared spectra of photoinduced species on hydroxylated titania surfaces. *J. Phys. Chem. B* **2000**, *104*, 9842–9850. [CrossRef]

35. Ramesha, G.K.; Brennecke, J.F.; Kamat, P.V. Origin of catalytic effect in the reduction of CO_2 at nanostructured TiO_2 films. *ACS Catal.* **2014**, *4*, 3249–3254. [CrossRef]

36. Lindan, P.; Harrison, N.; Gillan, M. Mixed dissociative and molecular adsorption of water on the rutile (110) surface. *Phys. Rev. Lett.* **1998**, *80*, 762–765. [CrossRef]

37. Bavykin, D.V.; Lapkin, A.A.; Plucinski, P.K.; Friedrich, J.M.; Walsh, F.C. Reversible Storage of Molecular Hydrogen by Sorption into Multilayered TiO_2 Nanotubes *J. Phys. Chem. B* **2005**, *109*, 19422–19427. [CrossRef] [PubMed]

38. Nosaka, A.Y.; Fujiwara, T.; Yagi, H.; Akutsu, H.; Nosaka, Y. Photocatalytic Reaction Sites at the TiO_2 Surface as Studied by Solid-State 1H NMR Spectroscopy. *Langmuir* **2003**, *19*, 1935–1937. [CrossRef]

39. Murrell, J.N.; Sorbie, K.S. New analytic form for the potential energy curves of stable diatomic states. *J. Chem. Soc. Faraday Trans.* **1996**, *92*, 2791–2798. [CrossRef]

40. Morcombe, C.R.; Zilm, K.W. Chemical shift referencing in MAS solid state NMR. *J. Magn. Reson.* **2003**, *162*, 479–486. [CrossRef]

41. Hayashi, S.; Hayamizu, K. Chemical Shift Standards in High-Resolution Solid-State NMR (1) 13C, 29Si, and 1H Nuclei. *Bull. Chem. Soc. Jpn.* **1991**, *64*, 685–687. [CrossRef]

Article

The Pros and Cons of Polydopamine-Sensitized Titanium Oxide for the Photoreduction of CO_2

Tongyao Wang, Ming Xia and Xueqian Kong *

Center for Chemistry of Novel & High-Performance Materials, and Department of Chemistry,
Zhejiang University, Hangzhou 310027, China; Wangtongyao@zju.edu.cn (T.W.); Xiam@zju.edu.cn (M.X.)
* Correspondence: kxq@zju.edu.cn; Tel.: +86-177-6715-0513

Received: 17 April 2018; Accepted: 15 May 2018; Published: 17 May 2018

Abstract: Photocatalytic reduction of CO_2 into fuels is a promising route to reduce greenhouse gas emission, and it demands high-performance photocatalysts that can use visible light in the solar spectrum. Due to its broadband light adsorption, polydopamine (PDA) is considered as a promising photo-sensitization material for semiconductor photocatalysts. In this work, titanium oxides have been coated with PDA through an in-situ oxidation polymerization method to pursue CO_2 reduction under visible light. We have shown that the surface coated PDA with a thickness of around 1 nm can enhance the photocatalytic performance of anatase under visible light to reduce CO_2 into CO. Assisted with additional UV-vis adsorption and photoluminescence characterizations, we confirmed the sensitization effect of PDA on anatase. Furthermore, our study shows that thicker PDA coating might not be favorable, as PDA could decompose under both visible and UV-vis light irradiations. ^{13}C solid-state nuclear magnetic resonance showed structural differences between thin and thick PDA coatings and revealed compositional changes of PDA after light irradiation.

Keywords: photocatalysis; CO_2 reduction; anatase; polydopamine; sensitization

---

1. Introduction

The large-scale consumption of fossil fuels rapidly increases atmospheric CO_2 concentration that leads to the consequence of climate change. Alongside technologies to capture and sequester CO_2 [1,2], photocatalytic reduction of CO_2 into fuels is a promising route to turn waste into resources [3,4]. Ultimately, the photocatalysts could turn into artificial leaves which utilize only solar light and minimal reagents to enable a sustainable carbon cycle. To reach such ambitious goal, semiconductor materials will play a very important role owing to their desired photocatalytic activity [5,6]. In 1979, Inoue et al. [7] initially reported semiconductor including TiO_2, WO_3, ZnO, CdS, GaP and SiC as photocatalysts on reduction of CO_2 into organic compounds such as methane, methanol, and formic acid. Encouraged by this pioneering work, various kinds of semiconductor photocatalysts (e.g., Fe_2O_3, Cu_2O, $BiVO_4$, Zn_2GeO_4 and g-C_3N_4) have been developed and tested, although overall their performance was far from satisfaction. The major reasons are their low chemical stability, rapid electron–hole recombination, low visible light adsorption and difficulty activating CO_2 [8,9].

TiO_2 is the most widely studied semiconductor photocatalyst due to its high chemical stability, low toxicity, commercial availability, and suitable energy levels of conduction and valance bands [10,11]. Three crystalline polymorphs of TiO_2 exist in nature, anatase, rutile and brookite, which have different levels of photocatalytic activities. Limited by its wide bandgap (e.g., 3.2 eV for anatase), however, TiO_2 can only absorb UV light of the solar spectrum which is only ~4% of the total solar energy. Various methods, including impurity doping, surface sensitization, oxygen vacancies, and the localized surface plasmon resonance (SPR) effect have been developed to improve the light-harvesting ability of TiO_2 under visible light which accounts for 45% of energy in the solar spectrum [12–20]. Among them, surface sensitization is a frequently-used method that improves visible adsorption and electron/hole separation by transferring the photo-excited electron in photosensitizer to the conduction band of TiO_2.

Compared to other photosensitizers such as pure organic dyes and organometallic complexes [15–20], polymer photosensitizers are more stable under photo-irradiation and in harsh environment. In this study, we focused on polydopamine (PDA), which is a polymer bioinspired by mussel adhesive proteins [21]. PDA not only is an effective surface-coating material, but also possesses unique optical and electronic properties such as broad-spectrum light adsorption and enhanced photoconductivity under the visible light irradiation [22–25]. Recently, it has been used to modify semiconductors such as TiO_2 [26–28], g-C_3N_4 [29,30], Cu_2O [31], and β-FeOOH [32] and metals such as Pd [33] and Ag [34] to improve their photocatalytic activities by the synergetic effect of π–π* transition. However, up to date, almost all the studies have only reported the photodegradation of organic dyes, i.e., the photo-oxidation reaction. PDA-TiO_2 inorganic–organic hybrid materials have not been applied to the photoreduction of CO_2. More importantly, the control of PDA sensitizer, e.g., the coating thickness and surface bonding structure is critical for the resulting photocatalytic performance. Fundamental understanding of the surface chemistry in such composite systems, however, is still rather limited.

In this work, we evaluated the photocatalytic performance of anatase-PDA (A@PDA) under UV-vis or visible light. It was found that the surface sensitizing by PDA can lead to an improved CO_2 conversion into CO. Our results suggest PDA could be a promising photosensitizer for anatase targeted for CO_2 reduction applications. However, we also noticed that, although the increased thickness of PDA apparently increases the CO production, part of the CO increase could come from the photo-degradation of PDA itself as indicated by solid state NMR (SSNMR) and thermogravimetric analysis (TGA) measurements before and after reaction. SSNMR also suggests the chemical structure surface coated PDA could be quite different from the bulk as the aliphatic portions significantly increase. Our experimental work shows that, for the purpose of surface sensitization, balancing the thickness of surface layer is extremely important. It is recommended that the photocatalytic performance (especially about the organic products) should be evaluated judiciously and careful characterization of photo-stability is necessary.

2. Results and Discussions

2.1. Physical and Optical Properties

X-ray diffraction (XRD) was performed to examine the crystal structure of A@PDA. As the results shown in Figure 1a, A@PDA shows a typical pattern of anatase, while the feature of disordered PDA is not observed due to limited thickness of the surface layer. To confirm the successful PDA coating on anatase, we carried out Raman spectroscopy and transmission electron microscopy (TEM) measurements. As shown in Figure 1b, besides the four main Raman vibration modes of anatase, i.e., 141 cm^{-1} (E_g), 395 cm^{-1} (B_{1g}), 516 cm^{-1} (A_{1g}) and 638 cm^{-1} (E_g) [35,36], A@PDA also showed two broad Raman bands at around 1354 cm^{-1} and 1585 cm^{-1} which match those of pure PDA. These broad bands originate from linear stretching of C–C bonds within the rings and in-plane stretching of the aromatic rings, similar to the D and G bands of carbon nanomaterials such as graphene and graphite [37]. Moreover, TEM images (Figure 1c) of A@PDA samples show a thin surface layer of PDA whose thickness ranges from 1 nm to 5 nm depending on the amount of dopamine hydrochloride precursor added. The presence of PDA may also be confirmed by ^{13}C SSNMR, as is mentioned below.

UV-vis diffraction reflectance spectroscopy measurement (Figure 1d) shows that the light adsorption of A@PDA is in general the superposition of A-Ti and PDA. Our calculation shows that the A-Ti has a band-gap of 3.06 eV which is in agreement with typical anatase (Figure S1).

Time-resolved photoluminescence (PL) decay (Figure 1e) is employed to study the charge carrier recombination behavior of A@PDA with different thickness of PDA. In these A@PDA samples, the fluorescence life time is increased versus anatase itself which means a longer life time of excited state that potentially benefit electron transfer for CO_2 reduction.

We also compared the specific surface area and porosity of A@PDA with anatase with N_2 adsorption measurements (Figure S2). According to the classification of International Union of Pure and Applied Chemistry (IUPAC) [38], A@PDA displays a type IV isotherm, suggesting the existence of mesopore structure. The specific surface area of A@PDA is somewhat larger than anatase, and this could contribute to a higher density of active sites (Table 1).

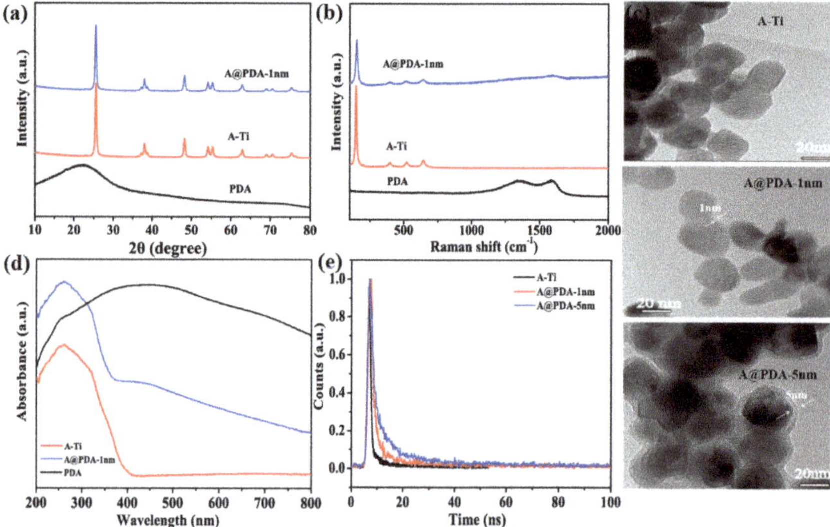

Figure 1. (**a**) XRD pattern and (**b**) Raman spectra of A@PDA-1 nm, A-Ti and PDA; (**c**) TEM images of A-Ti, A@PDA-1 nm, and A@PDA-5 nm; (**d**) UV-vis diffraction reflectance spectra of A-Ti, A@PDA-1 nm, and PDA; and (**e**) PL decay spectra of A-Ti, A@PDA-1 nm, and A@PDA-5 nm.

Table 1. Specific surface area, mean pore size and pore volume of the photocatalysts.

Photocatalyst	S_{BET} (m^2/g)	Mean Pore Size (nm)	Pore Volume (cm^3/g)
A-Ti	58.2	21.5	0.31
A@PDA-1 nm	60.1	23.7	0.36

2.2. Physical and Optical Properties

We first evaluated the photocatalytic activity of A@PDA under visible light using the UV cut-off filter of the Xenon light. The PDA coated anatase shows an increased production of CO per gram of catalyst compared to anatase (Figure 2a). Reaction under such conditions i.e., visible light, no extra sacrificial agents, and a reduced pressure of CO_2 is a viable pathway for the practical application of CO_2 reduction. This again proves the sensitization effect of PDA which works through either direct electron injection [26,27], or more likely a sequential electron injection mechanism, as there is no new band in the adsorption spectrum [28].

In this work, we also tested rutile (R@PDA-1 nm) and brookite (B@PDA-1 nm) using the same PDA sensitization strategy. However, the CO_2 reduction performance of R@PDA-1 nm and B@PDA-1 nm is worse than the pure TiO_2 phases (Figures S3 and S4). It might be due to the mismatch of the excited state of PDA with the CB edge of rutile and brookite [39–41].

On the other hand, under the UV-vis range of Xenon light, the performance of anatase improves and reaches about the same level as A@PDA-1 nm (Figure 2b). By increasing the thickness of PDA, the CO production of A@PDA can be increased further. As the photo-generated electrons have a limited migration range, the photo-sensitization should take place at the shallow interface between PDA and anatase. It is not clear whether such increased CO production for the thicker PDA coating of A@PDA is due to the sensitization effect or other mechanisms such as photo-degradation of PDA itself. We therefore performed a series of control experiments and further characterized the A@PDA samples after photocatalytic reactions.

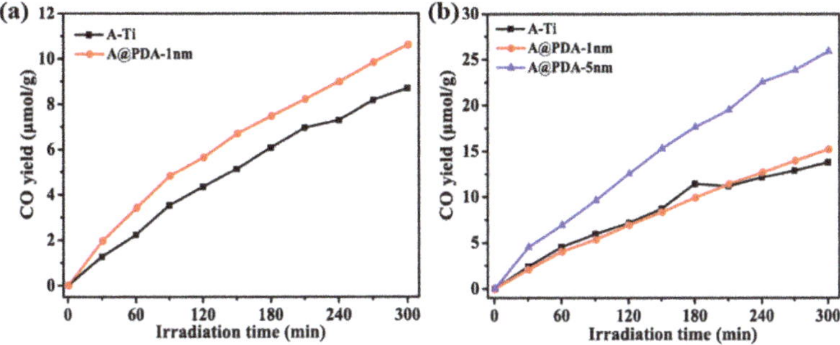

Figure 2. The comparison of CO yield for A-Ti and A@PDA-1 nm under visible light (**a**) and for the A-Ti, A@PDA-1 nm, and A@PDA-5 nm under UV-vis light (**b**) in CO_2.

2.3. Photo-Stability Evaluation

In the literature [42,43], PDA is generally regarded as a stable compound under light irradiation and mild temperature. However, Proks et al. [44] reported that, under elevated temperatures, PDA produces a continuous evolution of CO_2 and they observed that some aliphatic moieties convert into unsaturated species. Bearing this in mind, we were cautious about the extra CO generated by the A@PDA samples, even though it is mostly unexpected in the literature. Control experiments under UV vis light and pure nitrogen show that both A@PDA samples can still generate CO without the reagent of CO_2 (Figure 3a). The A@PDA with thicker coating generates significantly more CO than the thin one. As PDA is the only carbon source in the system, this inevitably proved that the surface covered PDA is not that stable under UV-vis light irradiation as people would expect. We then took a further step to study whether pure PDA itself can withstand light irradiation. As shown in Figure 3b, when the light was turned off, the PDA sample generated minimal CO suggesting it is relatively stable thermally. However, when light was on, no matter UV-vis or only visible light, extra CO was generated. This clearly indicates pure PDA also undergoes photo-degradation.

Further characterization was performed to analyze the structural evolution of A@PDA and PDA after light irradiation. Although no significant changes were detected in the TEM or XPS measurements for the PDA (Figures S5 and S6), clear changes were observed in TGA (Figure 3c) and SSNMR spectra (Figure 4a–c). In the TGA experiments, the A@PDA samples after light irradiation gave less weight loss, suggesting photo-degradation eliminated part of the uncrosslinked monomers or oligomers [45]. The difference in weight-loss of the A@PDA-1 nm sample is much less before and after light irradiation which in good agreement with control experiments observed in Figure 3a.

Qualitatively, ^{13}C SSNMR provides a consistent story that PDA and A@PDA-5 nm sample changed noticeably but not quite for the A@PDA-1 nm sample. More importantly, SSNMR can offer much more details on the changes in chemical structures [46]. In accordance to the literature [47–49], we assigned the resonances between 30 and 60 ppm to the carbon atoms of partially saturated five-member rings and aliphatic CH_2–CH_2–N segments; the signals spanning from 110 to 150 ppm to the aromatic species; and the peaks at between 170 and 180 ppm to the quinones or carboxylate groups. For the PDA sample, the signals of aliphatic carbon (30–50 ppm), the protonated arene (114 ppm) and catecholic groups (145 ppm) decreased after UV-vis light irradiation (Figure 4a), whereas the spectrum of the A@PDA-1 nm sample did not vary much after irradiation despite that the signal-to-noise ratio is low due to the limited amount of surface coating (Figure 4b). In contrast, the signals of aliphatic carbon decreases noticeably for the A@PDA-5 nm sample, indicating a thicker layer of PDA is prone to photo-decomposition (Figure 4c). These findings agree with other measurements and provide more chemical insights on how the PDA degrades under light. We reason that the self-degradation of PDA is the primary effect under such low oxygen concentrations [50]. Interestingly, ^{13}C SSNMR also suggests that the chemical compositions of PDA between sample A@PDA-1 nm and A@PDA-5 nm are somewhat different as the A@PDA-1 nm seems to have more aliphatic portions. This could be due to either the disordered nature of the tightly bound surface structures or different polymerization mechanisms induced by the TiO_2.

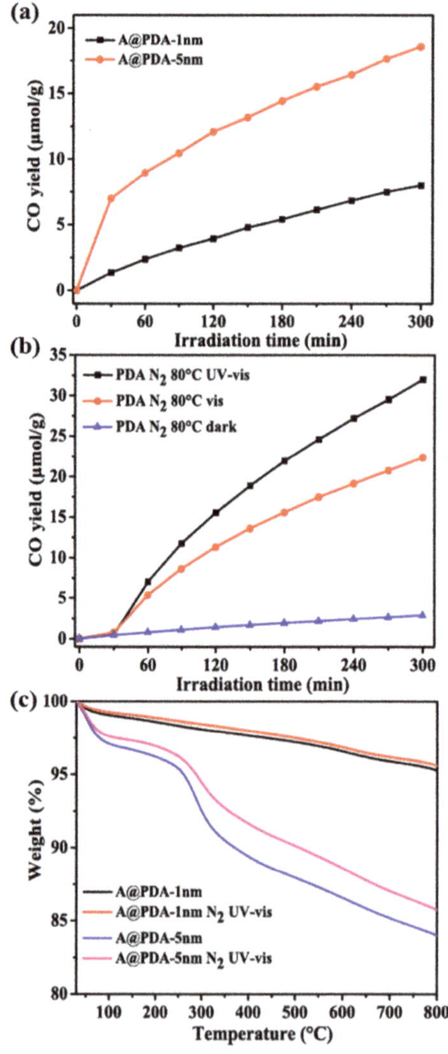

Figure 3. (a) The control experiments of A@PDA-1 nm and A@PDA-5 nm under UV-vis light in N_2; (b) CO yields for the PDA in N_2 under UV-vis light and visible light and in dark; and (c) TGA for the pristine A@PDA-1 nm and A@PDA-5 nm and them after illuminating under UV-vis light in N_2 for 6 h.

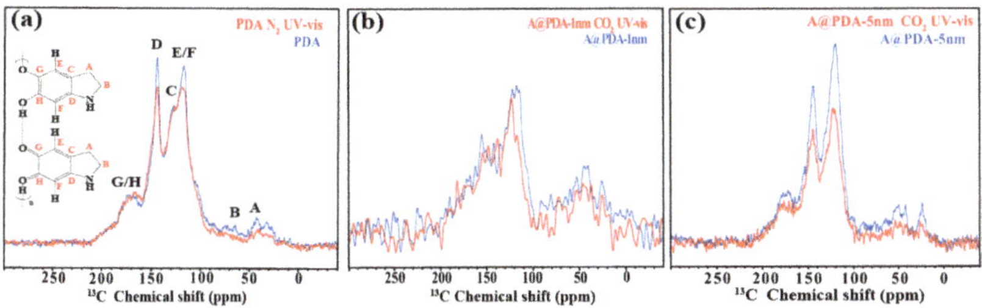

Figure 4. ^{13}C cross-polarization total suppression of spinning sidebands (CP-TOSS) spectrum of: pristine PDA and PDA after illuminating under UV-vis light in the presence of N_2 for 11 h (**a**); pristine A@PDA-1 nm and A@PDA-1 nm after illuminating in CO_2 for 11 h (**b**); and pristine A@PDA-5 nm and A@PDA-5 nm after illuminating in CO_2 for 11 h under UV-vis light (**c**).

2.4. Proposed Reaction Mechanisms

The mechanism of CO_2 photoreduction with H_2O using PDA-sensitized TiO_2 under visible light is illustrated in Figure 5. According to the literature [28], the lowest unoccupied molecular orbital (LUMO) and the highest occupied molecular orbital (HOMO) of PDA are ca. -1.2 and 0.4 eV, respectively. The conduction band (CB) of anatase TiO_2 is ca. -0.5 eV and valence band (VB) of TiO_2 is at 2.56 eV. Under irradiation, the electrons on HOMO of surface-coated PDA are excited to LUMO by adsorbing a considerable amount of visible light. These photoelectrons are then transferred to the conduction band of TiO_2, assisting the reduction of CO_2 into CO.

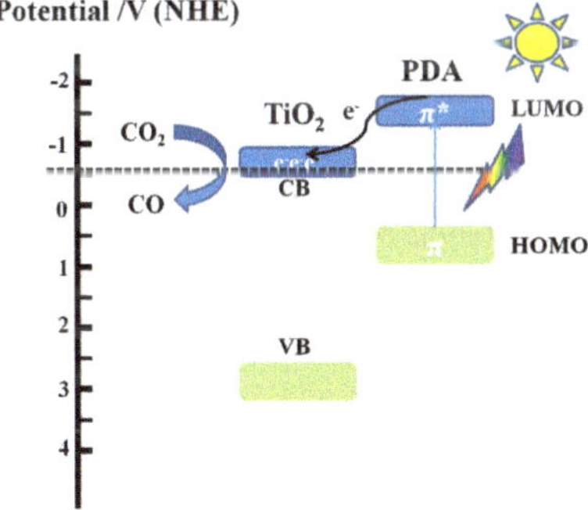

Figure 5. Illustration of the possible reaction mechanism for photoreduction of CO_2 with H_2O under visible light over TiO_2@PDA.

3. Experimental Section

3.1. Materials and Methods

Anatase TiO_2 nano-powder (25 nm), named as A-Ti, and Dopamine hydrochloride were purchased from Aladdin Bio-chem Technology Co. Ltd. (Shanghai, China). Hexamethylenetetramine was purchased from Sinopharm Chemical Reagent Co. Ltd. (Shanghai, China). The anatase powder was calcinated at $450\,^{\circ}C$ for 3 h to improve the crystallinity and remove the organic impurity on the surface of the TiO_2.

3.1.1. Preparation of TiO_2@PDA Composites

The preparation of TiO_2-PDA composites followed the procedures reported in the literature [26] with some modifications. First, 0.2 g of A-Ti and a calculated amount of dopamine hydrochloride were mixed with 50 mL of deionized water and the suspension was sonicated for 15 min before 0.4 g of hexamethylenetetramine was added. The above mixture was kept stirring for 3 h at 90 °C. The resulting product was obtained after centrifugation, washed thoroughly with water and dried overnight at 80 °C in an oven. The as-prepared sample is denoted as A@PDA. For the comparison, we prepared different thickness of PDA on the anatase with added dopamine hydrochloride varying from 0.02 g to 0.12 g, which resulted in samples A@PDA-1 nm and A@PDA-5 nm.

3.1.2. Preparation of PDA

Polydopamine was prepared according to the method reported in the literature [51]. First, 0.75 mL ammonia aqueous solution (NH_4OH, 28–30%) was added to the mixture of ethanol (40 mL) and deionized water (90 mL) under mild stirring at room temperature for 30 min. Second, 0.5 g of dopamine hydrochloride was dissolved in deionized water (10 mL) and then added to the mixture above under mild stirring. The mixture was kept for 30 h at room temperature. The resulting product was collected, washed thoroughly with water and dried overnight at 80 °C in an oven.

3.2. Characterizations

X-ray diffraction (XRD) patterns of the materials were collected on an Ultima IV X-ray diffractometer (Rigaku, Tokyo, Japan) with Cu Kα irradiation in the 2θ range of 10–80°. Raman spectra were recorded on a Labor Raman HR-800 (JobinYvon, Longjumeau, France) with an Ar^+ laser excitation at 514 nm. Transmission electron microscopy (TEM) (Hitachi HT7700, Tokyo, Japan) was used to obtain the morphologies of the materials. The UV-vis diffuse reflectance spectra of the samples over the range of 200–800 nm were obtained on a Shimadzu UV-2600 UV-vis spectrophotometer with an integration sphere diffuse reflectance attachment using $BaSO_4$ as the reference. The time-resolved photoluminescence (PL) decay measurements for the solid samples were investigated on an Edinburgh FLSP920 spectrometer, the laser device of 340 nm was used to get the laser beam. The surface area and porosity were analyzed by nitrogen adsorption at 77K on a gas adsorption apparatus (MicrotracBEL, BELSORP-max, Osaka, Japan). Thermogravimetric analysis (TGA) was performed in N_2 using a SDT Q600 V20.9 Build 20 thermogravimetric analyzer (TA Instruments, New Castle, DE, USA) from 30 °C to 800 °C with a heating rate of 10 °C min^{-1}. X-ray photoelectron spectroscopy (XPS) measurements were conducted on an Esclab MARK II spectrometer (VG Scientific, West Sussex, UK). The solid-state NMR (SSNMR) experiments were performed on a Bruker 14.1T magnet in 3.2-mm ZrO_2 rotors. The ^{13}C Cross-Polarization Total Suppression of Spinning Sidebands (CP-TOSS) NMR spectra were measured at a spinning rate of 10 kHz. The ^{13}C chemical shifts were referenced to adamantane at 38.4 ppm. A CP contact time was set to 2 ms. A total of 10,240 scans were accumulated for each spectrum. All experiments were conducted at room temperature.

3.3. Photocatalytic Reactions

CO_2 photocatalytic reduction in the presence of water was performed with a homemade stainless-steel photocatalytic reactor. The total volume of the reactor was about 415 mL. A 300 W Xe arc lamp (PL-X300D, Beijing Precise Technology Co. Ltd., Beijing, China) equipped with a UV cut-off filter ($\lambda \geq 400$ nm) was used as the UV-vis light source and visible light source and it was positioned ~7.5 cm above the quartz boat. The UV-vis light intensity measured using a radiometer was ca. 220 mW/cm^2 and the visible light intensity was ca. 150 mW/cm^2. In a typical experiment, 30 mg solid photocatalyst was placed in a quartz boat inside the closed quartz tube. A culture dish containing 4 mL water was placed at the side of the quartz tube, which supplies water vapor to the system. Before light irradiation, the reactor was sealed and was subjected to vacuum degassing at room temperature and then backfilled the mixture of N_2 and CO_2 (v/v 90/10) until the pressure rose slightly above 1 bar. The temperature of the reactor was maintained at 80 °C. The gas products from the reactor were analyzed at a 30-min interval using an online gas chromatography (Fuli GC-9790, Wenling, China) equipped with a flame ionized detector (FID) and methanizer.

4. Conclusions

In summary, core–shell structured TiO_2@PDA samples were prepared and they exhibited improved CO yield under visible light, confirming the sensitization effect of PDA. On the other hand, even though thicker PDA coating may lead to higher CO yield, part of the CO product could come from the photodegradation of PDA itself, as indicated by SSNMR and TGA. To achieve better CO_2 reduction performance, further investigations on polymer sensitized semiconductors are needed. Meanwhile, the photo-stability of the organic components has to be carefully evaluated and optimized.

Supplementary Materials: The following are available online at http://www.mdpi.com/2073-4344/8/5/215/s1, Figure S1: The plot of $(\alpha h\nu)^{1/2}$ versus $h\nu$ for the bandgap calculation of A-Ti, Figure S2: (**a**) Nitrogen adsorption and desorption isotherms; and (**b**) pore size distribution curves of A-Ti and A@PDA-1 nm, Figure S3: CO yield comparison of R-Ti and R@PDA-1 nm under Visible light (**a**) and UV-vis light (**b**) for the photoreduction of CO_2, Figure S4: CO yield comparison of B-Ti and B@PDA-1 nm under Visible light (**a**) and UV-vis light (**b**) for the photoreduction of CO_2, Figure S5: TEM images of : (**a**) pristine PDA; and (**b**) PDA after illuminating under UV-vis light in N_2 for 6 h, Figure S6: XPS spectra of PDA and PDA after illuminating under UV-vis light in N_2 for 6 h: (**a**) C1s; (**b**) N1s; and (**c**) O1s.

Author Contributions: X.K. and T.W. conceived and designed the experiments; T.W. performed the experiments; T.W. analyzed the data; M.X. contributed reagents/materials/analysis tools; and T.W. and X.K. wrote the paper.

Acknowledgments: This work was financially supported by National Natural Science Foundation of China (No. 21573197) and State Key Laboratory of Chemical Engineering No. SKL-ChE-16D03. The authors are grateful to the School of Materials Science and Engineering of Zhejiang University for providing the characterizations of time-resolved PL decay in this paper.

Conflicts of Interest: The authors declare no conflict of interest.

References

1. D'Alessandro, D.M.; Smit, B.; Long, J.R. Carbon Dioxide Capture: Prospects for New Materials. *Angew. Chem. Int. Ed.* **2010**, *49*, 6058–6082. [CrossRef] [PubMed]

2. Yaumi, A.L.; Abu Bakar, M.Z.; Hameed, B.H. Recent advances in functionalized composite solid materials for carbon dioxide capture. *Energy* **2017**, *124*, 461–480. [CrossRef]

3. Lewis, N.S.; Nocera, D.G. Powering the planet: Chemical challenges in solar energy utilization. *Proc. Natl. Acad. Sci. USA* **2006**, *103*, 15729–15735. [CrossRef] [PubMed]

4. Habisreutinger, S.N.; Schmidt-Mende, L.; Stolarczyk, J.K. Photocatalytic Reduction of CO_2 on TiO_2 and Other Semiconductors. *Angew. Chem. Int. Ed.* **2013**, *52*, 7372–7408. [CrossRef] [PubMed]

5. Hoffmann, M.R.; Martin, S.T.; Choi, W.Y.; Bahnemann, D.W. Environmental applications of semiconductor photocatalysis. *Chem. Rev.* **1995**, *95*, 69–96. [CrossRef]

6. Zhou, Y.; Tu, W.G.; Zou, Z.G. New Materials for CO_2 Photoreduction. In *Photocatalysis: Fundamentals and Perspectives*; Schneider, J., Bahnemann, D., Ye, J.H., Puma, G.L., Dionysiou, D.D., Eds.; The Royal Society of Chemistry: Cambridge, UK, 2016; pp. 318–322. ISBN 978-1-78262-041-9.

7. Inoue, T.; Fujishima, A.; Konishi, S.; Honda, K. Photoelectrocatalytic reduction of carbon dioxide in aqueous suspensions of semiconductor powders. *Nature* **1979**, *277*, 637–638. [CrossRef]

8. Li, X.; Wen, J.Q.; Low, J.X.; Fang, Y.P.; Yu, J.G. Design and fabrication of semiconductor photocatalyst for photocatalytic reduction of CO_2 to solar fuel. *Sci. China Mater.* **2014**, *57*, 70–100. [CrossRef]

9. Chang, X.X.; Wang, T.; Gong, J.L. CO_2 photo-reduction: Insights into CO_2 activation and reaction on surfaces of photocatalysts. *Energy Environ. Sci.* **2016**, *9*, 2177–2196. [CrossRef]

10. Carp, O.; Huisman, C.L.; Reller, A. Photoinduced reactivity of titanium dioxide. *Prog. Solid State Chem.* **2004**, *32*, 33–177. [CrossRef]

11. Ma, Y.; Wang, X.L.; Jia, Y.S.; Chen, X.B.; Han, H.X.; Li, C. Titanium Dioxide-Based Nanomaterials for Photocatalytic Fuel Generations. *Chem. Rev.* **2014**, *114*, 9987–10043. [CrossRef] [PubMed]

12. Liu, G.H.; Hoivik, N.; Wang, K.Y.; Jakobsen, H. Engineering TiO_2 nanomaterials for CO_2 conversion/solar fuels. *Sol. Energy Mater. Sol. Cells* **2012**, *105*, 53–68. [CrossRef]

13. Das, S.; Daud, W. Photocatalytic CO_2 transformation into fuel: A review on advances in photocatalyst and photoreactor. *Renew. Sustain. Energy Rev.* **2014**, *39*, 765–805. [CrossRef]

14. Zhao, H.L.; Pan, F.P.; Li, Y. A review on the effects of TiO_2 surface point defects on CO_2 photoreduction with H_2O. *J. Materiomics* **2017**, *3*, 17–32. [CrossRef]

15. Ozcan, O.; Yukruk, F.; Akkaya, E.U.; Uner, D. Dye sensitized artificial photosynthesis in the gas phase over thin and thick TiO_2 films under UV and visible light irradiation. *Appl. Catal. B Environ.* **2007**, *71*, 291–297. [CrossRef]

16. Nguyen, T.V.; Wu, J.C.S.; Chiou, C.H. Photoreduction of CO_2 over Ruthenium dye-sensitized TiO_2-based catalysts under concentrated natural sunlight. *Catal. Commun.* **2008**, *9*, 2073–2076. [CrossRef]

17. Qin, G.H.; Zhang, Y.; Ke, X.B.; Tong, X.L.; Sun, Z.; Liang, M.; Xue, S. Photocatalytic reduction of carbon dioxide to formic acid, formaldehyde, and methanol using dye-sensitized TiO_2 film. *Appl. Catal. B Environ.* **2013**, *129*, 599–605. [CrossRef]

18. Won, D.I.; Lee, J.S.; Ji, J.M.; Jung, W.J.; Son, H.J.; Pac, C.; Kang, S.O. Highly Robust Hybrid Photocatalyst for Carbon Dioxide Reduction: Tuning and Optimization of Catalytic Activities of Dye/TiO_2/Re(I) Organic-Inorganic Ternary Systems. *J. Am. Chem. Soc.* **2015**, *137*, 13679–13690. [CrossRef] [PubMed]

19. Do, J.Y.; Tamilavan, V.; Agneeswari, R.; Hyun, M.H.; Kang, M. Synthesis and optical properties of TDQD and effective CO_2 reduction using a TDQD-photosensitized TiO_2 film. *J. Photochem. Photobiol. A Chem.* **2016**, *330*, 30–36. [CrossRef]

20. Huang, H.W.; Lin, J.J.; Zhu, G.B.; Weng, Y.X.; Wang, X.X.; Fu, X.Z.; Long, J.L. A Long-Lived Mononuclear Cyclopentadienyl Ruthenium Complex Grafted onto Anatase TiO_2 for Efficient CO_2 Photoreduction. *Angew. Chem. Int. Ed.* **2016**, *55*, 8314–8318. [CrossRef] [PubMed]

21. Lee, H.; Dellatore, S.M.; Miller, W.M.; Messersmith, P.B. Mussel-inspired surface chemistry for multifunctional coatings. *Science* **2007**, *318*, 426–430. [CrossRef] [PubMed]

22. Pezzella, A.; Iadonisi, A.; Valerio, S.; Panzella, L.; Napolitano, A.; Adinolfi, M.; d'Ischia, M. Disentangling Eumelanin "Black Chromophore": Visible Absorption Changes As Signatures of Oxidation State- and Aggregation-Dependent Dynamic Interactions in a Model Water-Soluble 5,6-Dihydroxyindole Polymer. *J. Am. Chem. Soc.* **2009**, *131*, 15270–15275. [CrossRef] [PubMed]

23. Liu, Y.L.; Ai, K.L.; Lu, L.H. Polydopamine and Its Derivative Materials: Synthesis and Promising Applications in Energy, Environmental, and Biomedical Fields. *Chem. Rev.* **2014**, *114*, 5057–5115. [CrossRef] [PubMed]

24. D'Ischia, M.; Napolitano, A.; Ball, V.; Chen, C.T.; Buehler, M.J. Polydopamine and Eumelanin: From Structure-Property Relationships to a Unified Tailoring Strategy. *Acc. Chem. Res.* **2014**, *47*, 3541–3550. [CrossRef] [PubMed]

25. Son, E.J.; Kim, J.H.; Kim, K.; Park, C.B. Quinone and its derivatives for energy harvesting and storage materials. *J. Mater. Chem. A* **2016**, *4*, 11179–11202. [CrossRef]

26. Mao, W.X.; Lin, X.J.; Zhang, W.; Chi, Z.X.; Lyu, R.W.; Cao, A.M.; Wan, L.J. Core-shell structured TiO_2@polydopamine for highly active visible-light photocatalysis. *Chem. Commun.* **2016**, *52*, 7122–7125. [CrossRef] [PubMed]

27. Zhou, X.S.; Jin, B.; Luo, J.; Xu, X.Y.; Zhang, L.L.; Li, J.J.; Guan, H.J. Dramatic visible light photocatalytic degradation due to the synergetic effects of TiO_2 and PDA nanospheres. *RSC Adv.* **2016**, *6*, 64446–64449. [CrossRef]

28. Kim, S.; Moon, G.H.; Kim, G.; Kang, U.; Park, H.; Choi, W. TiO_2 complexed with dopamine-derived polymers and the visible light photocatalytic activities for water pollutants. *J. Catal.* **2017**, *346*, 92–100. [CrossRef]

29. He, F.; Chen, G.; Yu, Y.G.; Zhou, Y.S.; Zheng, Y.; Hao, S. The synthesis of condensed C-PDA-g-C_3N_4 composites with superior photocatalytic performance. *Chem. Commun.* **2015**, *51*, 6824–6827. [CrossRef] [PubMed]

30. Yu, Z.X.; Li, F.; Yang, Q.B.; Shi, H.; Chen, Q.; Xu, M. Nature-Mimic Method To Fabricate Polydopamine/Graphitic Carbon Nitride for Enhancing Photocatalytic Degradation Performance. *ACS Sustain. Chem. Eng.* **2017**, *5*, 7840–7850. [CrossRef]

31. Zhou, X.S.; Jin, B.; Luo, J.; Gu, X.X.; Zhang, S.Q. Photoreduction preparation of Cu_2O@polydopamine nanospheres with enhanced photocatalytic activity under visible light irradiation. *J. Solid State Chem.* **2017**, *254*, 55–61. [CrossRef]

32. Zhang, C.; Yang, H.C.; Wan, L.S.; Liang, H.Q.; Li, H.Y.; Xu, Z.K. Polydopamine-Coated Porous Substrates as a Platform for Mineralized beta-FeOOH Nanorods with Photocatalysis under Sunlight. *ACS Appl. Mater. Interfaces* **2015**, *7*, 11567–11574. [CrossRef] [PubMed]

33. Xie, A.M.; Zhang, K.; Wu, F.; Wang, N.N.; Wang, Y.; Wang, M.Y. Polydopamine nanofilms as visible light-harvesting interfaces for palladium nanocrystal catalyzed coupling reactions. *Catal. Sci. Technol.* **2016**, *6*, 1764–1771. [CrossRef]

34. Feng, J.J.; Zhang, P.P.; Wang, A.J.; Liao, Q.C.; Xi, J.L.; Chen, J.R. One-step synthesis of monodisperse polydopamine-coated silver core-shell nanostructures for enhanced photocatalysis. *New J. Chem.* **2012**, *36*, 148–154. [CrossRef]

35. Ohsaka, T. Temperature dependence of the Raman spectrum in anatase TiO_2. *J. Phys. Soc. Jpn.* **1980**, *48*, 1661–1668. [CrossRef]

36. Choi, H.C.; Jung, Y.M.; Kim, S.B. Size effects in the Raman spectra of TiO_2 nanoparticles. *Vib. Spectrosc.* **2005**, *37*, 33–38. [CrossRef]

37. Ye, W.C.; Wang, D.A.; Zhang, H.; Zhou, F.; Liu, W.M. Electrochemical growth of flowerlike gold nanoparticles on polydopamine modified ITO glass for SERS application. *Electrochim. Acta* **2010**, *55*, 2004–2009. [CrossRef]

38. Thommes, M.; Kaneko, K.; Neimark, A.V.; Olivier, J.P.; Rodriguez-Reinoso, F.; Rouquerol, J.; Sing, K.S.W. Physisorption of gases, with special reference to the evaluation of surface area and pore size distribution (IUPAC Technical Report). *Pure Appl. Chem.* **2015**, *87*, 1051–1069. [CrossRef]

39. Magne, C.; Dufour, F.; Labat, F.; Lancel, G.; Durupthy, O.; Cassaignon, S.; Pauporte, T. Effects of TiO_2 nanoparticle polymorphism on dye-sensitized solar cell photovoltaic properties. *J. Photochem. Photobiol. A Chem.* **2012**, *232*, 22–31. [CrossRef]

40. Akimov, A.V.; Neukirch, A.J.; Prezhdo, O.V. Theoretical Insights into Photoinduced Charge Transfer and Catalysis at Oxide Interfaces. *Chem. Rev.* **2013**, *113*, 4496–4565. [CrossRef] [PubMed]

41. Lin, H.; Fratesi, G.; Selcuk, S.; Brivio, G.P.; Selloni, A. Effects of Thermal Fluctuations on the Structure, Level Alignment, and Absorption Spectrum of Dye-Sensitized TiO_2: A Comparative Study of Catechol and Isonicotinic Acid on the Anatase (101) and Rutile (110) Surfaces. *J. Phys. Chem. C* **2016**, *120*, 3899–3905. [CrossRef]

42. Phua, S.L.; Yang, L.P.; Toh, C.L.; Ding, G.Q.; Lau, S.K.; Dasari, A.; Lu, X.H. Simultaneous Enhancements of UV Resistance and Mechanical Properties of Polypropylene by Incorporation of Dopamine-Modified Clay. *ACS Appl. Mater. Interfaces* **2013**, *5*, 1302–1309. [CrossRef] [PubMed]

43. Yang, X.; Duan, L.; Cheng, X.J.; Ran, X.Q. Effect of polydopamine coating on improving photostability of poly(1,3,4-oxadiazole)s fiber. *J. Polym. Res.* **2016**, *23*. [CrossRef]

44. Proks, V.; Brus, J.; Pop-Georgievski, O.; Vecernikova, E.; Wisniewski, W.; Kotek, J.; Urbanova, M.; Rypacek, F. Thermal-Induced Transformation of Polydopamine Structures: An Efficient Route for the Stabilization of the Polydopamine Surfaces. *Macromol. Chem. Phys.* **2013**, *214*, 499–507. [CrossRef]

45. Wang, Z.H.; Tang, F.; Fan, H.L.; Wang, L.; Jin, Z.X. Polydopamine Generates Hydroxyl Free Radicals under Ultraviolet-Light Illumination. *Langmuir* **2017**, *33*, 5938–5946. [CrossRef] [PubMed]

46. Marchetti, A.; Chen, J.E.; Pang, Z.F.; Li, S.H.; Ling, D.S.; Deng, F.; Kong, X.Q. Understanding Surface and Interfacial Chemistry in Functional Nanomaterials via Solid-State NMR. *Adv. Mater.* **2017**, *29*. [CrossRef] [PubMed]

47. Adhyaru, B.B.; Akhmedov, N.G.; Katritzky, A.R.; Bowers, C.R. Solid-state cross-polarization magic angle spinning ^{13}C and ^{15}N NMR characterization of Sepia melanin, Sepia melanin free acid and Human hair melanin in comparison with several model compounds. *Magn. Reson. Chem.* **2003**, *41*, 466–474. [CrossRef]

48. Dreyer, D.R.; Miller, D.J.; Freeman, B.D.; Paul, D.R.; Bielawski, C.W. Elucidating the Structure of Poly(dopamine). *Langmuir* **2012**, *28*, 6428–6435. [CrossRef] [PubMed]

49. Liebscher, J.; Mrowczynski, R.; Scheidt, H.A.; Filip, C.; Hadade, N.D.; Turcu, R.; Bende, A.; Beck, S. Structure of Polydopamine: A Never-Ending Story? *Langmuir* **2013**, *29*, 10539–10548. [CrossRef] [PubMed]

50. Gaya, U.I.; Abdullah, A.H. Heterogeneous photocatalytic degradation of organic contaminants over titanium dioxide: A review of fundamentals, progress and problems. *J. Photochem. Photobiol. C: Photochem. Rev.* **2008**, *9*, 1–12. [CrossRef]

51. Ai, K.L.; Liu, Y.L.; Ruan, C.P.; Lu, L.H.; Lu, G.Q. Sp2 C-Dominant N-Doped Carbon Sub-micrometer Spheres with a Tunable Size: A Versatile Platform for Highly Efficient Oxygen-Reduction Catalysts. *Adv. Mater.* **2013**, *25*, 998–1003. [CrossRef] [PubMed]

 *catalysts*

Article

Photocatalytic Water Disinfection under Solar Irradiation by D-Glucose-Modified Titania

Agata Markowska-Szczupak [1,*], Paulina Rokicka [1], Kunlei Wang [2], Maya Endo [2], Antoni Waldemar Morawski [1] and Ewa Kowalska [2]

[1] Institute of Inorganic Technology and Environment Engineering,
West Pomeranian University of Technology in Szczecin, Pulaskiego 10,
70-322 Szczecin, Poland; rokicka16@poczta.onet.pl (P.R.); amor@zut.edu.pl (A.W.M.)

[2] Institute for Catalysis, Hokkaido University, N21, W10, Sapporo 001-0021, Japan;
kunlei@cat.hokudai.ac.jp (K.W.); m_endo@cat.hokudai.ac.jp (M.E.); kowalska@cat.hokudai.ac.jp (E.K.)

* Correspondence: Agata.Markowska@zut.edu.pl; Tel.: +48-91-449-42-30

Received: 18 July 2018; Accepted: 30 July 2018; Published: 1 August 2018

Abstract: Modified titania photocatalysts were synthesized by the pressure method using titanium(IV) oxide from Grupa Azoty Zakłady Chemiczne "Police" S.A., Police, Poland, and D-glucose solution. Characterization of obtained composites was performed by X-ray diffraction (XRD), X-ray photoelectron spectroscopy (XPS), elemental analysis, and measurements of zeta potential and specific surface area (SSA). The possibility of using glucose-titania composites as photocatalysts for simulated solar-assisted disinfection against gram-negative *Escherichia coli* and gram-positive *Stapchyloccocus epidermidis* bacteria were examined in two reaction systems, i.e., for suspended and immobilized photocatalysts (on the concrete). It was found that an increase in the D-glucose concentration, i.e., higher carbon content, led to a decrease in antibacterial properties. The sample obtained from 1% of D-glucose solution at 100 °C (TiO$_2$-1%-G-100) showed superior photocatalytic activity under UV-Vis irradiation toward both bacteria species. Water disinfection was more efficient for suspended photocatalyst than that for supported one, where complete disinfection was reached during 55–70 min and 120 min of irradiation, respectively. For the first time, it has been shown that titania modified with monosaccharides can be efficiently used for water disinfection, and the immobilization of photocatalyst on the concrete might be a prospective method for public water supplies.

Keywords: C/TiO$_2$; photocatalysis; solar radiation; disinfection; immobilized catalyst

1. Introduction

In recent years, the intensification of research directed to delivery new, ecological, and cost-effective disinfection methods that are focused on increased water safety has taken place. It seems that the photocatalytic oxidation with titanium(IV) oxide (TiO$_2$, titania) fulfills above conditions. Unfortunately, the majority of commercially available titania photocatalysts are the first-generation catalyst, which means that their activity occurs only under UV irradiation, i.e., with wavelengths (λ) shorter than 400 nm. Therefore, the amount of solar radiation reaching Earth's surface might be insufficient for photocatalyst activation. In order to enhance the photocatalytic activity of TiO$_2$ under solar radiation, what is essential from the economical point of view, different ways of titania modification have been widely studied, e.g., doping, surface modification, and heterojunction. Various physical and chemical strategies, and many kinds of modifiers, such as metals, metal oxides, and non-metals have been used for titania modification, particularly noble metals (e.g., Au, Ag) and nonmetals (e.g., S, C, N) [1–4]. Scientific studies on nonmetal modification started about two decades ago [5]. Among them, the aliphatic alcohols, urea, thiourea, calcium carbide, activated carbon, multi-wall carbon nanotubes, and graphene have been used. It has been shown that carbon modification of titania resulted in significant enhancement of photocatalytic activity under solar radiation and even under sole visible irradiation [5–9]. The main advantage of carbon modification of TiO_2 is that the simple and fast methods of preparation can be used. It has been proved that C/TiO_2 demonstrates better photocatalytic activity than un-modified TiO_2 for organic (including phenol and organic dyes) or inorganic (e.g., nitrogen oxides) compounds' degradation [6,8,10–12]. However, except our own research [13–15], only a few studies on disinfection using carbon-modified TiO_2 have been performed [16–18]. For example, titania modified with carbon from coconut shell, alcohols, and carbide presents better biocidal properties than un-modified titania, even under irradiation with visible light. Moreover, it should be mentioned that preliminary results on antibacterial properties for other modified titania samples (with graphene oxide) were quite promising [15]. Therefore, the antibacterial properties of sunlight-activated D-glucose-modified titania was evaluated in the present study.

2. Results

2.1. Characteristics of D-Glucose-Modified TiO$_2$

Starting material contained majority of amorphous titania (ca. 75.0%, data from producer), and thus had very high specific surface area (SSA) of 312 m^2 g^{-1}, whereas crystalline part was composed mainly of anatase (95.0% and 5% of rutile). Thermal treatment did not cause significant changes in the crystalline composition of titania, where only slight phase transition was noticed, i.e., from amorphous titania to anatase, resulting in a slight increase in anatase content to ca. 98% and a decrease in SSA from 312 m^2 g^{-1} to 266, 158, and 88 m^2 g^{-1} after annealing at 100 °C, 150 °C, and 200 °C, respectively. Similarly, the presence of D-glucose practically did not influence the crystalline composition of photocatalysts, and anatase phase was predominant form of titania also in those samples (97.6–98.1%), as shown in Table 1. The largest anatase crystallites and the smallest SSA were obtained for the sample that was prepared with the smallest content of D-glucose at the highest temperature (TiO$_2$-G-1%-200). An increase in annealing temperature caused an increase in anatase crystallite size and a decrease in SSA (Figure 1a). It is well known that thermal treatment results in increases in both crystalline and particle sizes (a decrease in SSA), due to transformation of amorphous titania (e.g., amorphous layer on nano-crystals) and particle sintering/aggregation, respectively. In contrast, an influence of D-glucose on surface properties indicates that D-glucose restrains crystal growing (crystallite size of anatase in un-modified samples annealed at 100 °C, 150 °C, and 200 °C were larger, i.e., 12.0, 16.3, and 21.9 nm, respectively). It has been reported that titania modifiers could influence crystal growth either positively or negatively. For example, Grzybowska et al. found an increase in SSA after surface modification of titania with calcium, tungsten, iron, and aluminum, suggesting that adsorbed modifiers on titania surface disturbed in particles' sintering [19]. Usually, a decrease in crystallite size corresponds to an increase in SSA. Contrary, D-glucose-modification resulted in a decrease in both crystallite size and SSA for samples annealed at 100 °C (with an increase in D-glucose content), whereas higher temperatures of annealing resulted in a significant increase in SSA. Therefore, it is proposed that photocatalysts that were annealed at lower temperature could be composed of fine titania nanoparticles (NPs) covered with a thin layer of adsorbed D-glucose, whereas for those annealed at higher temperatures the presence of carbon (decomposed D-glucose) inhibited significantly particles' aggregation, as also proved by a decrease and an increase in micropore volume, respectively (Figure 1c). Interestingly, in the case of modified samples annealed at higher temperatures, the largest values of SSA were obtained for those with middle content of D-glucose (5%), suggesting that smaller content (1%) was not sufficient to prevent particles' aggregation, whereas too large content (10%) could form a thick layer on titania surface, as confirmed by pore size distribution (a decrease in total pore volume and mesopore volumes, Figure 1b,d). The carbon content increased with an increase in D-glucose content, as shown in Table 1.

Table 1. Physicochemical properties of photocatalysts used in this study (commercial photocatalysts: starting TiO$_2$ (from Grupa Azoty Zakłady Chemiczne "Police" S.A., Police, Poland) and KRONOClean 7000), and starting TiO$_2$ modified with D-glucose.

Sample Name	Anatase Content [%] [a]	Anatase Crystallite Size [nm]	SBET [m^2/ g]	Carbon Content (wt %)	Zeta Potential ζ [mV]
Starting TiO$_2$	95.0	11.0	312	0.0	−24.13
KRONOClean7000	100.0	11.0	242	0.96	−17.71
TiO$_2$-G-1%-100	97.8	11.7	268	0.59	−13.49
TiO$_2$-G-5%-100	97.6	11.4	252	2.23	−25.28
TiO$_2$-G-10%-100	97.6	11.2	214	4.49	−26.49
TiO$_2$-G-1%-150	98.0	16.0	155	0.49	−17.04
TiO$_2$-G-5%-150	98.1	13.4	206	2.10	−20.24
TiO$_2$-G-10%-150	97.8	11.4	195	3.91	−23.70
TiO$_2$-G-1%-200	98.0	21.8	87	0.29	−16.80
TiO$_2$-G-5%-200	97.9	17.2	120	1.90	−18.72
TiO$_2$-G-10%-200	97.9	17.3	110	3.37	−23.34

[a] anatase content considering only crystalline forms (anatase + rutile).

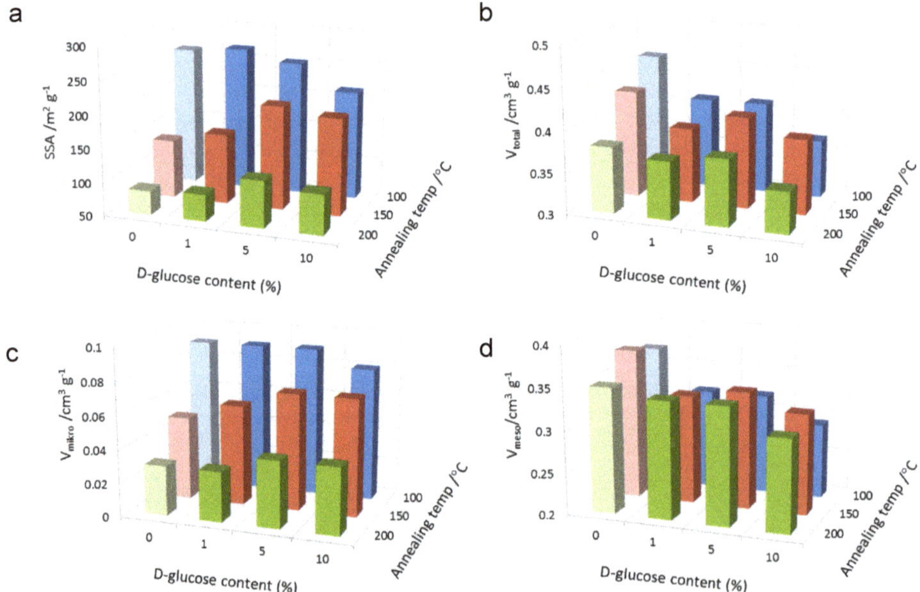

Figure 1. BET (Brunauer, Emmett and Teller) results for un-modified (D-glucose content: 0) and modified with D-glucose (1%, 5% and 10%): (**a**) specific surface area (SSA), (**b**) total pore volume (V_{total}), (**c**) micropore volume (V_{micro}), and (**d**) mesopore volume (V_{meso}).

However, an increase in annealing temperature resulted in a decrease in carbon content, especially for samples that were prepared at 200 °C (Table 1). This decrease in carbon content should be caused by thermal decomposition of D-glucose, which usually starts at ca. 165 °C [20]. Therefore, it is proposed that high pressure in an autoclave could accelerate D-glucose decomposition (at lower temperatures), similarly as reported for D-glucose instability during HPLC analysis [21]. Obviously, the presence of D-glucose on titania surface resulted in changes of zeta potential from −19.49, −21.09 and −21.44 mV for samples annealed at 100 °C, 150 °C, and 200 °C, respectively, to values shown in Table 1.

To investigate surface composition of titania after modification, X-ray photoelectron spectroscopy (XPS) analysis has been performed for three samples: starting TiO_2, TiO_2-100 (starting TiO_2 thermally treated at 100 °C), and TiO_2-G-1%-100 (starting TiO_2 modified with 1% of D-glucose and annealed at 100 °C; the most active sample), and the obtained results are shown in Figure 2 and Table 2. It was found that the surface of titania was enriched with oxygen as the ratio of oxygen to titanium exceeded the stoichiometric ratio (2) being 2.2, which is typical for various titania samples, due to adsorption of water and carbon dioxide on the titania surface. For example, oxygen enrichment of titania was reported for samples that were prepared by the laser ablation (O/Ti = 2.5) [22], the microemulsion (O/Ti = 4.6) [23], the hydrothermal (O/Ti = 5.2) [24] and the gas-phase (O/Ti = 7.7) [24] methods. Carbon is present in all titania samples (and also other oxides), mainly due to adsorption of carbon dioxide from air (forming bicarbonate and mono- and bidentate carbonate [25]), but also from precursors that are used for titania synthesis, such as titanium alcoholates. XPS analysis proved the surface modification of titania with D-glucose since the ratio of carbon to titanium increased by almost double after sample modification with 1 wt % of D-glucose. The banding energies of titanium, oxygen, and carbon were estimated by deconvolution of Ti $2p_{3/2}$, O 1s, and C 1s peaks, respectively, accordingly to published reports on XPS analysis of titania samples [26–31]. Titanium consisted mainly Ti^{4+}, and the reduced form of titanium (Ti^{3+}) did not exceed 2.2%. D-glucose presence and thermal treatment practically did not change form of titanium. Deconvolution of oxygen peak into three peaks at ca. 529.6 eV, 531.3 eV, and 533.3 eV, respectively, confirmed the co-existence of three forms of oxygen, i.e., (i) lattice oxygen in TiO_2, (ii) oxygen bound to: carbon (C=O), titanium in Ti_2O_3, and hydroxyl groups bound to two titanium atoms, (iii) hydroxyl groups that were bound to titanium or carbon (Ti-OH/C-OH). A significant increase in the content of oxygen in the form of hydroxyl groups (from 1.4 to 6.8) and an increase in the carbon content in the form of carbon bound to carbon (C-C) for TiO_2-G-1%-100 sample confirmed the presence of adsorbed D-glucose on the titania surface.

To examine photocatalytic activity of prepared samples, the generation of hydroxyl radicals under irradiation was performed, and the obtained data are shown in Figure 3. It was found that titania modification with glucose resulted in a significant enhancement of photocatalytic activity (more than twice), where the most active sample contained the smallest content of glucose (1 wt %) and was annealed at lowest temperature (100 °C), i.e., TiO_2-G-1%-100. An increase in annealing temperature resulted in a decrease in activity, which is probably due to glucose decomposition, as discussed above.

Table 2. Surface composition and fraction of oxidation states of Ti, O, and C from deconvolution of X-ray photoelectron spectroscopy (XPS) peaks.

Sample name	Content (at. %)			Ratio	
	Ti $2p_{3/2}$	O 1s	C 1s	O/Ti	C/Ti
Starting TiO_2	17.71	43.95	38.34	2.5	2.2
TiO_2-100	15.05	43.3	41.65	2.9	2.8
TiO_2-G-1%-100	12.1	36.06	51.84	3.0	4.3

Sample name	Ti $2p_{3/2}$ (%)		O 1s (%)			C 1s (%)		
	Ti^{4+}	Ti^{3+}	TiO_2	=O[a]	-OH[b]	C-C	C-OH	C=O
Starting TiO_2	97.9	2.1	47.2	50.0	2.8	58.7	30.1	11.2
TiO_2-100	98.1	1.9	45.0	53.6	1.4	54.7	37.2	8.1
TiO_2-G-1%-100	97.8	2.2	46.0	47.2	6.8	61.5	28.5	10.0

=O[a]: Ti-(OH)-Ti/Ti_2O_3/C=O; -OH[b]: Ti-OH/C-OH.

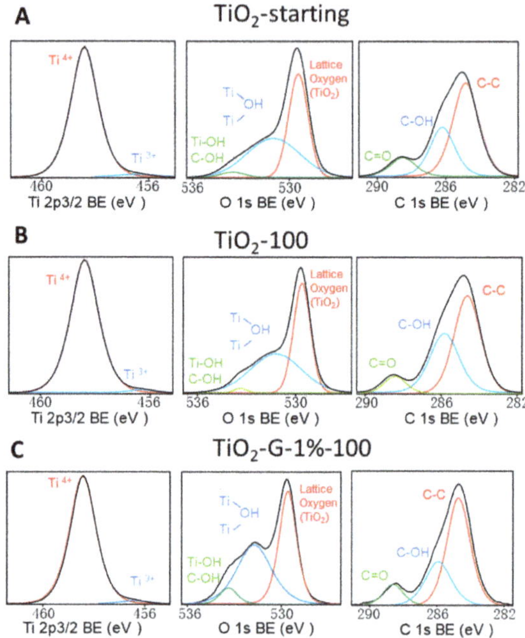

Figure 2. XPS results for Ti 2p$_{3/2}$, O 1s, and C 1s for starting TiO$_2$ (**A**; top), TiO$_2$-100 (**B**; middle), and TiO$_2$-G-1%-100 (**C**; bottom).

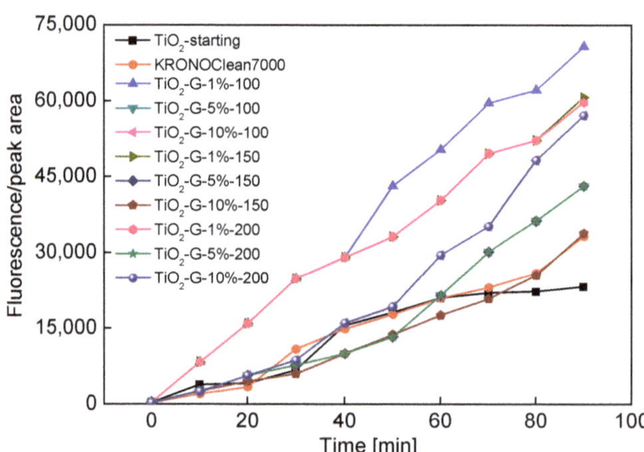

Figure 3. The amount of generated 2-hydroxyterephthalic acid (fluorescence intensity) during 90 min of UV-Vis irradiation.

2.2. Dose-Dependent Photocatalytic Inactivation of Escherichia coli and Staphylococcus epidermidis

The correlation between disinfection duration and dose of the glucose-modified photocatalyst with the highest photocatalytic activity (i.e., rate of hydroxyl radicals' formation), i.e., TiO$_2$-G-1%-100 sample, was examined. The fastest disinfection process under UV-Vis irradiation (after 70 min) was obtained for photocatalyst dose of 0.1 g $\times$ dm^{-3} (Figure 4a). It was found that both an increase and a decrease in TiO$_2$-G-1%-100 dose resulted in extension of disinfection duration. All of the differences were statistically significant (Table 3). Taken into account the obtained results, it was assumed that 0.1 g per dm^3 was an optimal dose for photocatalytic disinfection. Next, the influence of bacteria concentration (*E. coli* and *S. epidermidis*) on disinfection duration was studied, and the obtained data are shown in Figure 4b. The necessary duration of disinfection did not differ significantly for both bacteria solutions in the case of their low concentration from 1.5×10^4 CFU $\times$ cm^{-3} to 1.5×10^5 CFU $\times$ cm^{-3} (Table 3), reaching complete disinfection during 60 min of UV-Vis irradiation (Figure 4b). However, an increase in bacteria concentration resulted in a significant extension of necessary disinfection time, even up to 120 min for concentration of 1.5×10^8 CFU.

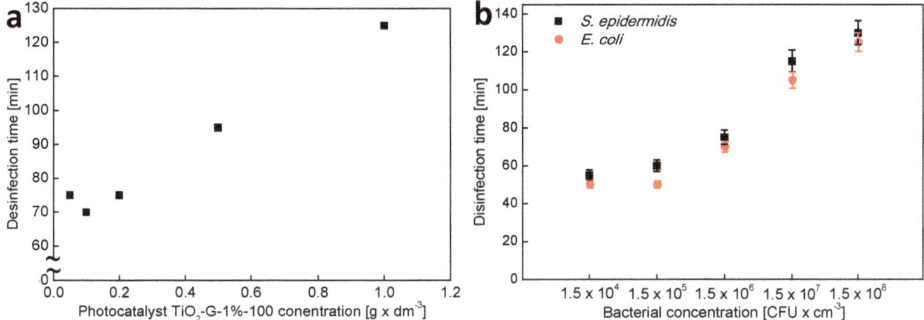

Figure 4. The correlations between disinfection time and: (**a**) photocatalyst TiO$_2$-G-1%-100 content, and (**b**) bacteria concentration.

Table 3. Statistical analysis for factors influencing *E. coli* disinfection time.

Photocatalyst TiO$_2$-G-1%-100 Concentration [g $\times$ dm^3]	0.05	0.1	0.2	0.5	1.0
0.05	-	SI	SI	*	***
0.1	SI	-	SI	*	***
0.2	*	*	*	-	***
0.5					
1.0	***	***	***	***	-
Initial concentration of *E. coli* [CFU $\times$ cm^3]	1.5×10^4	1.5×10^5	1.5×10^6	1.5×10^7	1.5×10^8
1.5×10^4	-	SI	SI	**	***
1.5×10^5	SI	-	SI	**	***
1.5×10^6	SI	SI	-	**	***
1.5×10^7	**	**	**	-	*
1.5×10^8	***	***	***	*	-
Initial concentration of *S. epidermidis* [CFU $\times$ cm^3]	1.5×10^4	1.5×10^5	1.5×10^6	1.5×10^7	1.5×10^8
1.5×10^4	-	SI	SI	**	***
1.5×10^5	SI	-	SI	**	***
1.5×10^6	SI	SI	-	**	***
1.5×10^7	**	**	**	-	*
1.5×10^8	***	***	***	*	-

* $p < 0.05$; ** $p < 0.01$; *** $p < 0.001$; SI—statistically insignificant.

2.3. Influenece of D-Glucose Content, Used for TiO$_2$ Modification at 100 °C, on Disinfection Properties

The influence of D-glucose content, used for titania modification at 100 °C, was investigated. It was found that disinfection duration strongly depended on the D-glucose content, as shown in Figure 5. TiO$_2$-G-1%-100 photocatalyst led to the complete inactivation of gram-negative bacteria *E. coli* within 70 min (Figure 5a) and gram-positive bacteria *S. epidermidis* within 85 min (Figure 6a) under UV-Vis irradiation. Scanning electron microscopy (SEM) revealed the significant change in bacterial morphology (e.g., shape, disruption of outer membranes, etc.) after photocatalytic process, as shown in Figures 5c and 6c. In contrast, all of the photocatalysts (commercial and modified) did not cause the change in bacterial number in dark conditions (Figures 5b and 6b).

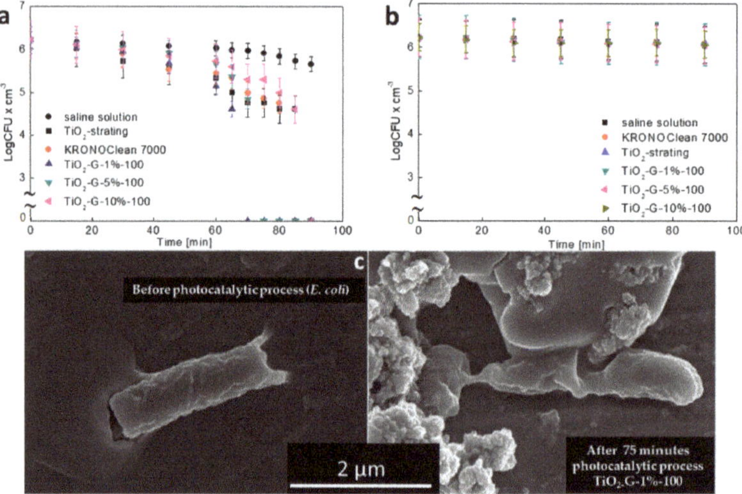

Figure 5. Antibacterial activity against *E. coli* of commercial and D-glucose-modified titania: (**a**) under UV-Vis irradiation, and (**b**) in dark conditions; and, (**c**) SEM images of *E. coli* before (left) and after 75 min of UV-Vis irradiation (right) in the presence of 0.1 g × dm^3 TiO$_2$-G-1%-100.

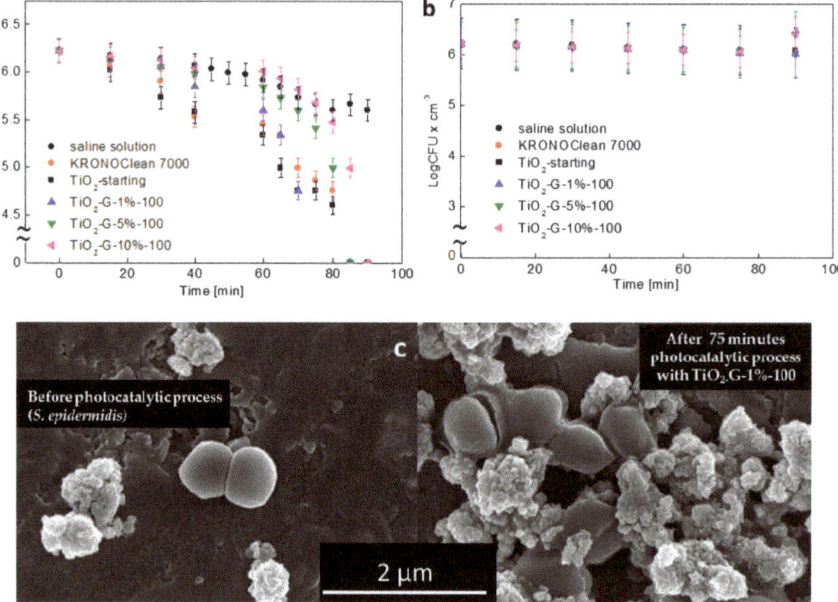

Figure 6. Antibacterial activity against *S. epidermidis* of commercial and D-glucose-modified titania: (**a**) under UV-Vis irradiation, and (**b**) in dark conditions; and, (**c**) SEM images of *S. epidermidis* before (left) and after 75 min of UV-Vis irradiation (right) in the presence of 0.1 g × dm^3 TiO$_2$-G-1%-100.

2.4. Influenece of D-Glucose Content, Used for TiO₂ Modification at 100 °C, on Enzymatic Activity

The enzymatic activity, i.e., catalase activity (CAT) and superoxide dismutase activity (SOD), secreted by *E. coli* and *S. epidermidis*, under UV-Vis irradiation and in the dark, was investigated for commercial and D-glucose-modified (at 100 °C) photocatalysts. The tests were performed before and after 30, 60, 90, and 120 min of process. The enzymatic activity was not observed from 90 min of process, and thus the obtained results were only presented till 60 min, as shown in Figures 7 and 8. It was demonstrated that without activation (dark conditions) all of the photocatalysts did not influence the CAT and SOD activity (Figures 7 and 8). However, 30-min UV-irradiation caused an increase in CAT and SOD enzyme activities, whereas next 30-min UV-Vis irradiation resulted in a decrease in those activities below the initial levels.

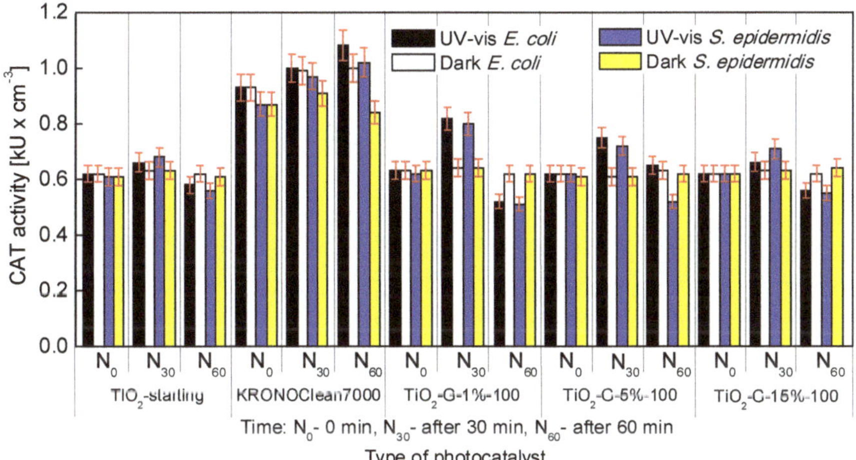

Figure 7. Influence of commercial and D-glucose-modified titania under UV-Vis irradiation and in dark conditions on *E. coli* and *S. epidermidis* catalase activity (CAT).

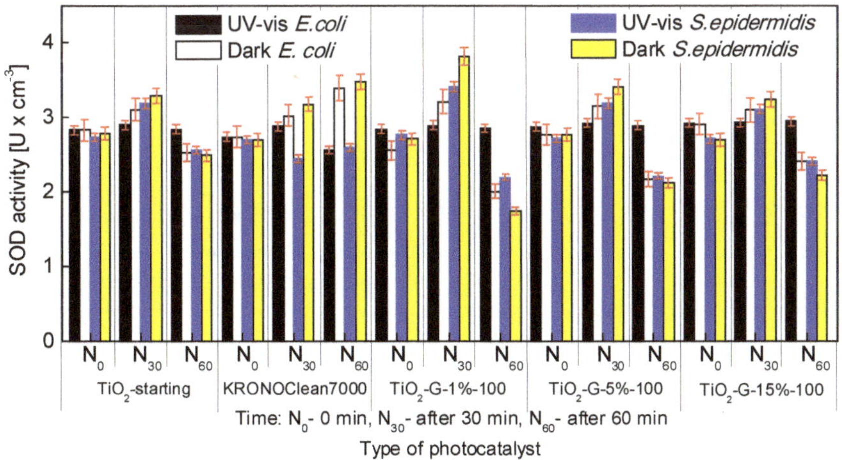

Figure 8. Influence of commercial and D-glucose-modified titania under UV-Vis irradiation and in dark conditions on *E. coli* and *S. epidermidis* superoxide dismutase activity (SOD).

2.5. Influenece of D-Glucose Content, Used for TiO$_2$ Modification at 100 °C, on Bacteria Mineralization

To study the probability of complete decomposition of bacterial cells (mineralization), the liberation of carbon dioxide during UV-Vis irradiation and in the dark, were investigated. It was found that, indeed, bacterial cells could be efficiently decomposed only under photocatalytic process (Figures 9a and 10a), whereas no changes in carbon dioxide concentration were noticed in the dark (Figures 9b and 10b). Moreover, the fastest evolution of carbon dioxide was observed for TiO$_2$-G-1%-100 sample with the lowest content of carbon among the modified samples, confirming its highest photocatalytic activity.

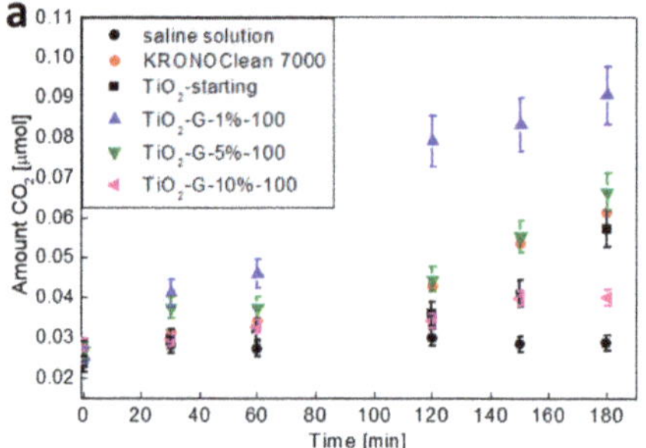
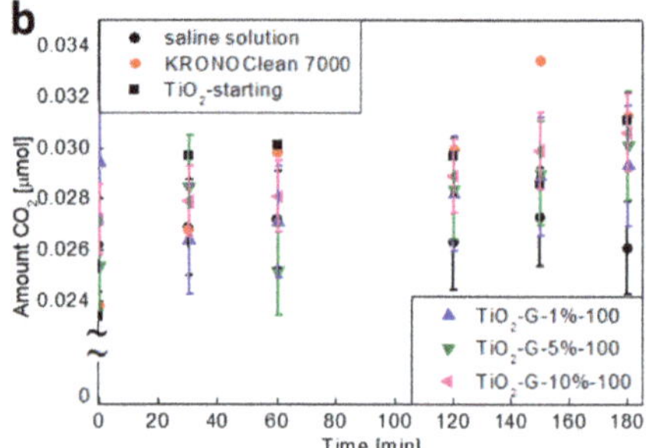

Figure 9. Evolution of CO$_2$ during mineralization of *E. coli* cells under: (**a**) UV-Vis irradiation, and (**b**) in the dark on commercial and D-glucose-modified titania.

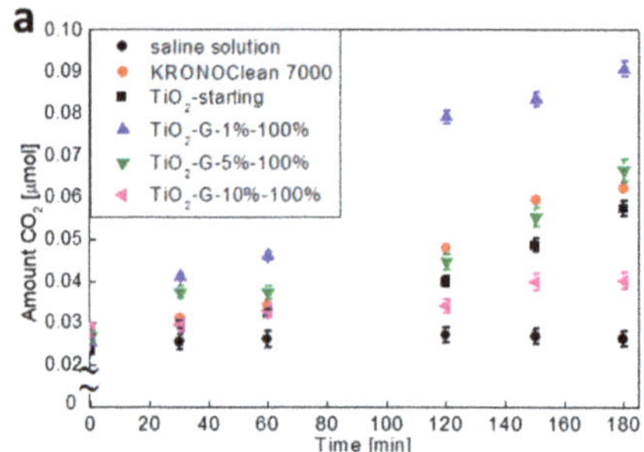
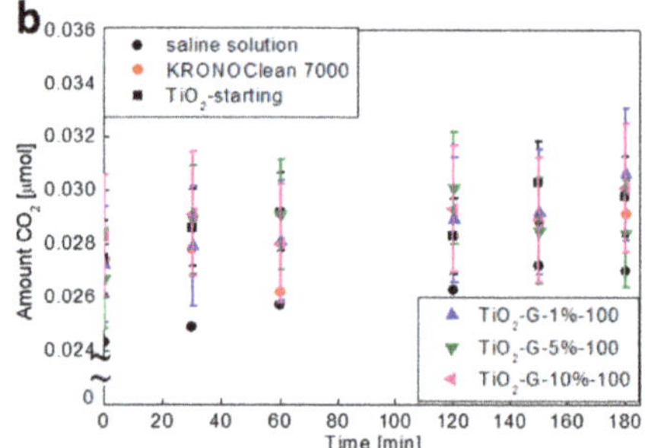

Figure 10. Evolution of CO$_2$ during mineralization of *S. epidermidis* cells under: (**a**) UV-Vis irradiation, and (**b**) in the dark on commercial and D-glucose-modified titania.

2.6. Commercialization of Antimicrobial Photocatalysts for Building Materials

To check the commercial feasibility of antimicrobial photocatalysts, cement mortars were supplemented with 10 wt % of TiO$_2$-G-1%-100 photocatalyst and tested for *E. coli* inactivation. It was clearly demonstrated that D-glucose-modified titania contributed to the additive benefits of the final product. Under UV-Vis irradiation the obtained concrete exhibited disinfection potential, as shown in Figure 11. The control experiments (concrete without photocatalyst) of *E. coli* inactivation did not show any significant differences in the number of bacteria after 120-min incubation in dark condition and under light irradiation. In contrast, in water being in contact with the plate, made of concrete supplemented with TiO$_2$-G-1%-100, the complete disinfection was achieved after only 120 min under UV-Vis irradiation (Figure 11).

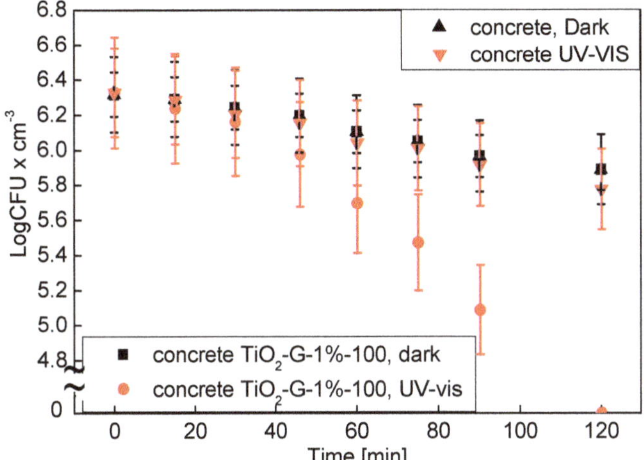

Figure 11. Antibacterial activity (against *E. coli*) of concrete supplemented with 10% TiO$_2$-G-1%-100 and reference samples (concrete without TiO$_2$) under UV-Vis irradiation and in dark condition.

3. Discussion

Antibacterial properties of titania have received a great attention by many researchers [2,14,18,32–34], and thus studies on TiO$_2$ disinfection under solar radiation are mainstream research and development [33]. Therefore, in this study it has been hypothesized that surface modification of titania by D-glucose could result in development of antimicrobial agent towards both gram-negative (*Escherichia coli*) and gram-positive (*Staphylococcus epidermidis*) bacteria under solar radiation. Physicochemical characteristics of D-glucose-modified titania at 100 °C showed that this group of samples has various advantageous, such as large SSA (214–268 m^2 × g^{-1}), large content of anatase (97.6–97.8%), and small size of anatase crystallites (11.2–11.7 nm). The photocatalytic activity of the D-glucose-modified TiO$_2$ was examined for the generation of •OH radicals under UV-Vis light irradiation. The most reactive photocatalyst, inducing the fastest *E. coli* and *S. epidermidis* inactivation, was obtained by titania modification with 1% of D-glucose solution at 100 °C (TiO$_2$-1%-100). The enhancement of photocatalytic activity (in the visible range of solar spectrum ($\lambda > 400$ nm)) has been reported for TiO$_2$ modified with glucose by Kim at al. [35]. According to these studies, glucose-adsorbed TiO$_2$ can form the charge transfer complex by ligand-to-metal charge transfer (LMCT) sensitization mechanism. Moreover, enhanced photocatalytic activity under UV irradiation could be achieved by the hindering of electron–hole recombination, since photogenerated holes could be trapped by hydroxyl groups [36]. Indeed, the enrichment of titania surface with hydroxyl groups was observed after its modification with D-glucose (TiO$_2$-1%-100), as shown in Table 2.

In the literature, there are many reports concerning the optimization procedures for photocatalyst and bacterial doses used for water disinfection. The results that were obtained by various teams differ significantly. For example, Herrmann [37] found that 2.5 g TiO$_2$ × dm^{-3} was the best value corresponding to the maximal photon absorption (all of the particles could be illuminated). However, Cho et al. [38] determined that the bactericidal activity of P25 at concentration of 1.0 g × dm^{-3} was two times more effective than that at 0.1 g × dm^{-3}, whereby increasing the TiO$_2$ concentration from 1.0 to 2.0 g × dm^{-3} did not enhance the inactivation efficiency. Therefore, it is proposed that for disinfection process less content of titania is required than that for decomposition of organic compounds. The presented results in this report have confirmed this hypothesis, but much less content of titania (0.1 g × dm^{-3}) than that by Cho et al. [38] appeared to be an optimal value.

The most effective concentration of bacteria (*E. coli* and *S. epidermidis*) was in the range from 1.5×10^4 CFU $\times$ cm^{-3} to 1.5×10^5 CFU $\times$ cm^{-3}, confirming the previous findings by Benabbou et al. [39]. Necessary time for complete inactivation of microorganisms was longer in the case of more concentrated bacterial suspension due to the shielding effect by bacteria cells. It was found that conditions of disinfection, such as photocatalyst content and bacterial concentration, have had a critical influence on the obtained results, and the optimal conditions must be experimentally determined.

Unfortunately, there are no reports on the antibacterial properties of glucose-modified titania prepared at temperatures lower than 300 °C, and thus the direct comparison of our results with others was impossible. The best antibacterial activity was obtained for the sample that was modified with 1% of D-glucose at 100 °C (TiO$_2$-1%-100). Under UV-Vis irradiation, the complete disinfection of *E. coli* and *S. epidermidis* was achieved after 70 and 75 min, respectively, whereas carbon-modified commercial photocatalyst KRONOClean 7000 was less efficient, needing 85 min and 90 min, respectively, of UV-Vis irradiation. Therefore, it is proposed that the form of carbon is crucial for antimicrobial properties, i.e., hydroxyl groups in glucose structure responsible for holes' trapping.

Antioxidant enzymes, secreted by bacteria, are considered to be important indicators for the oxidative stress. Catalase (CAT) and superoxide dismutase (SOD) are two commonly investigated intracellular enzymes [35,40]. CAT is an antioxidant enzyme catalyzing the decomposition of H$_2$O$_2$ to water and oxygen, whereas SOD is a metalloenzyme catalyzing the dismutation of $\bullet$O$_2$ into H$_2$O$_2$ and O$_2$. Accordingly, the loss of their activity accelerated the accumulation of reactive oxygen species (ROSs) and led to the deficiency of cell viability. On the basis of the obtained results, the two-step mechanism of response to the presence of photocatalyst, activated under UV-Vis irradiation, was described. It was found that the induced levels of these enzymes increased during the first 30 min, and then decreased (after 60 min). The changes in enzyme activity during the photocatalytic disinfection process indicated that defense capacity was overwhelmed by the rapidly created ROSs at the initial stage, and then suppressed gradually. The correlation between high amount of generated hydroxyl radicals (Figure 3) during 90 min of UV-Vis irradiation and the loss of enzymatic activity suggested that oxidative stress could act as an important pathway by which photocatalyst induced bacterial death. Interestingly, no significant differences in the oxidative stress response between two examined bacteria species was found. The protein nature of enzymes determines their quick and explicit reaction on various environmental factors whether naturally or artificially introduced, such as photocatalyst. Despite short lifetime of ROSs, generated on the surface of photoexcited photocatalyst, they may probably cause conformational changes (in secondary structure), damage of the active sites or irreversible loss of the activity.

It was found that carbon content was the main factor influencing antibacterial efficiency of D-glucose modified photocatalysts, and larger than 1% content of D-glucose was detrimental for activity. Similarly, it was reported that too high carbon content on the photocatalyst surface could block the active sites decreasing photocatalytic decomposition of organic compounds, such as phenol [41] and methylene blue [42]. The results of mineralization of the bacterial cells clearly indicated that the amount of CO$_2$, liberated from TiO$_2$-1%-100 suspension under UV-Vis irradiation, was 1.5 times higher than that from suspension containing commercial photocatalyst KRONOClean 7000 or starting TiO$_2$. Although an increase of CO$_2$ concentration proved the mineralization of bacteria cells, this did not guarantee water "microbial purity", as reported by Kacem at al. [43]. Photoexcited TiO$_2$ could cause only the partial mineralization of bacterial cells that could lead to bacteria transition to viable, but nonculturable (VBNC) state. This apparent dormant state, in which bacterial cells are metabolically active, maintains their pathogenic features, but bacteria cannot grow on culture media. Therefore, it is necessary to introduce genetic methods, based on the quantification of DNA (e.g., polymerase chain reaction PCR), which may justify the complete destruction of bacterial DNA.

The interaction between bacteria and photocatalyst was observed by SEM observations. It was found that modified samples with 1% of D-glucose showed stronger adhesion to the bacterial cells. The carbon content of 0.59 (TiO$_2$-G-1%-100) and enrichment of the titania surface with hydroxyl groups, resulted in a significant modification of surface properties, as clearly shown by zeta potential changes, and thus increased bacterial adhesion to the photocatalyst surface. It is very important finding since, according to the literature reports, the titania NPs with particle sizes larger than 8 nm cannot penetrate bacteria through the cell-wall porins of ca. 1.2-nm diameters [44]. Accordingly, the electrostatic interaction between photocatalyst and bacteria has a crucial significance for antibacterial activity and should be extensively investigated in future. It is well-known that gram-negative bacteria contain both anionic and zwitterionic phospholipids, whereas gram-positive bacteria contain predominantly anionic lipids. When considering the observation that was made by Clogston and Patri [45], that NPs with a zeta potential between -10 and $+10$ mV were considered as neutral, whereas NPs with zeta potentials that were lower than -30 mV were strongly anionic, and thus cationic particles would attract negatively charged bacteria (e.g., *E. coli* ζ ~35 mV). The obtained results confirmed these findings since the best antibacterial properties were found for TiO_2-G-1%-100 photocatalyst with the least negative zeta potential of -13.49 mV. Deterioration of antibacterial properties was observed for photocatalyst characterized by more negative ζ (<-20 mV), due to worse contact between bacterial cells and photocatalyst NPs. It is important to notice that bacteria cells are much larger than NPs of photocatalysts (approx. 180 times bigger), and thus stronger adsorption of photocatalyst on bacteria is advisable for better interaction. An attractive feature of this mechanism is that it can be used to design antimicrobial agents with selective toxicity against certain organisms.

The present work fulfilled previous studies on disinfection process. First time, the application of D-glucose-modified photocatalyst in a cement mortar for water disinfection was presented. The complete disinfection was achieved during 120-min contact of contaminated water with a small concrete plate, made of cement mortar and supplemented with 10 wt % of TiO_2-G-1%-100 photocatalyst. According to the obtained results, it would be possible to use solar energy to activate the D-glucose-modified photocatalyst, and thus disinfect water in a concrete container used for storage of drinking water or fish farms (e.g., trout ponds). Last mentioned application should improve health conditions in fish farms since the microbial purity of water is the most important factor affecting performance in aquaculture production systems.

4. Materials and Methods

4.1. Preparation of C/TiO$_2$

The intermediate product, taken directly from the production line of titanium(IV) oxide, from the Grupa Azoty Zakłady Chemiczne "Police" S.A., Police, Poland), was used as the starting material in the process of photocatalysts preparation. Titanium dioxide was treated with D-glucose (Chempur, Piekary Śląskie, Poland) solutions of various concentrations (1%, 5%, 10%) at various annealing temperature (100 °C, 150 °C, and 200 °C) in a pressure autoclave BHL-800 (Berghof, Eningen, Germany).

4.2. Characterisation of Photocatalyst

The characteristic of commercial and obtained photocatalyst was studied by measurement of specific surface area (SSA) by the N_2 adsorption-desorption method on a Quadrasorb SI analyzer (Quantachrome Instruments, Boynton Beach, FL, USA). The crystalline phase and crystal structures of prepared samples were identified by X-Ray Diffraction analysis (PANalytical Empyrean X-ray diffractometer, Malvern Panalytical Ltd., Melvern, UK) using Cu Kα radiation ($\lambda = 1.54056$ Å). Total carbon amount in the samples was calculated using CN628 elemental analyzer (LECO Corporation, St Joseph, MI, USA). Chemical composition of the surface (chemical state and content of titanium, oxygen, and carbon) was estimated by XPS JEOL JPC-9010MC with MgKα X-ray (JEOL, Tokyo, Japan). The zeta potential was determined by using ZetaSizerNanoSeries ZS (Malvern Instruments, Melvern, UK).

4.3. Antimicrobial Properties of Dispersed C/TiO$_2$

D-glucose-modified titania, starting TiO$_2$ (Grupa Azoty Zakłady Chemiczne "Police" S.A., Police, Poland), commercial photocatalyst KRONOClean 7000 (International Inc., Leverkusen, Germany) were dispersed in 10 cm^{-3} *Escherichia coli* (*E. coli*) K12 ATCC29425 or *Staphylococcus epidermidis* (*S. epidermidis*) ACCT 49461 suspension in a sterile sodium chloride aqueous solution (8.5 g dm^{-3}) in a glass test tube, and irradiated with lamp emitting artificial solar light (250 W OSRAM) under continuous stirring (using a magnetic stirrer at speed of 250 rpm) at 37 °C. The distance between the solution and the light source was fixed at ca. 15 cm. The radiant flux was monitored with a Radiation Intensity Meter LB901/WCM3 & PD204AB cos. sensor meter. The photocatalyst and bacterial cell concentration was adjusted, according to the results described in Section 2.2. The control experiments in darkness and for NaCl solution were also performed. Serial dilutions (10^{-1} to 10^{-6}) were prepared in saline solution (0.9%). The samples were placed on Plate Count Agar (PCA, BTL Polska Sp. z.o.o, Warszawa, Poland). The plates were incubated for 24 h at 37 °C and then colony forming unit (CFU $\times$ cm^{-3}) was counted. Statistical analysis of obtained results was conducted using Excel spreadsheet. The comparisons among the means and the statistical significance of differences were evaluated by Tukey's test. Microscopic bacteria specimen were prepared according to general protocol and dehydrated by graded series of ethanol (30, 50, 70, 90, 95, and 99.5% (*v*/*v*)). Samples were sputtered with gold for 60 s from three-directions and observed by SEM (SU8020 UHR FE-SEM, Hitachi, Tokyo, Japan). The bacteria mineralization rate was calculated as amounts of liberated CO$_2$ (gas chromatography; GC 8610C; SRI Instrument Inc., Torrance, CA, USA) after 180 min irradiation or under dark condition. The catalase (CAT) and superoxide dismutase (SOD) in bacterial suspension were determined before photocatalytic process (N$_0$) and after 30 (N$_{30}$) and 60 (N$_{60}$) min of irradiation and in dark conditions. Catalase activity was determined spectrophotometrically at 25 °C (1-cm cuvette, spectrophotometer F-2500, Hitachi, Tokyo, Japan) by monitoring the change in absorbance A (λ = 374 nm), according protocols that are given in [46] and own modification. One unit (U) of activity was defined as the amount of enzyme that catalyses the oxidation of 1 mmol H$_2$O$_2$ in 1 min under the assay conditions. Superoxidase dismutase activity was determined spectrophotometrically at 25 °C (1-cm cuvette, spectrophotometer F-2500, Hitachi, Tokyo, Japan) by monitoring the change in absorbance A (λ = 420 nm), according to [47] and own modification. One unit of SOD activity inhibits the rate of pyrogallol oxidation at 25 °C by 50% in 1 min.

4.4. Antimicrobial Properties of Employed in Concrete TiO$_2$-G-1%-100

The bactericidal properties of cement mortars supplemented with modified titania (TiO$_2$-G-1%-100) were examined in an own-designed photoreactor according patented procedure [48]. Cement mortar Fix M-15 Kreisel (Kreisel Technika Budowlana Sp. z.o.o., Poznań, Poland) was used to prepare reference concrete plates. The same material was used to prepare concrete plated with 10% weight of TiO$_2$-G-1%-100. The antimicrobial activity was evaluated against *Escherichia coli* K12 (ATCC 25922) under artificial solar light (UV-Vis) and in dark condition. The bacterial concentration was calculated as CFU per milliliter according the procedure described in Section 4.3.

5. Conclusions

Titania modification with cheap carbon source, i.e., D-glucose, resulted in the appearance of high antimicrobial activity against both gram-positive and gram-negative bacteria. It is thought that strong interaction between bacterial cells and photocatalyst surface are responsible for inactivation and complete decomposition of bacteria cells. It is proposed that D-glucose-modified titania photocatalyst is highly prospective material, and may be commercially used for cheap and effective water disinfection and wastewater treatment since its immobilization on a concrete results in the efficient removal of bacteria under solar radiation.

Author Contributions: A.M.-S. conceived and designed the experiments, and wrote the paper; P.R. performed the experiments; M.E. analyzed the data; K.W. performed XPS analysis and edited figures; A.W.M. corrected the manuscript; and E.K. analyzed the data and wrote the paper. All authors read and approved the final manuscript.

Acknowledgments: A.M.-S. acknowledges Institute for Catalysis, Hokkaido University for research stay in a framework of Fusion Emergent Research project by the MEXT program of Integrated Research Consortium on Chemical Sciences (IRCCS).

Conflicts of Interest: The author declares no conflict of interest.

References

1. Rengifo-Herrera, J.A.; Mielczarski, E.; Mielczarski, J.; Castillo, N.C.; Kiwi, J.; Pulgarin, C. Escherichia coli inactivation by N, S co-doped commercial TiO_2 powders under UV and visible light. *Appl. Catal. B Environ.* **2008**, *84*, 448–456. [CrossRef]

2. Kowalska, E.; Mahaney, O.O.P.; Abe, R.; Ohtani, B. Visible-light-induced photocatalysis through surface plasmon excitation of gold on titania surfaces. *Phys. Chem. Chem. Phys.* **2010**, *12*, 2344. [CrossRef] [PubMed]

3. Liu, G.; Han, C.; Pelaez, M.; Zhu, D.; Liao, S.; Likodimos, V.; Ioannidis, N.; Kontos, A.G.; Falaras, P.; Dunlop, P.S.M.; et al. Synthesis, characterization and photocatalytic evaluation of visible light activated c-doped TiO_2 nanoparticles. *Nanotechnology* **2012**, *23*, 294003. [CrossRef] [PubMed]

4. Asahi, R.; Morikawa, T.; Irie, H.; Ohwaki, T. Nitrogen-doped titanium dioxide as visible-light-sensitive photocatalyst: Designs, developments, and prospects. *Chem. Rev.* **2014**, *114*, 9824–9852. [CrossRef] [PubMed]

5. Asahi, R.; Morikawa, T.; Ohwaki, T.; Aoki, K.; Taga, Y. Visible-light photocatalysis in nitrogen-doped titanium oxides. *Science* **2001**, *293*, 269–271. [CrossRef] [PubMed]

6. Palanivelu, K.; Im, J.-S.; Lee, Y.-S. Carbon doping of TiO_2 for visible light photo catalysis—A review. *Carbon Lett.* **2007**, *8*, 214–224. [CrossRef]

7. Wu, G.; Nishikawa, T.; Ohtani, B.; Chen, A. Synthesis and characterization of carbon-doped TiO_2 nanostructures with enhanced visible light response. *Chem. Mater.* **2007**, *19*, 4530–4537. [CrossRef]

8. Wu, Z.; Dong, F.; Zhao, W.; Wang, H.; Liu, Y.; Guan, B. The fabrication and characterization of novel carbon doped TiO_2 nanotubes, nanowires and nanorods with high visible light photocatalytic activity. *Nanotechnology* **2009**, *20*, 235701. [CrossRef] [PubMed]

9. Zhong, J.; Chen, F.; Zhang, J. Carbon-deposited TiO_2: Synthesis, characterization, and visible photocatalytic performance. *J. Phys. Chem. C* **2010**, *114*, 933–939. [CrossRef]

10. Xie, C.; Yang, S.; Shi, J.; Niu, C. Highly crystallized C-doped mesoporous anatase TiO_2 with visible light photocatalytic activity. *Catalysts* **2016**, *6*, 117. [CrossRef]

11. Morawski, A.W.; Janus, M.; Tryba, B.; Inagaki, M.; Kałucki, K. TiO_2-anatase modified by carbon as the photocatalyst under visible light. *Rendus Rendu Chim.* **2006**, *9*, 800–805. [CrossRef]

12. Sakthivel, S.; Kisch, H. Daylight photocatalysis by carbon-modified titanium dioxide. *Angew. Chem. Int. Ed.* **2003**, *42*, 4908–4911. [CrossRef] [PubMed]

13. Janus, M.; Markowska-Szczupak, A.; Kusiak-Nejman, E.; Morawski, A.W. Disinfection of *E. coli* by carbon modified TiO_2 photocatalysts. *Environ. Prot. Eng.* **2012**, *38*, 90–97. [CrossRef]

14. Wanag, A.; Rokicka, P.; Kusiak-Nejman, E.; Markowska-Szczupak, A.; Morawski, A.W. TiO_2/glucose nanomaterials with enhanced antibacterial properties. *Mater. Lett.* **2016**, *185*, 264–267. [CrossRef]

15. Wanag, A.; Rokicka, P.; Kusiak-Nejman, E.; Kapica-Kozar, J.; Wrobel, R.J.; Markowska-Szczupak, A.; Morawski, A.W. Antibacterial properties of TiO_2 modified with reduced graphene oxide. *Ecotoxicol. Environ. Saf.* **2018**, *147*, 788–793. [CrossRef] [PubMed]

16. Li, Y.; Ma, M.; Wang, X.; Wang, X. Inactivated properties of activated carbon-supported TiO_2 nanoparticles for bacteria and kinetic study. *J. Environ. Sci.* **2008**, *20*, 1527–1533. [CrossRef]

17. McEvoy, J.G.; Cui, W.; Zhang, Z. Degradative and disinfective properties of carbon-doped anatase-rutile TiO_2 mixtures under visible light irradiation. *Catal. Today* **2013**, *207*, 191–199. [CrossRef]

18. Sun, D.-S.; Kau, J.-H.; Huang, H.-H.; Tseng, Y.-H.; Wu, W.-S.; Chang, H.-H. Antibacterial properties of visible-light-responsive carbon-containing titanium dioxide photocatalytic nanoparticles against anthrax. *Nanomaterials* **2016**, *6*, 237. [CrossRef] [PubMed]

19. Grzybowska, B.; Słoczyński, J.; Grabowski, R.; Samson, K.; Gressel, I.; Wcisło, K.; Gengembre, L.; Barbaux, Y. Effect of doping of TiO_2 support with altervalent ions on physicochemical and catalytic properties in oxidative dehydrogenation of propane of vanadia–titania catalysts. *Appl. Catal. A Gen.* **2002**, *230*, 1–10. [CrossRef]

20. örsi, F. Kinetic studies on the thermal decomposition of glucose and fructose. *J. Therm. Anal.* **1973**, *5*, 329–335. [CrossRef]

21. Slimestad, R.; Vågen, I.M. Thermal stability of glucose and other sugar aldoses in normal phase high performance liquid chromatography. *J. Chromatogr. Coruña* **2006**, *1118*, 281–284. [CrossRef] [PubMed]

22. Siuzdak, K.; Sawczak, M.; Klein, M.; Nowaczyk, G.; Jurga, S.; Cenian, A. Preparation of platinum modified titanium dioxide nanoparticles with the use of laser ablation in water. *Phys. Chem. Chem. Phys.* **2014**, *16*, 15199–15206. [CrossRef] [PubMed]

23. Zielińska-Jurek, A.; Kowalska, E.; Sobczak, J.W.; Lisowski, W.; Ohtani, B.; Zaleska, A. Preparation and characterization of monometallic (au) and bimetallic (ag/au) modified-titania photocatalysts activated by visible light. *Appl. Catal. B Environ.* **2011**, *101*, 504–514. [CrossRef]

24. Wei, Z.; Janczarek, M.; Endo, M.; Balčytis, A.; Nitta, A.; Mendez Medrano, M.G.; Colbeau-Justin, C.; Juodkazis, S.; Ohtani, B.; Kowalska, E. Noble metal-modified faceted anatase titania photocatalysts: Octahedron versus decahedron. *Appl. Catal. B Environ.* **2018**, *237*, 574–587. [CrossRef]

25. Baltrusaitis, J.; Schuttlefield, J.; Zeitler, E.; Grassian, V.H. Carbon dioxide adsorption on oxide nanoparticle surfaces. *Chem. Eng. J.* **2011**, *170*, 471–481. [CrossRef]

26. Fleisch, T.H.; Mains, G.J. Reduction of copper oxides by uv radiation and atomic hydrogen studied by xps. *Appl. Surf. Sci.* **1982**, *10*, 51–62. [CrossRef]

27. Akhavan, O.; Ghaderi, E. Self-accumulated ag nanoparticles on mesoporous TiO_2 thin film with high bactericidal activities. *Surf. Coat. Technol.* **2010**, *204*, 3676–3683. [CrossRef]

28. Kruse, N.; Chenakin, S. Xps characterization of Au/TiO_2 catalysts: Binding energy assessment and irradiation effects. *Appl. Catal. A Gen.* **2011**, *391*, 367–376. [CrossRef]

29. Rossnagel, S.M.; Sites, J.R. X-ray photoelectron spectroscopy of ion beam sputter deposited SiO_2, TiO_2, and Ta_2O_5. *J. Vac. Sci. Technol. A Vac. Surf. Films* **1984**, *2*, 376–379. [CrossRef]

30. Jensen, H.; Soloviev, A.; Li, Z.; Søgaard, E.G. Xps and ftir investigation of the surface properties of different prepared titania nano-powders. *Appl. Surf. Sci.* **2005**, *246*, 239–249. [CrossRef]

31. Yu, J.; Zhao, X.; Zhao, Q. Effect of surface structure on photocatalytic activity of TiO_2 thin films prepared by sol-gel method. *Thin Solid Films* **2000**, *379*, 7–14. [CrossRef]

32. Lonnen, J.; Kilvington, S.; Kehoe, S.C.; Al-Touati, F.; McGuigan, K.G. Solar and photocatalytic disinfection of protozoan, fungal and bacterial microbes in drinking water. *Water Res.* **2005**, *39*, 877–883. [CrossRef] [PubMed]

33. Wang, L.; Hu, C.; Shao, L. The antimicrobial activity of nanoparticles: Present situation and prospects for the future. *Int. J. Nanomed.* **2017**, *12*, 1227–1249. [CrossRef] [PubMed]

34. Maletić, M.; Vukčević, M.; Kalijadis, A.; Janković-Častvan, I.; Dapčević, A.; Laušević, Z.; Laušević, M. Hydrothermal synthesis of TiO_2/carbon composites and their application for removal of organic pollutants. *Arab. J. Chem.* **2016**. [CrossRef]

35. Kim, G.; Lee, S.-H.; Choi, W. Glucose–TiO_2 charge transfer complex-mediated photocatalysis under visible light. *Appl. Catal. B Environ.* **2015**, *162*, 463–469. [CrossRef]

36. Dong, F.; Wang, H.; Wu, Z. One-step "green" synthetic approach for mesoporous c-doped titanium dioxide with efficient visible light photocatalytic activity. *J. Phys. Chem. C* **2009**, *113*, 16717–16723. [CrossRef]

37. Herrmann, J. Heterogeneous photocatalysis: Fundamentals and applications to the removal of various types of aqueous pollutants. *Catal. Today* **1999**, *53*, 115–129. [CrossRef]

38. Cho, M.; Chung, H.; Choi, W.; Yoon, J. Linear correlation between inactivation of *E. coli* and OH radical concentration in TiO_2 photocatalytic disinfection. *Water Res.* **2004**, *38*, 1069–1077. [CrossRef] [PubMed]

39. Benabbou, A.K.; Derriche, Z.; Felix, C.; Lejeune, P.; Guillard, C. Photocatalytic inactivation of escherischia coli. *Appl. Catal. B Environ.* **2007**, *76*, 257–263. [CrossRef]

40. Kumar, A.; Pandey, A.K.; Singh, S.S.; Shanker, R.; Dhawan, A. Engineered ZnO and TiO2 nanoparticles induce oxidative stress and dna damage leading to reduced viability of escherichia coli. *Free. Radic. Biol. Med.* **2011**, *51*, 1872–1881. [CrossRef] [PubMed]

41. Wanag, A.; Morawski, A.W.; Kapica-Kozar, J.; Kusiak-Nejman, E. Photocatalytic performance of thermally prepared TiO$_2$/C photocatalysts under artificial solar light. *Micro Nano Lett.* **2016**, *11*, 202–206. [CrossRef]

42. Cui, Y.; Li, H.; Hong, W.; Fan, S.; Zhu, L. The effect of carbon content on the structure and photocatalytic activity of nano-bi2wo6 powder. *Powder Technol.* **2013**, *247*, 151–160. [CrossRef]

43. Kacem, M.; Bru-Adan, V.; Goetz, V.; Steyer, J.P.; Plantard, G.; Sacco, D.; Wery, N. Inactivation of escherichia coli by TiO2-mediated photocatalysis evaluated by a culture method and viability-qpcr. *J. Photochem. Photobiol. A Chem.* **2016**, *317*, 81–87. [CrossRef]

44. Lin, X.; Li, J.; Ma, S.; Liu, G.; Yang, K.; Tong, M.; Lin, D. Toxicity of TiO2 nanoparticles to escherichia coli: Effects of particle size, crystal phase and water chemistry. *PLoS ONE* **2014**, *9*, e110247. [CrossRef] [PubMed]

45. Clogston, J.D.; Patri, A.K. Zeta potential measurement. *Methods Mol. Biol.* **2011**, *697*, 63–70. [CrossRef] [PubMed]

46. Hadwan, M.H.; Abed, H.N. Data supporting the spectrophotometric method for the estimation of catalase activity. *Data Brief* **2016**, *6*, 194–199. [CrossRef] [PubMed]

47. Marklund, S.; Marklund, G. Involvement of the superoxide anion radical in the autoxidation of pyrogallol and a convenient assay for superoxide dismutase. *Eur. J. Biochem.* **1974**, *47*, 469–474. [CrossRef] [PubMed]

48. Markowska-Szczupak, A.; Rokicka, P.; Janus, M.; Kusiak Najman, E.; Morawski, A.W. Method for Testing Antibacterial Properties of Materials Made from Cements Containing Titanium Dioxide. Patent Number 416,872, 19 April 2016.

 *catalysts*

Article

Synergistic Effect of Cu$_2$O and Urea as Modifiers of TiO$_2$ for Enhanced Visible Light Activity

Marcin Janczarek [1,2,*], Kunlei Wang [1] and Ewa Kowalska [1]

[1] Institute for Catalysis, Hokkaido University, N21, W10, Sapporo 001-0021, Japan; kunlei@cat.hokudai.ac.jp (K.W.); kowalska@cat.hokudai.ac.jp (E.K.)

[2] Department of Chemical Technology, Gdansk University of Technology, Narutowicza Str. 11/12, 80-233 Gdansk, Poland

* Correspondence: mjancz@pg.edu.pl; Tel.: +48-58-3472352

Received: 14 March 2018; Accepted: 1 June 2018; Published: 6 June 2018

Abstract: Low cost compounds, i.e., Cu$_2$O and urea, were used as TiO$_2$ modifiers to introduce visible light activity. Simple and cheap methods were applied to synthesize an efficient and stable nanocomposite photocatalytic material. First, the core-shell structure TiO$_2$–polytriazine derivatives were prepared. Thereafter, Cu$_2$O was added as the second semiconductor to form a dual heterojunction system. Enhanced visible light activity was found for the above-mentioned nanocomposite, confirming a synergistic effect of Cu$_2$O and urea (via polytriazine derivatives on titania surface). Two possible mechanisms of visible light activity of the considered material were proposed regarding the type II heterojunction and Z-scheme through the essential improvement of the charge separation effect.

Keywords: photocatalysis; nanocomposites; heterojunction; Cu$_2$O; urea; polytriazine; Z-scheme

1. Introduction

Titanium dioxide (TiO$_2$, titania) is widely recognized as an efficient, stable, and green photocatalytic material (long term stability, chemical inertness and corrosion resistance). Therefore, its application potential in photocatalysis is still growing and presently focuses on emergency areas, such as environmental remediation (water treatment and air purification), renewable energy processes (i.e., photocurrent generation, water splitting for hydrogen production, conversion of CO$_2$ to hydrocarbons) and self-cleaning surfaces [1–5]. However, the application of titania is still limited to regions with a high intensity of solar radiation due to its wide bandgap (ca. 3.0 to 3.2 eV). The following strategies of titania doping, modification, semiconductor coupling, and dye sensitization can be applied to incorporate visible light absorption to TiO$_2$ [6–8]. The nature of electron transfer between TiO$_2$ and other materials has been intensively studied and recognized as the origin of the high performances of TiO$_2$-based composites [9]. The photocatalytic activity of TiO$_2$ systems depend on the following properties, such as particle size, surface area, crystal phase, morphology, uncoordinated surface sites, defects in the lattice, and degree of crystallinity. Design of TiO$_2$ composite structures based on the heterojunction between titania and other semiconducting materials can improve many of these properties. Moreover, this strategy can create and tune other properties, such as mid-band-gap electronic states, which can be responsible for the intensification of charge separation or incorporation of a red shift to the absorption spectrum [7,9].

Copper oxides (Cu$_2$O, CuO) have been intensively studied as titania modifiers due to their intrinsic p-type configuration. Cu$_2$O and CuO are inexpensive semiconductors with band gap energies of 2.1–2.2 eV and 1.2–1.7 eV, respectively. This fact makes both materials promising for research directed to solar energy utilization [10–12]. If the electronic properties of both oxides are compared, CuO has a significantly smaller band gap than Cu$_2$O, and thus can absorb more photons, but the positions of CB and VB for CuO are insufficient to catalyze the generation of hydroxyl and superoxide radicals, which are the primary initiators for the photocatalytic oxidation of organic compounds [13,14]. Therefore, the photocatalytic oxidation of organic compounds over Cu_2O/TiO_2 composites has been of particular interest, especially to introduce visible light activity to TiO_2-based systems [15–19]. For example, Liu et al. prepared Cu_2O/TiO_2 composites where titania was in the form of nanosheets with exposed {001} facets. They reported visible light photocatalytic activity for Cu_2O/TiO_2 nanosheets three times higher than that for nitrogen-doped titania nanosheets. Owing to the type II heterojunction between Cu_2O and TiO_2, an efficient charge separation is observed and visible light induced electron transfer from Cu_2O to TiO_2 occurs, resulting in visible light photocatalytic performance of the system [17]. Furthermore, Cu_2O possesses promising application potential because of its very good antipathogenic properties, even better than that of metallic copper [20–23].

Apart from p-type semiconducting metal oxides as composite junctions for titania, a promising option to prepare efficient and low cost visible light-active material based on TiO_2 is the application of organic compounds like urea as a modifier. Urea-derived titania materials were initially recognized as nitrogen "doped" TiO_2 photocatalysts [24–27]. Subsequently, the presence of nitridic and amidic species or nitrogen species with several oxidation states of nitrogen were suggested [25,26]. Finally, Mitoraj and Kisch proposed another explanation for the nature of urea-modification of titania including both nitrogen and carbon originating from urea as the elements building the chemical structure responsible for visible light-sensitization of TiO_2 [28–30]. They reported that thermal processing of urea with titania at 400 °C produced poly(amino-tri-s-triazine) derivatives (shell) covalently attached to the semiconductor (core), i.e., a unique example of inorganic with an organic (polytriazine as a crystalline layer) semiconductor connected through the Ti–N–C bond. Titania acts as a catalyst in this urea transformation. Condensation between the triazine amino and titania hydroxy groups forms Ti–N bonds. Considered chemical structures arise from condensed aromatic s-triazine compounds containing melem and melon (trimer of melem) units, which form a visible light absorbing semiconducting organic layer coupled with titania. Compared to unmodified TiO_2, the prepared photocatalyst may exhibit a band-gap narrowing. It has been proposed that the absorption shoulder in the visible region corresponding to this material is a charge-transfer band enabling an optical transfer from polytriazine component to titania [29]. In contrast, it is known that prolonged heating of urea at 550 °C produces polycondensed s-triazines with a graphitic structure (g-C_3N_4), which was recognized as a separate metal-free polymeric n-type semiconductor (band gap energy equal to 2.7 eV) with the possibility to create heterojunctions with other semiconductors, e.g., titania [31–34]. Therefore, thermal treatment of a TiO_2/urea composite at higher temperature results in the formation of a TiO_2/g-C_3N_4 heterojunction, whereas at lower temperatures, novel electronic states responsible for visible light activity located near valence band of titania originating from the presence of poly(amino-s-triazine) shell are formed [6].

Herein, we demonstrate that titania with a polytriazine layer originated from urea, and modified with Cu_2O (Cu_2O/PTr–TiO_2) becomes an efficient visible light photocatalyst, significantly more active than that of single modified titania, i.e., PTr–TiO_2 or Cu_2O/TiO_2. The preparation method is economically and practically attractive by using low cost components such as urea and Cu_2O and simple synthetic operations, i.e., lower temperatures of preparation in comparison with g-C_3N_4 synthesis (400 v. 550 °C). The important issue is the probable proposition of the explanation of the synergistic behavior of two different modifiers with corresponding visible light activation mechanisms. There are only two studies describing combinations of g-C_3N_4 with Cu_2O [35] and additionally with TiO_2 [36] showing improvements of visible light photocatalytic activity in comparison to single components (However, dyes have been used as test compounds, and thus origin of visible activity could not be unequivocally decided, i.e., dye sensitization could not be excluded [37,38]). Moreover, polycondensed s-triazines of graphitic structure were used, but not the non-graphitic forms obtained at lower temperature as proposed by Mitoraj and Kisch [29]. To the best of our knowledge, this report is the first study for this type of photocatalytic material.

2. Results and Discussion

2.1. Preparation Conditions and Visible Light Photocatalytic Efficiency

Commercially available P25 titania was used as a base to prepare the final photocatalytic material (Cu_2O/PTr–TiO_2). The first step of preparation was to obtain PTr–TiO_2 with urea as a modifier (preparation details in Materials and Methods). "PTr" symbolizes the polytriazine shell. Different ratios between urea and P25 were investigated: 0.5:1, 1:1, 2:1 and 3:1 to find the material with the best photocatalytic efficiency under visible light irradiation. Samples with different shades of yellow color were obtained (more intensive yellow represents a higher amount of urea). 2-propanol oxidation to acetone was selected as the reaction system to test visible light photocatalytic activity ($\lambda > 455$ nm). The sample prepared at 1:1 urea–P25 ratio possessed the highest photocatalytic activity (measured as the produced acetone amount) and was chosen for further research and marked as PTr–TiO_2 (see Figure 1). It was found that higher amounts of urea were not advantageous for the photocatalytic activity of the final product probably due to the detrimental influence of some not converted products of urea thermal transformation (excess of urea) on the semiconducting character of the polytriazine shell structure, as previously reported [29].

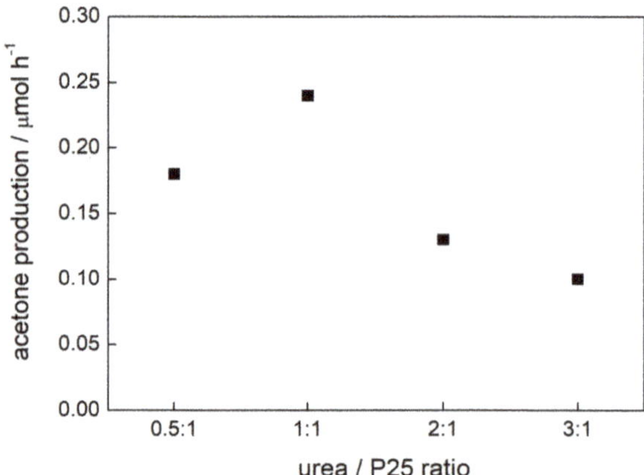

Figure 1. Visible light photocatalytic activity of the samples prepared with corresponding urea/P25 ratio.

The second step in the photocatalyst preparation was the physical mixing of PTr–TiO_2 with Cu_2O by using different contents of Cu_2O. Figure 2 shows the evidence of the advantageous role of the addition of cuprous oxide to the PTr–TiO_2 sample. The optimum Cu_2O content was found to be in the range of ca. 5 wt %, whereas 10 wt % and larger amounts were detrimental for photocatalytic efficiency. Detrimental influence of modifiers at their larger contents on photocatalytic activity is not surprising, and has already been reported for various systems. There are two main reasons for this behavior, i.e., (i) a shielding effect as Cu_2O blocks PTr–TiO_2 for photon absorption; and (ii) competitive adsorption as the oxygen and/or organic compounds (here 2-propanol) could not adsorbed directly on the titania surface, occupied by its modifier. Based on the obtained results, 5 wt % was chosen the Cu_2O content in the Cu_2O/PTr–TiO_2 sample.

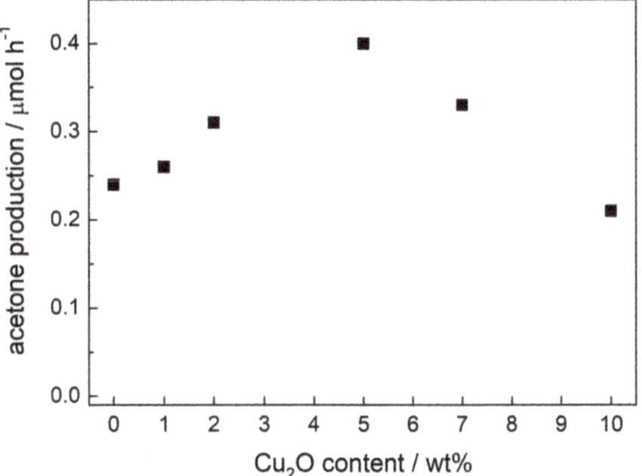

Figure 2. Visible light photocatalytic activity of the samples prepared with different Cu_2O content.

2.2. Characterization of Samples

P25 is a well-known mixed-phase titania photocatalyst containing two crystalline phases: anatase (86.4%) and rutile (13.6%), and amorphous titania (exact composition using NiO as the internal standard was estimated previously showing the anatase/rutile/amorphous ratio to be 76–80/13–15/6–11 (P25 is not a uniform sample) [39]) with a 21 nm crystallite size of anatase (determined from XRD) and 59.1 m^2/g specific surface area. After the thermal treatment with urea, the properties of the PTr–TiO$_2$ sample (1:1 urea–P25 ratio) changed slightly to a 23 nm crystallite size of anatase and 62.3 m^2 g^{-1} specific surface area, but the crystalline phase content remained unchanged. It is proposed that an increase in crystallite size should result from amorphous titania conversion to anatase during calcination, whereas an increase in specific surface area comes from the adsorbed organic layer of poly(amino-tri-*s*-triazine) derivatives. Cu_2O, which was used as a component of the Cu_2O–PTr–TiO$_2$ mixture, was characterized by a 65 nm crystallite size and 23 m^2 g^{-1} specific surface area. Therefore, the results stated that the PTr–TiO$_2$ crystallites were almost three times smaller than the Cu_2O ones. These values were confirmed by XRD analysis of the Cu_2O/PTr–TiO$_2$ sample. Figure 3 shows the XRD diffractogram of Cu_2O/PTr–TiO$_2$ with peaks corresponding to anatase, rutile, and Cu_2O. An aromatic system of carbon nitrides should appear at around 27.4°, but a strong peak of rutile overlaps it. It was reported that in the case of a small fraction of polycondensed polytriazines, this peak would not be detected [29].

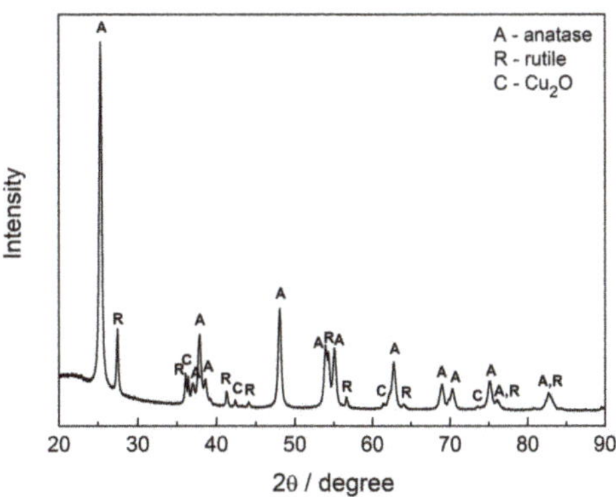

Figure 3. XRD diffractogram of Cu_2O/PTr–TiO$_2$ sample.

The diffuse reflectance spectra of the PTr–TiO$_2$ and Cu$_2$O/PTr–TiO$_2$ samples (Figure 4) show the strong red shift originated from the absorption properties of the polytriazine layer on P25. The visible light absorption effect was stronger than that for similar samples reported by Mitoraj and Kisch [28,30]. However, a different type of titania was used as the base for synthesis in that study. Therefore, it was proposed that the content of the surface hydroxyl groups on the surfaces of the various TiO$_2$ samples were responsible for better/worse ability of polytriazine layer formation (the issue of finishing the process of formation polytriazine layer by the reaction of amino groups of the relevant intermediates with titania surface OH groups [30]). To check this hypothesis, reference experiments were performed for own-prepared decahedral anatase particles (DAP) [40] modified with urea using the same conditions as for P25. Interestingly, it was observed that the DAP sample after this modification practically did not change, i.e., the color was still white and the visible light activity was not observed. Accordingly, it is proposed that the highly crystalline titania samples with low surface area (DAP as an example) possess too low number of surface OH groups to successfully introduce the polytriazine layer on their surfaces. It is also possible that surface defects, e.g., oxygen vacancy, play a crucial role in polytriazine layer formation since the DAP (well-crystallized faceted titania) possess an extremely low content of such defects (clarification of this phenomenon for other titania samples is under study). Mitoraj and Kisch proposed that the visible light absorption of urea-modified samples with a yellow color originated from the presence of poly(tri-*s*-triazine) centered intra-bandgap levels which may form (depending on PTr concentration) a narrow energy band in titania [29]. Additionally, in the case of the Cu$_2$O/PTr–TiO$_2$ photocatalyst, more intensive light absorption in the range 450–600 nm, as the consequence of the inherent light absorption properties of Cu$_2$O, was observed (Figure 3). The Cu$_2$O/PTr–TiO$_2$ sample had a reddish-yellow color.

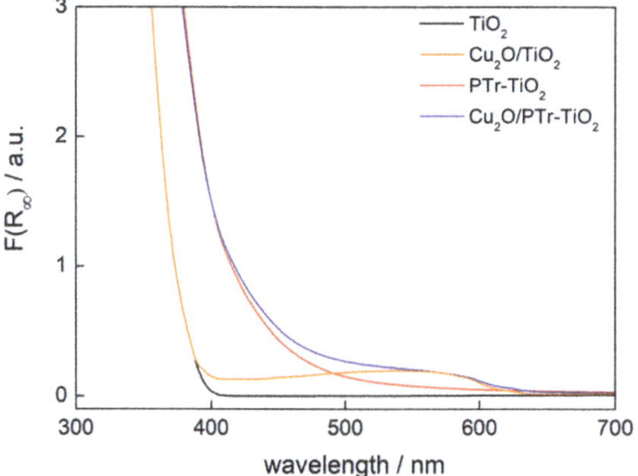

Figure 4. Diffuse reflectance spectra of TiO$_2$, PTr–TiO$_2$, Cu$_2$O/PTr–TiO$_2$. The Kubelka–Munk function F(R$_\infty$) is equivalent to absorbance.

XPS analysis of PTr–TiO$_2$ and Cu$_2$O/PTr–TiO$_2$ samples showed the presence of nitrogen 1s with binding energies of 399.1 and 400.5 eV, which indicates the presence of C=N–C and (–C=N–)$_x$ bonds, respectively, being in agreement with previously reported values [29,41,42]. XPS results showing the fractions of oxidation states of titanium, oxygen and copper are shown in Table 1. For samples modified with urea (PTr–TiO$_2$ and Cu$_2$O/PTr–TiO$_2$), the content of oxygen related to hydroxyl groups bound to titanium and carbon was lower than for bare TiO$_2$ due to the fact that surface hydroxyl groups participate in poly-*s*-triazine layer formation. In contrast, the content of Ti^{3+} increased after modification, which could confirm the participation of surface defects in the formation of PTr–TiO$_2$ (as discussed above for modified faceted titania (DAP)). It is proposed that thermal treatment could increase the formation of Ti^{3+}.

Therefore, it is also possible that a decrease in the content of hydroxyl groups after titania modification could result from their replacement by poly-*s*-triazine, similar to titania surface modification with nanoparticles of noble metals [43]. Moreover, samples with the poly-*s*-triazine layer also had a higher carbon content (higher C/Ti ratio). These results confirmed the presence of an organic layer on the titania surface consisting of poly-*s*-triazine derivatives, as reported previously [29].

Table 1. XPS analysis for TiO_2, PTr–TiO_2 and Cu_2O/PTR–TiO_2 samples including fraction of oxidation states of Ti, O, and Cu from the deconvolution of XPS peaks of Ti $2p_{3/2}$, O 1s and Cu $2p_{3/2}$.

Samples	Ti $2p_{3/2}$ (%)		O 1s (%)			Ratio		Valent State (%)		
	Ti^{4+}	Ti^{3+}	TiO_2 [a]	Ti-OH [b]	Ti-OH [c]	O/Ti	C/Ti	Cu^{2+}	Cu^+	Cu(0)
TiO_2	98.5	1.5	57.9	26.7	15.4	2.6	3.5	–	–	–
PTr-TiO_2	97.2	2.8	62.3	23.7	14.0	2.5	4.1	–	–	–
Cu_2O/PTr-TiO_2	97.0	3.0	61.8	24.1	14.1	2.5	4.3	0.9	99.1	–

[a] Oxygen in the TiO_2 crystal lattice; [b] Ti–(OH)–Ti, Ti_2O_3, C=O; [c] Ti–OH, C–OH.

To confirm the presence of the organic layer and possible formation of the Cu_2O/PTr–TiO_2 heterojunction, scanning transmission electron microscopy (STEM) observation was carried out. It was found that Cu_2O existed in both crystalline and amorphous forms (large aggregates of very fine NPs). Moreover, two kinds of crystalline structures were noticed, i.e., large crystals of 100–150 nm and fine faceted nanocrystals of 10–30 nm (cubic and decahedral). For the Cu_2O/PTr–TiO_2 sample, titania particles were covered with a 5–10-nm layer of polytriazine located nearby larger Cu_2O crystallites, as clearly shown in Figure 5 (respective SEM and TEM modes of the same image). Moreover, fine amorphous Cu_2O interconnecting PTr–TiO_2 particles was also observed. Interestingly, it was found that long-term electron beam during STEM observations could destroy the polytriazine layer around the titania since new nanostructures were formed (gravitationally-formed honey-like tails).

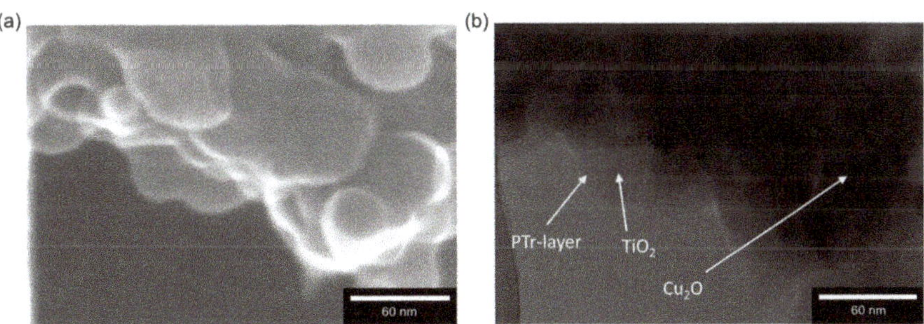

Figure 5. STEM images of Cu_2O/PTr–TiO_2 sample in SEM (**a**) and TEM (**b**) modes.

2.3. Enhanced Visible Light Photocatalytic Activity as a Synergistic Effect of Two Modifiers

Figure 6 shows the results of the photocatalytic activity tests of five photocatalysts: TiO_2, Cu_2O, Cu_2O/TiO_2, PTr-TiO_2, and Cu_2O/PTr–TiO_2 in the visible light-induced 2-propanol oxidation. The highest activity was observed for Cu_2O/PTr–TiO_2 with 1.2 µmol of acetone formation after 3 h of irradiation. The Cu_2O/TiO_2 and PTr–TiO_2 samples were characterized by significantly lower photoactivity: 0.18 and 0.72 µmol of acetone, respectively. The unmodified photocatalysts TiO_2 (P25) and Cu_2O were almost inactive in this reaction system. The improvement of visible light photocatalytic performance for the Cu_2O/PTr–TiO_2 sample in comparison to the Cu_2O/TiO_2 and PTr–TiO_2 photocatalysts is evidence of the synergistic effect of urea (poly(tri-*s*-triazine)) and Cu_2O modifiers for enhancing the visible light photocatalytic activity of titania. The photocatalytic activity test for the Cu_2O/PTr–TiO_2 sample was extended to 8 h to check the basic stability of the material. After 8 h of irradiation, a linear course of acetone production was still observed.

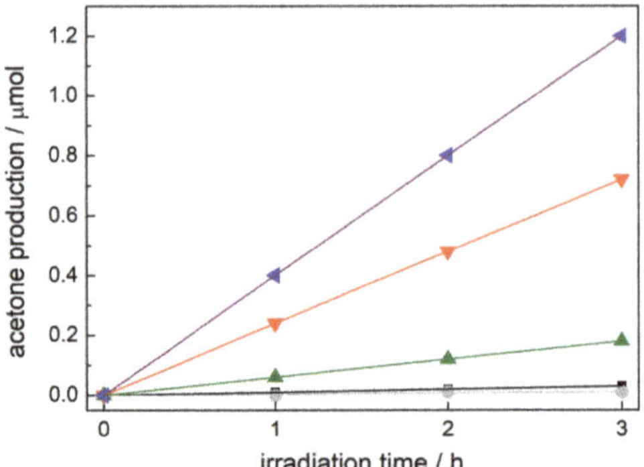

Figure 6. Visible light-induced ($\lambda > 455$ nm) 2-propanol oxidation in the presence of (■) TiO$_2$, (●) Cu$_2$O, (▲) Cu$_2$O/TiO$_2$, (▼) PTr-TiO$_2$, and (◄) Cu$_2$O/PTr-TiO$_2$.

A lack of visible light activity for a single Cu$_2$O, in spite of its good visible light absorption properties, can be explained by a strong photocorrosion effect and fast charge recombination [44,45]. It was reported that the coupling of cuprous oxide with titania reduced these detrimental effects by the heterojunction mechanism (type II) between the p- and n-type semiconductors introducing visible light activity and enhancing stability [15–19,46].

The PTr–TiO$_2$ photocatalyst possesses significantly higher visible light activity for 2-propanol oxidation than Cu$_2$O/TiO$_2$. The core-shell structure of titania–poly(tri-*s*-triazine) derivatives changes photoabsorption and electronic properties in comparison to the unmodified titania. Figure 7 shows a way to determine band gap energy and absorption onset for the PTr–TiO$_2$ sample. The band gap narrowing occurred from 3.17 eV for bare TiO$_2$ (P25) to 2.88 eV for PTr-TiO$_2$. The absorption onset is useful to determine the location of surface states in PTr–TiO$_2$ originating from the presence of an organic modifier (shaded area in the Figures 8 and 9) [6].

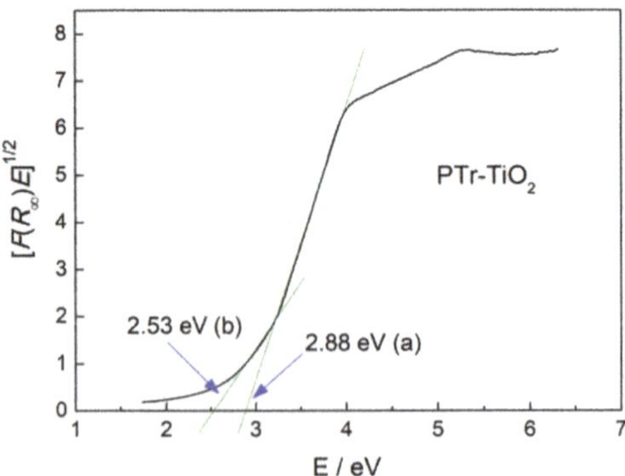

Figure 7. Plot of transformed Kubelka–Munk function vs. energy of light for the PTr–TiO$_2$ sample. Determination of band gap energy (**a**) and absorption onset (**b**).

Figures 8 and 9 illustrate the propositions of explanation for enhanced visible light photocatalytic activity of Cu$_2$O/PTr–TiO$_2$, a photocatalytic system with two main composites. In the first mechanistic variant (Figure 7), two semiconductors: p-type Cu$_2$O and n-type titania (P25) with an organic sensitizer created a type II heterojunction. This system provided the optimum band positions for efficient charge carrier separation. Visible light photoexcited electrons were transferred from CB(Cu$_2$O) to

CB(PTr–TiO$_2$) and this transfer could occur due to the favorable energetics of the relative positions of both CBs, whereas holes were simultaneously transferred from VB(PTr–TiO$_2$) to VB(Cu$_2$O). The main consequence of such phenomenon is the separation of the photogenerated electrons and holes reducing the probability of recombination and increasing the lifetimes of the charge carriers [46]. Cu$_2$O has a valence band potential of 1.07 V [47]. To form hydroxyl radicals, the valence band potential of Cu$_2$O should be more positive than the following values at pH = 7 [7]:

$$h^+ + H_2O \rightarrow OH^\bullet + H^+, E_0 = 2.73 \text{ V (vs. NHE)} \tag{1}$$

$$h^+ + OH^- \rightarrow OH^\bullet, E_0 = 1.90 \text{ V (vs. NHE)} \tag{2}$$

Therefore, the formation of hydroxyl radicals may be thermodynamically unfavorable. It can be concluded that holes accumulated would be mostly consumed through the direct oxidation of 2-propanol. Electrons accumulated on CB(PTr–TiO$_2$) play a key role in the formation of reactive oxygen species. The properties of the PTr–TiO$_2$ component of the considered semiconductor composition by introduction of visible light activity into titania enabled the formation of this advantageous heterojunction functioning under visible light irradiation.

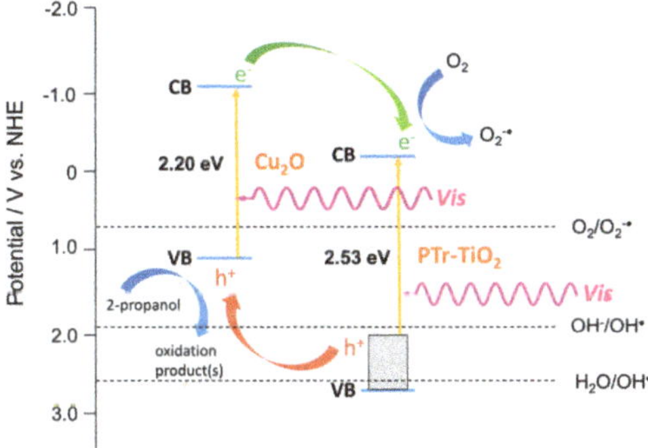

Figure 8. Energy diagram for the Cu$_2$O/PTr–TiO$_2$ photocatalytic system working under visible light, illustrating the coupling of two semiconductors as a type II heterojunction.

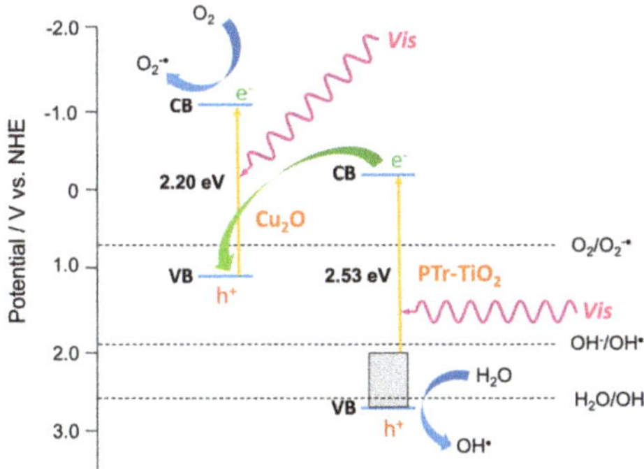

Figure 9. Energy diagram for Cu$_2$O/PTr–TiO$_2$ photocatalytic system working under visible light, illustrating the coupling of two semiconductors as a Z-scheme system.

Another mechanistic variant describing enhanced visible light activated Cu$_2$O/PTr–TiO$_2$ system relies on the Z-scheme concept [48,49]. As shown in Figure 9, photogenerated electrons in PTr–TiO$_2$

with a lower reduction ability recombined with the photogenerated holes in Cu_2O with lower oxidation ability. Therefore, electrons accumulated on $CB(Cu_2O)$ with a high reduction ability of holes accumulated on $VB(PTr–TiO_2)$ with a high oxidation ability can be maintained. The occurrence of the second mechanistic variant might be even more possible due to the existence of optimal Cu_2O content (Figure 2) and the low photocatalytic activity of the Cu_2O/TiO_2 system.

In the case of both mechanisms, the efficient charge separation induced by visible light irradiation was the main reason of the exceptional photocatalytic activity of the proposed system and provides the explanation of the synergistic role of the two considered types of titania modifiers. Clarification of which mechanistic variant is responsible for enhanced photocatalytic properties of the $Cu_2O/PTr–TiO_2$ system in the visible light is necessary. Further studies including action spectrum analysis, photoactivity tests in the presence of scavengers, and detailed characterization of these materials (and other photocatalysts prepared with different titania photocatalysts, i.e., varied by surface properties and/or content of oxygen defects), such as the estimation of the quasi–Fermi level, are along this line.

3. Materials and Methods

3.1. Preparation of $Cu_2O/PTr–TiO_2$

P25 (AEROXIDE® TiO_2 P25, Nippon Aerosil, Tokyo, Japan), urea (Wako Pure Chemicals, Osaka, Japan) and Cu_2O (Wako Pure Chemicals) were used for the study without purification. $PTr–TiO_2$ samples were prepared by the method based on the processes reported elsewhere [26,29]. Typically, 400 g of P25 powder and different amounts of urea corresponding to 0.5:1, 1:1, 2:1 and 3:1 (w/w) ratios were ground in an agate mortar, followed by calcination in air at 400 °C for 30 min. Powders were placed in the open test-tube with 15 cm of length. The resulting powders were washed five times with water to remove the excess of urea decomposition products and finally dried under air at 70 °C. In the second step, $PTr–TiO_2$ sample and Cu_2O powder (in different amounts: 1, 2, 5, 7, 10 wt %) were mixed thoroughly with 5 min of grinding.

3.2. Sample Characterization

The UV–Vis diffuse reflectance spectra (DRS) were recorded on JASCO V-670 (JASCO, Tokyo, Japan) equipped with PIN-757 integrating sphere using $BaSO_4$ as a reference. Gas-adsorption measurements of prepared titania samples were performed on a Yuasa Ionics Autosorb, 6AG (Yuasa Ionics, Osaka, Japan) surface area and pore size analyzer. Specific surface area (SSA) was calculated from nitrogen adsorption at 77 K using the Brunauer–Emmett–Teller equation. X-ray diffraction patterns (XRD) were collected using an X-ray diffractometer (Rigaku intelligent XRD SmartLab with a Cu target, Rigaku, Tokyo, Japan). X-ray photoelectron spectra (XPS) were recorded using a JEOL JPC-9010MC (JEOL, Tokyo, Japan) spectrometer with a MgKa X-ray source. Samples were also characterized by scanning transmission electron microscopy (STEM, HITACHI HD-2000, HITACHI, Tokyo, Japan).

3.3. Photocatalytic Reaction

The photocatalyst (50 mg) was suspended in an aqueous solution of 2-propanol (5 vol %, 5 mL) and photoirradiated (120 W-xenon lamp) with a Y48 cut-off filter mounted in the irradiation window, therefore the light of wavelengths >450 nm reached the suspension, which was under continuous magnetic stirring (1000 rpm) in a thermostated bath. Generated acetone was detected using GC–FID (Shimadzu GC-14B (Shimadzu, Kyoto, Japan) equipped with a flame ionization detector). Before the injection of the liquid sample to GC, the photocatalyst was separated using a filter (Whatman Mini-UniPrep, PVDF, Whatman, Maidstone, UK).

4. Conclusions

The results presented in this study clearly revealed that the application of both low cost- modifiers for titania: urea (first step of preparation) and Cu_2O (second step), significantly enhanced the visible light photocatalytic properties of TiO_2 in comparison to the single materials of urea-modified TiO_2 and Cu_2O-modified titania. Therefore, it is possible to describe this phenomenon as a synergistic effect of urea (more detailed: the presence of polytriazine layer on titania originated from the thermal treatment of urea) and Cu_2O. The type II heterojunction or Z-scheme systems formed by two semiconductors, p-type Cu_2O and n-type $PTr–TiO_2$, can be responsible for the improvement of photocatalytic activity through the intensification of visible light-induced charge separation and subsequent reduction of the electron-hole recombination effect. A preparation of composite photocatalysts based on titania and low-cost materials as modifiers, such as a precursor of the organic sensitizer of titania and metal oxide, to prepare the efficient and stable photocatalytic system operating in the visible light within heterojunction principles is a very promising direction for a wider practical application of photocatalysis given the preference for concomitantly cheap and efficient solutions.

Author Contributions: M.J. conceived, designed, performed the experiments and characterizations, interpreted the data, and wrote the paper. E.K. performed STEM, interpreted the data, and corrected the manuscript. K.W. performed STEM experiments. All authors read and approved the final manuscript.

Acknowledgments: M.J. acknowledges Hokkaido University for guest lecturer position (2016–2017).

Conflicts of Interest: The authors declare no conflict of interest.

References

1. Bahnemann, D.W. Photocatalytic water treatment: Solar energy applications. *Solar Energy* **2004**, *77*, 445–459. [CrossRef]

2. Chen, X.B.; Shen, S.H.; Guo, L.J., Mao, S.S. Semiconductor-based photocatalytic hydrogen generation. *Chem. Rev.* **2010**, *110*, 6503–6570. [CrossRef] [PubMed]

3. Li, K.; Peng, B.S.; Peng, T.Y. Recent advances in heterogeneous photocatalytic CO_2 conversion to solar fuels. *ACS Catal.* **2016**, *6*, 7485–7527. [CrossRef]

4. Pietron, J.J.; DeSario, P.A. Review of roles for photonic crystals in solar fuels photocatalysis. *J. Photonics Energy* **2016**, *7*, 012007. [CrossRef]

5. Ragesh, P.; Ganesh, V.A.; Nair, S.V.; Nair, A.S. A review on 'self-cleaning and multifunctional materials'. *J. Mater. Chem. A* **2014**, *2*, 14773–14797. [CrossRef]

6. Kisch, H. Semiconductor photocatalysis—Mechanistic and synthetic aspects. *Angew. Chem. Int. Ed.* **2013**, *52*, 812–847. [CrossRef] [PubMed]

7. Schneider, J.; Matsuoka, M.; Takeuchi, M.; Zhang, J.; Horiuchi, Y.; Anpo, M.; Bahnemann, D.W. Understanding TiO_2 photocatalysis: Mechanisms and materials. *Chem. Rev.* **2014**, *114*, 9919–9986. [CrossRef] [PubMed]

8. Pelaez, M.; Nolan, N.T.; Pillai, S.C.; Seery, M.K.; Falaras, P.; Kontos, A.G.; Dunlop, P.S.M.; Hamilton, J.W.J.; Byrne, J.A.; O'Shea, K.; et al. A review on the visible light active titanium dioxide photocatalysts for environmental applications. *Appl. Catal. B Environ.* **2012**, *125*, 331–349. [CrossRef]

9. Dahl, M.; Liu, Y.; Yin, Y. Composite titanium dioxide nanomaterials. *Chem. Rev.* **2014**, *114*, 9853–9889. [CrossRef] [PubMed]

10. Bhanushali, S.; Ghosh, P.; Ganesh, A.; Cheng, W.L. 1D copper nanostructures: Progress, challenges and opportunities. *Small* **2015**, *11*, 1232–1252. [CrossRef] [PubMed]

11. Clarizia, L.; Spasiano, D.; Di Somma, I.; Marotta, R.; Andreozzi, R.; Dionysiou, D.D. Copper modified-TiO_2 catalysts for hydrogen generation through photoreforming of organics. A short review. *Int. J. Hydrogen Energy* **2014**, *39*, 16812–16831. [CrossRef]

12. Janczarek, M.; Kowalska, E. On the origin of enhanced photocatalytic activity of copper-modified titania in the oxidative reaction systems. *Catalysts* **2017**, *7*, 317. [CrossRef]

13. Nguyen, M.A.; Bedford, N.M.; Ren, Y.; Zahran, E.M.; Goodin, R.C.; Chagani, F.F.; Bachas, L.G.; Knecht, M.R. Direct Synthetic Control over the Size, Composition, and Photocatalytic Activity of Octahedral Copper Oxide Materials: Correlation Between Surface Structure and Catalytic Functionality. *ACS Appl. Mater. Interfaces* **2015**, *7*, 13238–13250. [CrossRef] [PubMed]

14. Deng, X.L.; Wang, C.G.; Shao, M.H.; Xu, X.J.; Huang, J.Z. Low-temperature solution synthesis of CuO/Cu_2O nanostructures for enhanced photocatalytic activity with added H_2O_2: Synergistic effect and mechanism insight. *RSC Adv.* **2017**, *7*, 4329–4338. [CrossRef]

15. Yang, L.; Luo, S.; Li, Y.; Xiao, Y.; Kang, Q.; Cai, Q. High efficient photocatalytic degradation of p-nitrophenol on a unique Cu_2O/TiO_2 p-n heterojunction network catalyst. *Environ. Sci. Technol.* **2010**, *44*, 7641–7646. [CrossRef] [PubMed]

16. Chu, S.; Zheng, X.M.; Kong, F.; Wu, G.H.; Luo, L.L.; Guo, Y.; Liu, H.L.; Wang, Y.; Yu, H.X.; Zou, Z.G. Architecture of $Cu_2O@TiO_2$ core-shell heterojunction and photodegradation for 4-nitrophenol under simulated sunlight irradiation. *Mater. Chem. Phys.* **2011**, *129*, 1184–1188. [CrossRef]

17. Liu, L.; Gu, X.; Sun, C.; Li, H.; Deng, Y.; Gao, F.; Dong, L. In situ loading of ultra-small Cu_2O particles on TiO_2 nanosheets to enhance the visible-light photoactivity. *Nanoscale* **2012**, *4*, 6351–6359. [CrossRef] [PubMed]

18. Liu, L.M.; Yang, W.Y.; Li, Q.; Gao, S.A.; Shang, J.K. Synthesis of Cu_2O nanospheres decorated with TiO_2 nanoislands, their enhanced photoactivity and stability under visible light illumination, and their post-illumination catalytic memory. *ACS Appl. Mater. Interfaces* **2014**, *6*, 5629–5639. [CrossRef] [PubMed]

19. Xiong, L.B.; Yang, F.; Yan, L.L.; Yan, N.N.; Yang, X.; Qiu, M.Q.; Yu, Y. Bifunctional photocatalysis of TiO_2/Cu_2O composite under visible light: Ti^{3+} in organic pollutant degradation and water splitting. *J. Phys. Chem. Solids.* **2011**, *72*, 1104–1109. [CrossRef]

20. Qiu, X.; Miyauchi, M.; Sunada, K.; Minoshima, M.; Liu, M.; Lu, Y.; Li, D.; Shimodaira, Y.; Hosogi, Y.; Kuroda, Y.; et al. Hybrid Cu_xO/TiO_2 nanocomposites as risk-reduction materials in indoor environments. *ACS Nano* **2012**, *6*, 1609–1618. [CrossRef] [PubMed]

21. Hans, M.; Erbe, A.; Mathews, S.; Chen, Y.; Solioz, M.; Mucklich, F. Role of copper oxides in contact killing of bacteria. *Langmuir* **2013**, *29*, 16160–16166. [CrossRef] [PubMed]

22. Duan, W.; Zheng, M.; Li, R.; Wang, Y. Morphology transformation of Cu_2O by adding TEOA and their antibacterial activity. *J. Nanopart. Res.* **2016**, *18*, 342. [CrossRef]

23. Lee, Y.J.; Kim, S.; Park, S.H.; Park, H.; Huh, Y.D. Morphology-dependent antibacterial activities of Cu_2O. *Mater. Lett.* **2011**, *65*, 818–820. [CrossRef]

24. Nosaka, Y.; Matsushita, M.; Nishino, J.; Nosaka, A.Y. Nitrogen-doped titanium dioxide photocatalysts for visible response prepared by using organic compounds. *Sci. Technol. Adv. Mater.* **2005**, *6*, 143–148. [CrossRef]

25. Bacsa, R.; Kiwi, J.; Ohno, T.; Albers, P.; Nadtochenko, V. Preparation, testing and characterization of doped TiO_2 active in the peroxidation of biomolecules under visible light. *J. Phys. Chem. B* **2005**, *109*, 5994–6003. [CrossRef] [PubMed]

26. Kisch, H.; Sakthivel, S.; Janczarek, M.; Mitoraj, D. A low-band gap, nitrogen-modified titania visible-light photocatalyst. *J. Phys. Chem. C* **2007**, *111*, 11445–11449. [CrossRef]

27. Beranek, R.; Neumann, B.; Sakthivel, S.; Janczarek, M.; Dittrich, T.; Tributsch, H.; Kisch, H. Exploring the electronic structure of nitrogen-modified TiO_2 photocatalysts through photocurrent and surface photovoltage studies. *Chem. Phys.* **2007**, *339*, 11–19. [CrossRef]

28. Mitoraj, D.; Kisch, H. The nature of nitrogen-modified titanium dioxide photocatalysts active in visible light. *Angew. Chem. Int. Ed.* **2008**, *47*, 9975–9978. [CrossRef] [PubMed]

29. Mitoraj, D.; Kisch, H. On the mechanism of urea-Induced titania modification. *Chem. Eur. J.* **2010**, *16*, 261–269. [CrossRef] [PubMed]

30. Mitoraj, D.; Kisch, H. Surface modified titania visible light photocatalyst powders. *Solid State Phenom.* **2010**, *162*, 49–75. [CrossRef]

31. Dong, G.; Zhang, Y.; Pan, Q.; Qiu, J. A fantastic graphitic carbon nitride (g-C_3N_4) material: Electronic structure, photocatalytic and photoelectronic properties. *J. Photochem. Photobiol. C Photochem. Rev.* **2014**, *20*, 33–50. [CrossRef]

32. Liu, J.; Zhang, T.; Wang, Z.; Dawson, G.; Chen, W. Simple pyrolysis of urea into graphitic carbon nitride with recyclable adsorption and photocatalytic activity. *J. Mater. Chem.* **2011**, *21*, 14398–14401. [CrossRef]

33. Lee, S.C.; Lintang, H.O.; Yuliati, L. A urea precursor to synthesize carbon nitride with mesoporosity for enhanced activity in the photocatalytic removal of phenol. *Chem. Asian J.* **2012**, *7*, 2139–2144. [CrossRef] [PubMed]
34. Senthil, R.A.; Theerthagiri, J.; Selvi, A.; Madhavan, J. Synthesis and characterization of low-cost g-C$_3$N$_4$/TiO$_2$ composite with enhanced photocatalytic performance under visible-light rradiation. *Opt. Mater.* **2017**, *64*, 533–539. [CrossRef]
35. Peng, B.; Zhang, S.; Yang, S.; Wang, H.; Yu, H.; Zhang, S.; Peng, F. Synthesis and characterization of g-C$_3$N4/Cu$_2$O composite catalyst with enhanced photocatalytic activity under visible light irradiation. *Mater. Res. Bull.* **2014**, *56*, 19–24. [CrossRef]
36. Min, Z.; Wang, X.; Li, Y.; Jiang, J.; Li, J.; Qian, D.; Li, J. A highly efficient visible-light-responding Cu$_2$O-TiO$_2$/g-C$_3$N$_4$ photocatalyst for instantaneous discolorations of organic dyes. *Mater. Lett.* **2017**, *193*, 18–21. [CrossRef]
37. Kisch, H.; Macyk, W. Visible-light photocatalysis by modified titania. *Chem. Phys. Chem.* **2002**, *3*, 399–400. [CrossRef]
38. Yan, X.; Ohno, T.; Nishijima, K.; Abe, R.; Ohtani, B. Is methylene blue an appropriate substrate for a photocatalytic activity test? A study with visible-light responsive titania. *Chem. Phys. Lett.* **2006**, *429*, 606–610. [CrossRef]
39. Wang, K.; Wei, Z.; Ohtani, B.; Kowalska, E. Interparticle electron transfer in methanol dehydrogenation on platinum-loaded titania particles prepared from P25. *Catal. Today* **2018**, *303*, 327–333. [CrossRef]
40. Janczarek, M.; Kowalska, E.; Ohtani, B. Decahedral-shaped anatase titania photocatalyst particles: Synthesis in a newly developed coaxial-flow gas-phase reactor. *Chem. Eng. J.* **2016**, *289*, 502–512. [CrossRef]
41. Dementjev, A.P.; de Graaf, A.; van den Sanden, M.C.M.; Maslakov, K.I.; Naumkin, A.V.; Serov, A.A. X-Ray photoelectron spectroscopy reference data for identification of the C$_3$N$_4$ phase in carbon–nitrogen films. *Diam. Relat. Mater.* **2000**, *9*, 1904–1907. [CrossRef]
42. Guo, X.; Xie, Y.; Wang, X.; Zhang, S.; Hou, T.; Lv, S. Synthesis of carbon nitride nanotubes with the C$_3$N$_4$ stoichiometry via a benzene-thermal process at low temperatures. *Chem. Commun.* **2004**, 26–27. [CrossRef] [PubMed]
43. Janczarek, M.; Wei, Z.; Endo, M.; Ohtani, B.; Kowalska, E. Silver- and copper-modified decahedral anatase titania particles as visible light-responsive plasmonic photocatalyst. *J. Photonics Energy* **2017**, *7*, 012008. [CrossRef]
44. Bessekhouad, Y.; Robert, D.; Weber, J.-V. Photocatalytic activity of Cu$_2$O/TiO$_2$, Bi$_2$O$_3$/TiO$_2$ and ZnMn$_2$O$_4$/TiO$_2$ heterojunctions. *Catal. Today* **2005**, *101*, 315–321. [CrossRef]
45. Huang, L.; Peng, F.; Yu, H.; Wang, H. Preparation of cuprous oxides with different sizes and their behaviors of adsorption, visible-light driven photocatalysis and photocorrosion. *Solid State Sci.* **2009**, *11*, 129–138. [CrossRef]
46. Marschall, R. Semiconductor Composites: Strategies for Enhancing Charge Carrier Separation to Improve Photocatalytic Activity. *Adv. Funct. Mater.* **2014**, *24*, 2421–2440. [CrossRef]
47. Luna, A.L.; Valenzuela, M.A.; Colbeau-Justin, C.; Vazquez, P.; Rodriguez, J.; Avendano, J.R.; Alfaro, S.; Tirado, S.; Garduno, A.; De la Rosa, J.M. Photocatalytic degradation of gallic acid over CuO–TiO$_2$ composites under UV/Vis LEDs irradiation. *Appl. Catal. A Gen.* **2016**, *521*, 140–148. [CrossRef]
48. Li, H.; Tu, W.; Zhou, Y.; Zou, Z. Z-scheme photocatalytic systems for promoting photocatalytic performance: Recent progress and future challenges. *Adv. Sci.* **2016**, *3*, 1500389. [CrossRef] [PubMed]
49. Low, J.; Jiang, C.; Cheng, B.; Wageh, S.; Al-Ghamdi, A.A.; Yu, J. A review of direct Z-scheme photocatalysts. *Small Methods* **2017**, *1*, 1700080. [CrossRef]

 *catalysts*

Article

Highly Active TiO₂ Microspheres Formation in the Presence of Ethylammonium Nitrate Ionic Liquid

Anna Gołąbiewska [1], Micaela Checa-Suárez [1,2], Marta Paszkiewicz-Gawron [1], Wojciech Lisowski [3], Edyta Raczuk [1], Tomasz Klimczuk [4], Żaneta Polkowska [5], Ewelina Grabowska [1], Adriana Zaleska-Medynska [1] and Justyna Łuczak [6,*]

[1] Faculty of Chemistry, University of Gdansk, 80-308 Gdansk, Poland; anna.golabiewska@ug.edu.pl (A.G.); micaela.checa@epn.edu.ec (M.C.-S.); marta.paszkiewicz@phdstud.ug.edu.pl (M.P.-G.); edyta.raczuk@gmail.com (E.R.); ewelina.grabowska@ug.edu.pl (E.G.); adriana.zaleska@ug.edu.pl (A.Z.-M.)
[2] Faculty of Civil and Environmental Engineering, Escuela Politécnica Nacional, Quito 170525, Ecuador
[3] Institute of Physical Chemistry, Polish Academy of Sciences, 01-224 Warsaw, Poland; wlisowski@ichf.edu.pl
[4] Faculty of Applied Physics and Mathematics, Gdansk University of Technology, 80-233 Gdansk, Poland; t.klimczuk@gmail.com
[5] Department of Analytical Chemistry, Faculty of Chemistry, Gdansk University of Technology, 80-233 Gdańsk, Poland; zanpolko@pg.gda.pl
[6] Department of Chemical Technology, Faculty of Chemistry, Gdansk University of Technology, 80-233 Gdansk, Poland
* Correspondence: justyna.luczak@pg.gda.pl; Tel.: +48-58-347-13-65

Received: 14 May 2018; Accepted: 13 June 2018; Published: 11 July 2018

Abstract: Spherical microparticles of TiO₂ were synthesized by the ionic liquid-assisted solvothermal method at different reaction times (3, 6, 12, and 24 h). The properties of the prepared photocatalysts were investigated by means of UV-VIS diffuse-reflectance spectroscopy (DRS), Brunauer–Emmett–Teller (BET) surface area measurements, scanning electron microscopy (SEM), X-ray diffraction analysis (XRD), and X-ray photoelectron spectroscopy (XPS). The results indicated that the efficiency of the phenol degradation was related to the time of the solvothermal synthesis, as determined for the TiO₂_EAN(1:1)_24h sample. The microparticles of TiO₂_EAN(1:1)_3h that formed during only 3 h of the synthesis time revealed a really high photoactivity under visible irradiation (75%). This value increased to 80% and 82% after 12 h and 24 h, respectively. The photoactivity increase was accompanied by the increase of the specific surface area, thus the poresize as well as the ability to absorb UV-VIS irradiation. The high efficiency of the phenol degradation of the ionic liquid (IL)–TiO₂ photocatalysts was ascribed to the interaction between the surface of the TiO₂ and ionic liquid components (carbon and nitrogen).

Keywords: ionic liquids; ionic liquid-assisted solvothermal reaction; reaction time; titanium dioxide; heterogeneous photocatalysis; visible light

---

1. Introduction

A number of recent studies have explored the applications of photosensitized titanium dioxide photocatalysts inter alia in solar cells, energy storage, hydrogen production via the water-splitting process, and the photocatalytic degradation of organic pollutants for water/air purification [1–5]. Expected low exploitation costs, prevalence in utility, and safety has created the motivation for intensified research in the field of solar to chemical energy conversion. The photoactive performance of the pristine TiO₂ nano- and micro-particles is limited by its 3.2 eV bandgap to the absorption of UV light. As the UV light range constitutes the total radiant energy in only as diminutive an amount as 5%, harvesting a greater range of the solar spectrum is considered vital for achieving the significant effectiveness of the photocatalysis [5,6]. The extension of the TiO_2 spectral response range to absorb the photons under visible (43%) or/and near infrared (49% of solar spectrum energy) irradiation is crucial for this purpose. The TiO_2 photocatalytic properties can be controlled and their optical response can be expanded to absorb photons under visible or/and near infrared light irradiation through the alteration of the TiO_2 bandgap via morphology engineering [7–10].

It was found that, after excitation, the electrons and holes propagate to the nano- and micro-particles surface where they react with electron acceptors and donors, respectively [6,9,11,12]. Additionally, it has been noted that a slower recombination rate and a larger surface area accounted for more active adsorption/desorption reactions and the surface transfer of photoexcited electrons [9,12,13], whereas the potential adverse effects that originate in the highly defective sites, typically developing with the growth of a large surface area, may be rectified by a higher crystallization of the particles [9].

Ionic liquids (ILs) have gained increasing attention in terms of their assistance in TiO_2 synthesis as solvents as well as spatial and, perhaps, band structuring agents. Their high viscosity, dielectric constant, and thus polarity and dispersal capacity are widely recognized as the properties responsible for the charge, steric, and viscous stabilization of small-sized slow-growing crystallites, as well as the hindrance of aggregation and agglomeration processes that are disadvantageous for photocatalytic [7,10,11,14–23].

The synthesized photocatalysts in the presence of ILs nano- and microparticles are characterized by a larger specific surface area, higher crystallization level, and less crystalline defects [17]. Hence, with the assistance of ILs, the formation of the particles of the beneficial surface reactivity is promoted, inducing a more effective photon absorption, trapping, and their migration to the surface. Moreover, an energetically simplified pathway of excited electrons through an ionic liquid's HOMO and LUMO orbitals [16], along with prolonged stabilized charge separation was revealed to result in the formation of a greater amount of reactive oxygen species (ROS) in the subsequent reaction of electrons and holes with oxygen and water, respectively. The generated reactive oxygen species ($\bullet$OH and $O_2\bullet^-$) are crucial reagents in the photodegradation of pollutants [16,17,24].

Notwithstanding, the direct relationship between the structure of the ILs and the size/morphology of the nano- and microparticles of the semiconductors, such as the TiO_2 photocatalysts, still remains ambiguous. Up until now, the following factors had been reported as predominant in effectuating the structure, and thus the activity of said particles, as follows: (1) IL anion type (the number of atoms it is composed o growth; apart from this, π–π stacking of imidazolium cations promotes the ILs role as templating agents); (3) cation–anion interaction energy where the frailer, the weaker the cation–anion interactions, the firmer the capping on growing the TiO_2 particles and the more efficient the inhibition/hampering of the unfavorable Ostwald ripening process; and (4) the type of overall interactions (π–π, van der Waals, coulomb and electrostatic forces, and hydrogen bonding) [15,16,21,25].

Furthermore, the proposed *fons et orgio* of the influences on the TiO_2 photoactivity are tenable, as follows: (1) doping with N, C, and F elements after ILs thermal decomposition; (2) directly sensitizing the TiO_2 particles; (3) affecting the transfer of photo-generated charges through the bulk of particles; and (4) favoring oxygen vacancies and Ti^{3+} species formation during synthesis [26–28]. However, an up to date kinetics of the formation of the TiO_2 particles during ionic liquid-assisted synthesis has not been presented and discussed.

In this regard, in this study, the TiO_2 particles were synthesized solvothermally, with the assistance of the selected ionic liquid, ethylammonium nitrate [EAN][NO_3], which is one of the earliest reported in the literature of protic ionic liquid [29,30].

Apart from prevailing in studies on the topics of the synthesis process and characterization of the IL-assisted TiO_2 microparticles [22,30–32], we focused on evaluating the functional properties of the obtained micro-particles, namely, their photoactivity. Alongside this, we strived to infer the mechanisms of the ILs assistance through exploring the influence of selected ILs, illustrated with the example of ethylammonium nitrate [EAN][NO$_3$] [29].

We placed an emphasis on the essentiality of (1) researching VIS-light induced photoactivity in comparison to the already researched UV-induced photoactivity [19,25,33–35]; and (2) conducting research in the presence of a model pollutant (2a), neutral in terms of the photosensitization of TiO$_2$, (2b) proposing a simple mechanism of degradation and mineralization, and (2c) with a low photoabsorption coefficient [27,36,37]. In contrast to the common choice of organic dyes (methylene blue, methyl orange, and rhodamine B) [7,10,13,18,19,25,38,39], we applied phenol. Another imperative was that we excluded the minor ones in facilitating the photocatalysis reaction active species, while exposing the substantial ones through the active species scavenger tests.

In this article, we present the results of the previously non-analyzed and non-reviewed comprehensively [EAN][NO$_3$]-assisted TiO$_2$ microparticles, where the photocatalytic effect on the degradation and mineralization of phenol in aqueous solution vastly exceeded our expectations, reaching as high as an 82% degradation rate in our tests, which are described below. Moreover, for the first time, we have examined the kinetics of the TiO$_2$ microsphere formation in the presence of ethylammonium nitrite ionic liquid.

2. Results

First of all, a set of samples with selected IL to titanium (IV) butoxide (TBOT) molar ratios were synthesized and characterized, taking into account the surface area and photoactivity. The sample labeling and the amount of ILs to the precursor used during the preparation procedure, as well as the specific surface area, pore volume of the obtained photocatalysts, and their photocatalytic activity under VIS irradiation are given in Table 1. On the basis of the photocatalytic effect, we chose the sample with the highest activity in order to examine the kinetics of the TiO$_2$ microsphere formation in the presence of ethylammonium nitrite ionic liquid, which is shown in Figure 1. As presented in Table 1, the effect of the solvothermal synthesis duration (3, 6, 12, and 24 h) on the surface properties as well as the photoactivity for the IL:TBOT ratio equaled to 1:1 were also investigated.

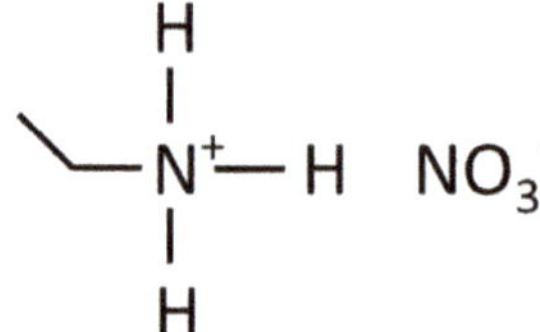

Figure 1. The structure of ethylammonium nitrite [EAN][NO$_3$] ionic liquid (IL).

2.1. The BET Surface Area and SEM Analysis

The results listed in Table 1 revealed the influence of the ionic liquid content on the specific surface area and the pore volume of the synthesized samples. All of the samples presented a higher specific surface area in comparison with the TiO$_2$ synthesized without IL (184 m^2·g^{-1}), and also than that reported for the commercially available P25 (50 m^2·g^{-1}) [10,40]. The values of the specific surface area for TiO$_2$ prepared in the presence of [EAN][NO$_3$] ranged from 190 m^2·g^{-1} (sample prepared with the lowest IL:TBOT molar ratio, 1:10) to 233 m^2·g^{-1} for the sample obtained with IL:TBOT, with a molar ratio of 1:2. In this regard, the direct relation between the amount of the IL in the reaction mixture and the specific surface area for these photocatalysts was detected. However, a further increase in the IL content taken for the synthesis (up to IL:TBOT molar ratio of 1:1) resulted in a decrease of the pore volume, thus, the specific surface area was due, most probably, to the overloading of the TiO$_2$ surface with organic salt. This might indicate that the ILs work like a designer agent of the microstructures' physical and structural properties.

The scanning electron microscopy images of the pure TiO_2 and IL-assisted TiO_2 particles obtained for various molar ratios of IL:TBOT in the presence of [EAN][NO$_3$] are presented in Figure 2. The pristine TiO_2 exhibited a smooth surface with an average size from 0.5–4 µm. In the case of the IL–TiO_2 samples, it showed that the sample with the molar ratio 1:1 and the sample with a low amount of IL (1:8) presented a spherical structure and did not change in relation to the pristine TiO_2.

Table 1. Characteristics of the TiO_2 particles obtained by ethylammonium nitrite [EAN][NO$_3$] assisted solvothermal synthesis. IL—ionic liquid; TBOT—titanium (IV) butoxide.

Sample	Time of the Synthesis	Molar Ratio (IL:TBOT)	Specific Surface Area ($m^2 \cdot g^{-1}$)	Pore Volume ($cm^3 \cdot g^{-1}$)	Phenol Degradation Efficiency under 60 min of VIS Irradiation (%)	Rate of Phenol Degradation under Visible Light ($\lambda > 420$) [$\mu mol \cdot dm^{-3} \cdot min^{-1}$]
TiO$_2$_pristine	24	-	184	0.069	7	0.18
TiO$_2$_EAN(1:10)_24h	24	1:10	190	0.093	28	1.01
TiO$_2$_EAN(1:8)_24h	24	1:8	211	0.102	19	0.55
TiO$_2$_EAN(1:5)_24h	24	1:5	216	0.105	33	1.11
TiO$_2$_EAN(1:3)_24h	24	1:3	216	0.105	36	1.17
TiO$_2$_EAN(1:2)_24h	24	1:2	233	0.113	41	1.32
TiO$_2$_EAN(1:1)_24h	24	1:1	221	0.108	82	3.12
TiO$_2$_EAN(1:1)_3h	3	1:1	239	0.12	75	2.28
TiO$_2$_EAN(1:1)_6h	6	1:1	207	0.101	75	2.38
TiO$_2$_EAN(1:1)_12h	12	1:1	209	0.102	80	2.53

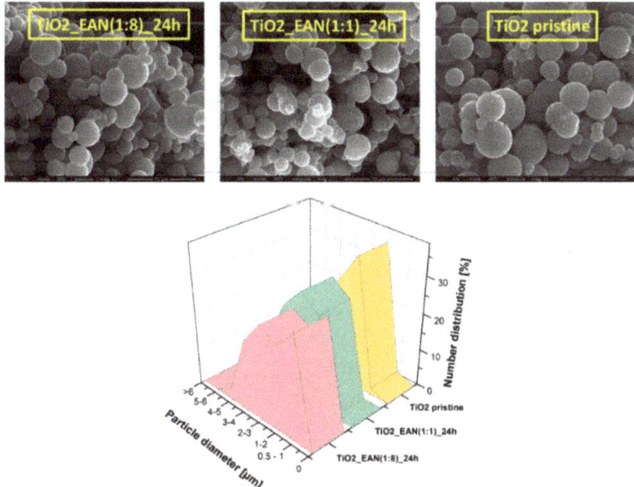

Figure 2. SEM images of TiO_2 obtained from IL-assisted solvothermal synthesis, TiO$_2$_EAN(1:8)_24h, TiO$_2$_EAN(1:1)_24h, and reference TiO_2.

The experiments of the TiO$_2$_EAN(1:1)_24h sample preparation performed at different time regimes revealed that 3 h of solvothermal synthesis was enough to obtain the TiO_2 microspheres (see SEM images, presented in Figure 3 below). However, the reaction yield was relatively low (only 19%) and increased with the increasing reaction time (38% for 6 h, 65% for 12 h, and 93% for 24 h). Moreover, the elongation of the synthesis time resulted, at first, in the decrease of the specific surface area from $239 \ m^2 \cdot g^{-1}$ (after 3 h) to $207 \ m^2 \cdot g^{-1}$ (after 6 h), and then in the enlargement of the specific surface area to $221 \ m^2 \cdot g^{-1}$ (after 24 h).

An explanation of this observation may be found in the SEM images of the microparticles prepared with the same amount of substrates (molar ratio of TBOT to ILs equaling 1:1), however, with different times of thermal treatment (3, 6, 12, and 24 h) presented in Figure 3. In all of the introduced variations, the majority of the particles lay within the scope of 1–3 μm for 3 h (48%), 2–3 μm for 6 (24%) and 24 h (28%), and 3–4 μm for 12 h (19%). The percentage contribution of the particles with a diameter above 4 μm was rarely reached and never exceeded 15% for all of the obtained samples. In the image sequences presented below, we noted that the sample subjected to thermal treatment for 3 h was mainly composed of a large number of small spherical particles with a range of 1–3 μm. Nonetheless, between the particles synthesized in 6, 12, and 24 h regimes, less of a difference was recognized. In this regard, the high surface area of the particles prepared within 3 h may be related to a higher contribution of the smallest particles.

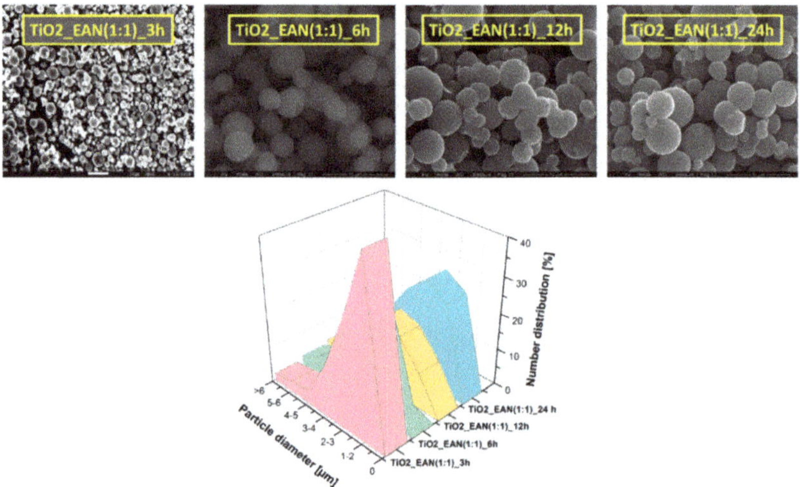

Figure 3. SEM images of TiO$_2$_EAN(1:1) obtained in different synthesis times of 3, 6, 12, and 24 h.

2.2. XRD Analysis

The PXRD patterns for the series of TiO$_2$_EAN are presented in Figures 4 and 5. All of the patterns looked similar as the samples only contained the anatase (TiO$_2$) phase. The open circles represent the experimental data points, a solid red line is a profile fitting (LeBail method), and the vertical bars mark the positions of the expected Bragg reflections for the used model (I 41/a m d, s.g. #141). The LeBail fit given lattice parameters for TiO$_2$ are gathered in Table 2. The lattice parameters were similar and are close to those reported by Djerdj and Tonejc [41]. The PXRD reflections were broad, which indicated a small crystallite size estimated to be between 5.0 and 6.5 nm. However, no correlation was observed between the amount of ionic liquid taken for the synthesis nor for the preparation time and the crystallite size.

Table 2. Lattice parameters and average crystallite size of the TiO$_2$_EAN photocatalysts.

Sample	a (Å)	c (Å)	d (Å)
TiO$_2$_EAN(1:1)_3h	3.8051(3)	9.554(2)	65
TiO$_2$_EAN(1:1)_6h	3.7907(7)	9.504(3)	50
TiO$_2$_EAN(1:1)_12h	3.7892(6)	9.507(3)	55
TiO$_2$_EAN(1:1)_24h	3.7890(6)	9.502(3)	60
TiO$_2$_EAN(1:2)_24h	3.7899(7)	9.499(3)	55
TiO$_2$_EAN(1:3)_24h	3.7926(8)	9.506(4)	60
TiO$_2$_EAN(1:5)_24h	3.7814(11)	9.472(6)	65
TiO$_2$_EAN(1:8)_24h	3.7953(9)	9.483(5)	55
TiO$_2$_EAN(1:10)_24h	3.7942(12)	9.476(7)	55

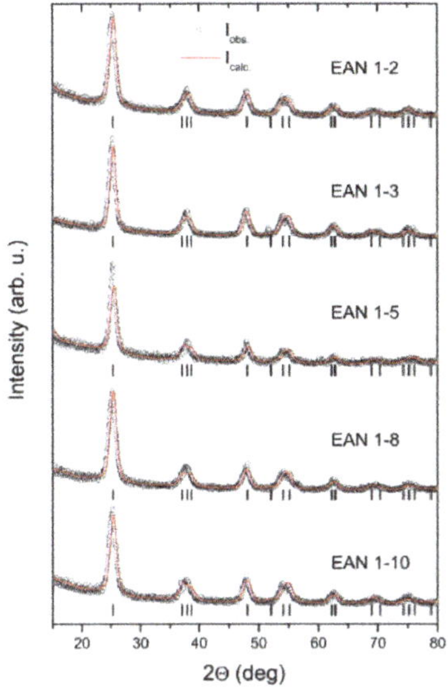

Figure 4. XRD pattern of TiO$_2$_EAN prepared with different IL–titanium (IV) butoxide (TBOT) molar ratios. A solid line is a profile fit to the experimental data (open circles). The Bragg reflections are marked by vertical bars.

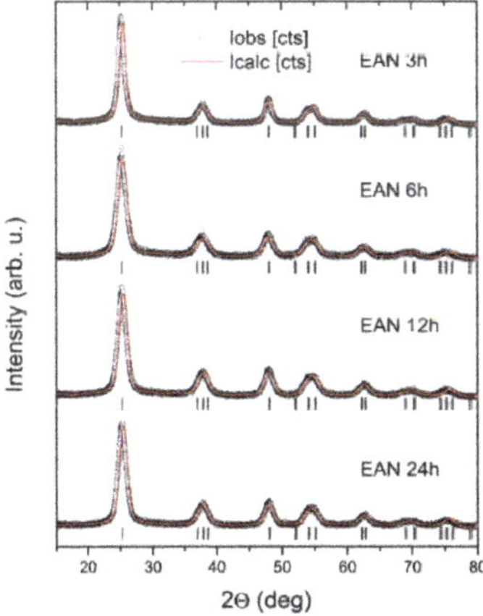

Figure 5. XRD pattern of the TiO$_2$_EAN prepared with the IL–TBOT molar ratio of 1:1, synthesized at different synthesis times. A solid line is a profile fit to the experimental data (open circles). The Bragg reflections are marked by vertical bars.

2.3. X-ray Photoelectron Spectroscopy (XPS) Analysis

The elemental surface composition of the selected IL–TiO$_2$ specimens, evaluated by XPS, is shown in Table 3. Titanium, oxygen, carbon, and nitrogen were detected and the corresponding high-resolution (HR) XPS spectra of Ti 2p, O 1s, C 1s, and N 1s are presented in Figure 6. The chemical character of the elements is identified in the deconvoluted spectra in Figure 6 and Table 3. The Ti 2p, O 1s, and C 1s spectra exhibited features that were characteristic of the IL–TiO$_2$ specimens [16,42]. The N 1s signal at 400 eV is commonly interpreted as the surface C–N bond. Only for the TiO$_2$_EAN(1:1) sample thermally treated for 24 h was the additional signal that appeared at about 402.8 eV observed, which may be assigned to the oxidized nitrogen surface species.

The XPS data collected in Table 3 revealed that the surface chemical composition of the TiO$_2$_EAN(1:1)_24h samples were different than the sample of the lower amount of TiO$_2$_EAN(1:8)_24h. For the last one, the contribution of the Ti$^{(3+)}$ fraction was about 30% smaller and the nitrogen amount was evidently lower than for the TiO$_2$_EAN(1:1)_24h sample. These observations confirmed a higher amount of IL on the TiO$_2$_EAN(1:1)_24h sample surface (an analogous relation was also observed for the carbon atom). In the series of TiO$_2$_EAN(1:1) samples, differing by the time of thermal treatment, we noted a systematic decrease in the oxygen of the Ti–O$_{surf}$ fraction, with an increasing time of the thermal treatment from 6 to 24 h, and a significantly smaller surface concentration of nitrogen for the 24 h synthesized sample (Table 3). Moreover, for the last sample, the oxidized form of the nitrogen species appeared in addition to the main nitrogen N–C surface fraction (Figure 6). These observations indicated the surface transformation of the IL-assisted TiO$_2$ with the prolonged time of thermal treatment.

To elucidate the effect of the phenol-degradation processing on the chemical composition of the TiO$_2$_EAN photocatalyst, we analyzed both the sample after three cycles of photocatalytic processing and the same sample washed with deionized water. The results are compared in Figure 6 and Table 3. One can see that the nitrogen atom concentration was similar for both of the samples. However, the water washed sample exhibited a significantly larger surface amount of titanium faction Ti$^{(3+)}$ and a relatively higher contribution of $-$OH surface species (see O 1s and C 1s fractions in Table 3).

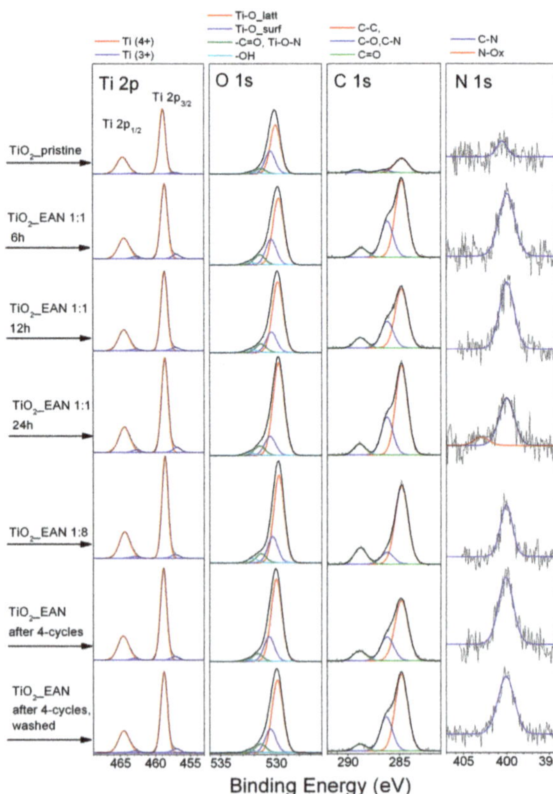

Figure 6. High resolution X-ray photoelectron spectroscopy (XPS) spectra of elements detected in the surface layer of the [EAN][NO$_3$]-modified TiO$_2$ particles.

Table 3. Elemental composition (in at. %) and chemical character of titanium, oxygen, carbon, and nitrogen states in the surface layer of [EAN][NO$_3$]-modified TiO$_2$ particles, evaluated by X-ray photoelectron spectroscopy (XPS) analysis.

Sample Label	Σ Ti (at. %)	Ti 2p$_{3/2}$ Fraction (%)		Σ O (at. %)	O 1s Fraction (%)				Σ C (at. %)	C 1s Fraction (%)			Σ N (at. %)	N 1s Fraction (%)	
		Ti$^{(4+)}$ 458.7 $\pm$ 0.1 eV	Ti$^{(3+)}$ 457.0 $\pm$ 0.1 eV		Ti-O$_{latt}$ 529.9 $\pm$ 0.1 eV	Ti-O$_{sur}$ 530.5 $\pm$ 0.1 eV	–Ti–O–N, –C=C 531.5 $\pm$ 0.1 eV	–OH 532.4 $\pm$ 0.1 eV		"A" C–C 284.8 eV	"B" C–OH, C–N 286.2 $\pm$ 0.1 eV	"C" –C=O, 288.9 $\pm$ 0.1 eV		N–C 399.9 $\pm$ 0.1 eV	N–O$_x$ 402.8 $\pm$ 0.1 eV
TiO$_2$_pristine	29.44	97.59	2.41	66.27	62.35	28.60	6.59	2.46	4.15	72.27	13.46	14.27	0.14	100	0
TiO$_2$_EAN 1:1_6h	23.15	93.37	6.63	61.97	63.55	24.13	9.54	2.78	14.35	63.03	29.19	7.78	0.53	100	0
TiO$_2$_EAN 1:1_12h	24.00	94.06	5.94	63.51	68.33	19.89	8.50	3.28	11.94	61.53	27.49	10.99	0.54	100	0
TiO$_2$_EAN 1:1_24h	23.37	93.64	6.36	63.00	74.56	15.33	7.72	2.39	13.24	64.56	27.42	8.02	0.38	84.89	5.11
TiO$_2$_EAN 1:8_24h	24.32	95.44	4.56	63.80	68.52	20.58	7.25	3.65	11.61	73.05	11.18	15.77	0.29	100	0
TiO$_2$_EAN_4-cycles	24.61	94.47	4.78	63.77	70.46	21.06	6.75	1.73	11.14	64.12	25.22	10.66	0.49	100	0
TiO$_2$_EAN_4-cycles (washed)	23.53	93.10	6.13	62.35	67.42	21.84	8.15	2.59	13.64	65.63	28.52	5.84	0.48	100	0

2.4. UV-VIS Spectrum

The UV-VIS adsorption spectra of the TiO_2 synthesized with various molar ratios of [EAN][NO$_3$] to TBOT are presented in Figure 7. Pristine TiO_2 was only photoactive under the UV region ($\lambda < 400$ nm). The addition of the ionic liquid to the TiO_2 synthesis environment increased the absorption range of IL–TiO_2, being noticeably photoactive at above 420 nm. The absorption properties of the samples prepared with IL were superior in comparison with the pristine TiO_2 in the visible light range, whereas all of the samples showed a similar UV absorption. Generally, a higher IL amount used for synthesis resulted in the enhancement of the VIS light absorption by the IL–TiO_2 photocatalysts.

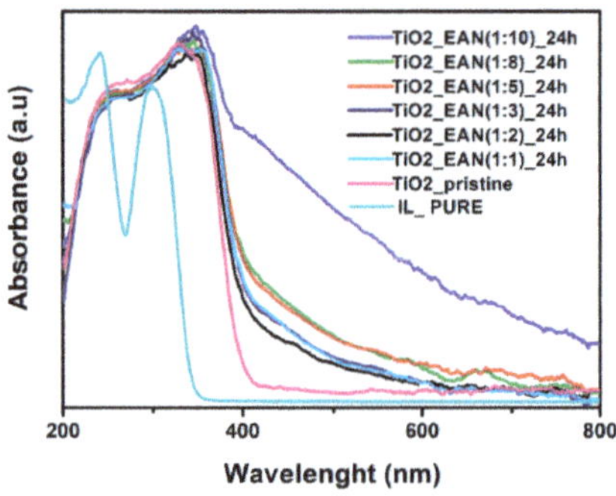

Figure 7. UV-VIS adsorption spectra of TiO_2 synthesized using various molar ratios of [EAN][NO$_3$] to TBOT.

Interesting results were also obtained for the experiments where the influence of the reaction time was taken into account (Figure 8). When increasing the reaction time, the enhancement of the visible light absorption by IL–TiO_2 was observed. Thereby, the sample prepared during 24 h was characterized by a significantly broader absorption as well as the highest red shift of the absorption edge; thus, there was a higher effectiveness in creating the electron–hole pairs.

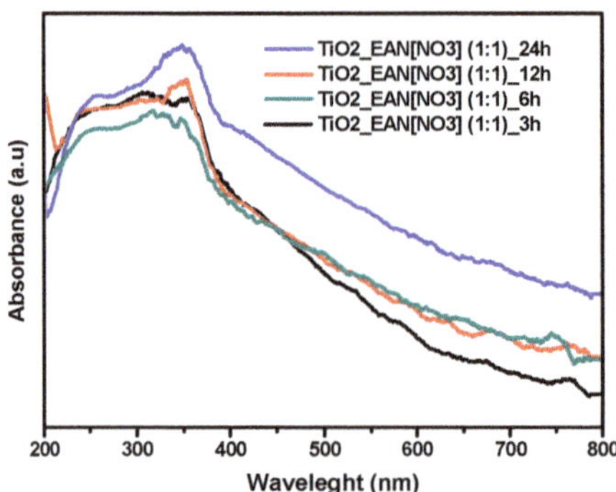

Figure 8. UV-VIS adsorption spectra of TiO_2_EAN(1:1) photocatalyst prepared in 3, 6, 12, and 24 h.

According to the literature, a significant increase in the absorption of visible light by TiO_2 can be related with the presence of carbon atoms in the sample [43–47]. As the carbon species exist mainly on the TiO_2 surface rather than occupying the lattice of TiO_2, the visible light response of the samples was attributed to the formation of the inter band C2p states [48]. However, our observations revealed that a correlation between the content of the IL used for the sample preparation and the apparent enhanced absorption of visible light was not observed. Additionally, based on the XPS analysis, a similar amount of at. % of carbon was observed for the samples containing small amounts of IL, as well as the sample with a higher IL content.

2.5. Photocatalytic Activity of IL–TiO₂ in Phenol Decomposition Model Reaction

The photocatalytic activity of the IL–TiO$_2$ samples was evaluated by the degradation of the phenol model compound under visible light ($\lambda > 420$ nm) irradiation. The obtained results are summarized in Table 1 and presented in Figure 9. As above-mentioned, before illumination, the solution was stirred for 30 min in the dark to establish a molecular adsorption equilibrium. Pristine TiO$_2$ synthesized by the solvothermal method, without addition of the ionic liquid, was used as the reference sample. For comparison purposes, the photocatalytic activity result of P25 TiO$_2$ under UV-VIS irradiation was also shown.

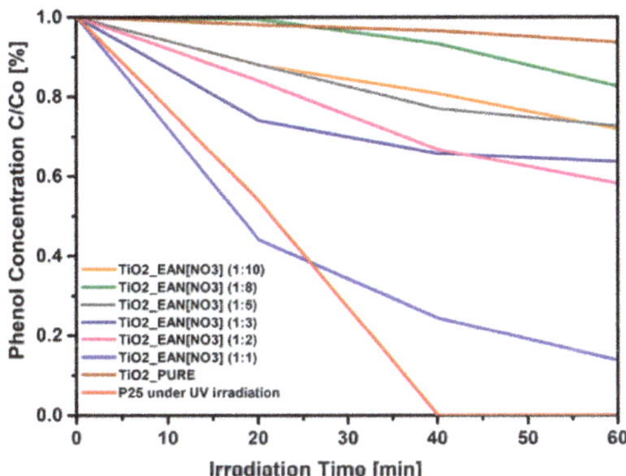

Figure 9. Efficiency of phenol degradation under visible light ($\lambda > 420$ nm) in the presence of TiO$_2$ prepared in [EAN][NO$_3$] and under UV-VIS light for P25.

It was found that the TiO$_2$_EAN microparticles exhibited a higher photoefficiency than the pristine TiO$_2$, which is also consistent with the higher BET specific surface areas and enhanced optical properties. After 60 min of the photocatalytic process, 7% of the phenol was degraded using unmodified TiO$_2$, however, the TiO$_2$_EAN efficiency was higher, and was strongly influenced by the amount of ILs that were used during the synthesis. For example, up to 82% of the phenol was degraded in the presence of photocatalyst TiO$_2$_EAN(1:1)_24h, where the molar ratio of [EAN][NO$_3$] to TBOT taken for synthesis was 1:1. This value was about 5.5 times higher when compared to the pristine TiO$_2$, indicating its excellent photocatalytic activity. This observation was correlated with the UV-VIS adsorption spectra, where the TiO$_2$_EAN(1:1)_24h sample showed the highest extension of the absorption edge to the visible light region. It was found that the samples prepared with IL:TBOT molar ratios of 1:10 and 1:8 revealed the lowest photoactivity among the photocatalysts that were obtained in the presence of ethylammonium nitrate IL. In this regard, the photodegradation efficiency increased with the increase in the quantity of ionic liquid taken to synthesis (thereby, the IL present at the TiO$_2$ surface).

In addition, the efficiency of the phenol total mineralization by the sample with the highest photoactivity was also determined. The total organic carbon measurements confirmed 20% of the total mineralization of the phenol after 60 min of irradiation in a presence of the TiO$_2$_[EAN](1:1)_24h sample that was performed under VIS irradiation and 48% under UV-VIS irradiation.

Moreover, the efficiency of the phenol degradation was also related with the time of the solvothermal synthesis, as determined for the TiO$_2$_EAN(1:1) sample. The microparticles of TiO$_2$_EAN(1:1)_3h formed during only 3 h of the synthesis time revealed a really high photoactivity under visible irradiation at 75%. This value increased to 80% and 82% after 12 h and 24 h, respectively. The photoactivity increase was accompanied by an increase in the specific surface area, thus the pore sizes as well as the ability to absorb UV-VIS irradiation. Additionally, based on the XPS measurements, it was concluded that the increase in the visible light absorption and the enhancement of the photocatalytic activity may be related to the highest quantity of the carbon and $Ti^{(+3)}$ defects at the TiO_2 surface.

To investigate the degradation/regeneration capacity and the structural stability during the entire process, we performed stability tests for the sample characterized by the best photocatalytic activity under VIS light irradiation. In the stability tests, the same sample was repeatedly used in the phenol photodegradation reaction three times. As shown in Figure 10, a significant drop in the phenol removal, from 84% to 33%, was found.

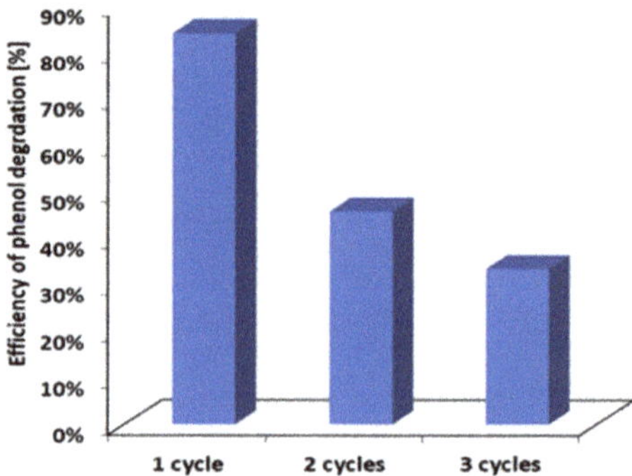

Figure 10. Phenol removals in the photodegradation using the TiO_2 microspheres in a cycled mode.

In order to further clarify the possible mechanism of phenol degradation, reactive species trapping tests were designed. Controlled photoactivity experiments using different radical scavengers (ammonium oxalate as a scavenger for h^+, $AgNO_3$ as a scavenger for e^-, benzoquinone as a scavenger for $O_2\bullet^-$ radical species, and tert-butyl alcohol as a scavenger for $\bullet OH$ species) were carried out similarly to the above described photocatalytic degradation process. The only exception was that the radical scavengers were added to the reaction system. The addition of ammonium oxalate and $AgNO_3$ had a weak inhibition efficiency of the phenol degradation, indicating that h^+ and e^- had a negligibly small effect on the mechanism of photocatalytic degradation. The photocatalytic conversion fell by approximately half when the tert-butyl alcohol (TBA) as the scavenger for the hydroxyl radicals was used. As shown in Figure 11, when benzoquinone as the trapping agent of $O_2\bullet^-$ was added into the phenol solution under visible irradiation, the photodegradation of phenol significantly declined to about 7%. These results clearly suggest that the photocatalytic degradation of phenol under VIS irradiation in the presence of $TiO_2\_EAN(1:1)$ was mainly intimate, with the photogenerated superoxide radical species. Secondly, the photogenerated OH radicals were also involved in the decomposition of phenol.

To explain what might have contributed to the increase in the photoactivity of TiO_2_ILs, the decomposition level of the ionic liquid cations was investigated using chromatography techniques. It was demonstrated that the ethylammonium nitrate ionic liquid was degraded in 97% after 24 h of the solvothermal reaction. Therefore, it is most likely that TiO_2 could be doped with nitrogen and carbon, and/or surface-modified by carbon species. However, based on the XPS analysis, the Ti–N interactions between the released nitrogen atoms, resulting from the IL's thermal decomposition, and the TiO_2 matrix, have not been observed. Thus, although we could expect the incorporation of nitrogen and carbon atoms in a crystalline lattice of TiO_2, the performed analysis did not confirm it. Nevertheless, it should be remembered that because of the low ionic liquid content on the TiO_2 surface, a low level of XPS detection (d.l. = 0.1 at. %) could affect the results. According to the literature, it could be generally stated that if TiO_2 particles grow in the presence of the N and C precursors and under elevated temperature conditions, usually, the N and C atoms are incorporated into the crystal lattice of the semiconductors [49–51].

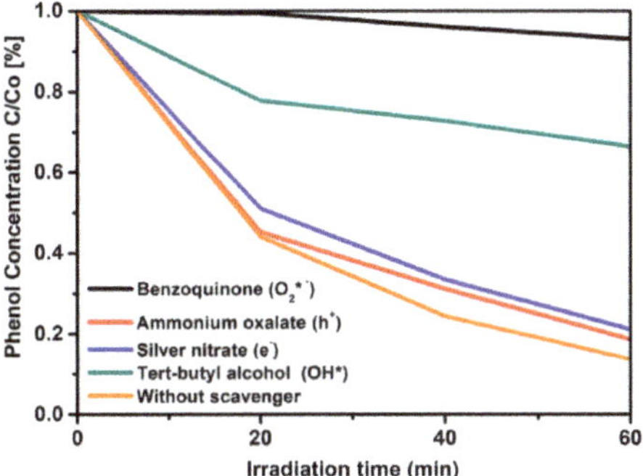

Figure 11. Effect of the addition of different radical scavengers on the phenol degradation over TiO_2_EAN(1:1)_24h composites under visible light irradiation. Initial pH 6.5, 20 mg/L, light intensity 3 mW/cm^2.

To identify the origin of visible light-induced activity, the phenol degradation rate was plotted depending on the surface area values as well as the carbon and nitrogen content in the surface layer. The data presented in Figure 12 suggests that the observed increase in photoactivity could be most interpreted by the presence of nitrogen in the surface layer. Although the origin of photoactivity is not clear at this moment, it is crucial to note that the preparation of highly active TiO_2 spheres was developed.

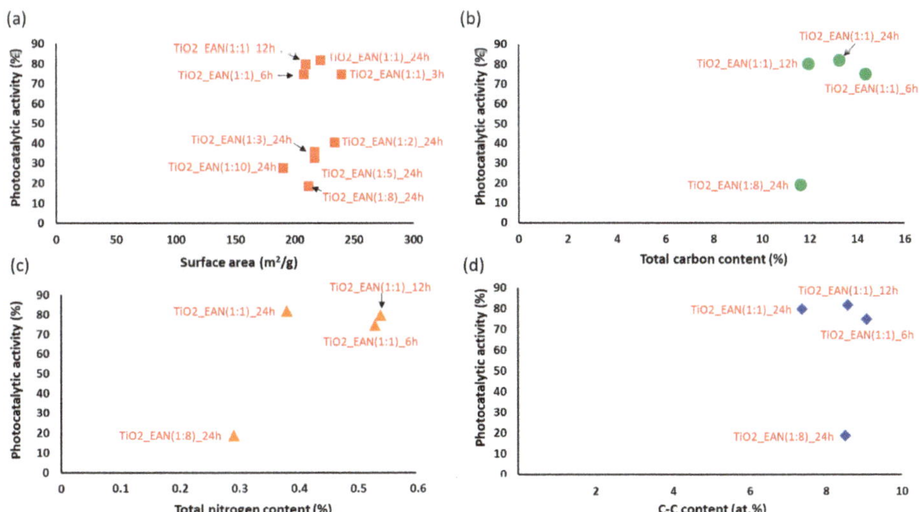

Figure 12. Dependence of the (a) surface area, (b) total carbon content, (c) total nitrogen content (d) C–C content of the photocatalytic activity.

3. Materials and Methods

3.1. Materials

For the aforementioned synthesis' purpose, the following reagents were applied: (1) titanium (IV) butoxide (TBOT) as the TiO_2 micro-particles direct precursor, and (2) 36% hydrochloric acid (HCl) as a pH stabilizer, sourced from Sigma-Aldrich; (3) anhydrous ethyl alcohol (99.8% ethanol) as the reaction medium (from POSCH S.A., Troine, Luxembourg); (4) ethylammonium nitrate (from IOLITEC, Heilbronn, Germany, purveyed with $\geq$97% of purity) as the assisting ionic liquid; and (5) deionized water, provided locally.

3.2. Preparation of IL-Assisted TiO_2 Particles

The preparation of the TiO_2 micro-particles was carried out emulating the method reported by Paszkiewicz et al. [16], summarized as follows: (1) TBOT was dispersed in ethanol through dropwise pouring under constant vigorous stirring; and (2) HCl, deionized water, and the due amount of IL—adequate to the applied molar ratio of TBOT to IL—were dissolved under unchanged conditions, to the point of attaining a pellucid solution. Afterwards, the ensuing mixture was transferred into an inner Teflon and an outer stainless steel autoclave, to be incubated at 180 °C for 24 h (for various TBOT to IL molar ratios) and for 6, 12, and 24 h (for 1:1 TBOT to IL molar ratio) for the purpose of the kinetics of the crystal growth evaluation. Subsequently, the autoclaves were cooled down to room temperature and the obtained precipitate was cleansed through numerous washings with deionized water and ethanol, and then dried at 60 °C for 4 h. The preparation ended with the 2 h calcination of plateau at 200 °C, which was reached at the 2 °C/min slope. For reference, the pristine TiO_2 was as-synthetized, with the exception of the IL presence.

3.3. Surface Properties Characterization

The Brunauer–Emmett–Teller (BET) surface area was calculated by the N_2 absorption–desorption isotherms at 77 K on a Micromeritics Gemini V200 Shimadzu analyzer (equipped with the VacPrep 061 Degasser) (Norcross, GA, USA). The morphology of the TiO_2 micro-particles was studied by scanning electron microscopy (SEM) analysis, performed under a Hitachi TM-1000 microscope (Tokyo, Japan). The chemistry of the surface was researched by X-ray photoelectron spectroscopy (XPS), the results were obtained with a PHI 5000 VersaProbeTM (ULVAC-PHI, Chigasaki, Japan) spectrometer of monochoromatic Al Kα radiation (h = 1486.6 eV). The phase purity of the samples was determined by powder X-ray diffraction (PXRD) using a PANalytical X'Pert Plus diffractometer (Almelo, The Netherlands) with Cu Kα radiation. To determine the unit cell parameters, the profile fits were performed on the powder diffraction data through the use of the HighScore program using Thompson$-$Cox$-$Hastings pseudo-Voigt peak shapes. The average crystallite size was calculated using the Scherrer equation.

The decomposition level of the ionic liquid cations was analyzed using a Dionex ICS 1100 liquid chromatograph. Deionized water containing 0.21% (v/v) of methanesulfonic acid (Sigma Aldrich, St. Louis, MO, USA) was used as a mobile phase. The separation was carried out isocratically using a Dionex (Sunnyvale, CA, USA) ION PAC, CS16 column (3 × 250 mm Dionex) at 35 °C. The flow rate was 0.36 mL min^{-1}. Each sample (before and after solvothermal reaction) was measured in triplicate. The decomposition level was calculated as follows:

$$\eta IL(\%) = 100 \times C_0 - C/C_0$$

where C_0 is the initial concentrations of the cations of the ILs; and C is the concentrations of the cations of the ILs after the solvothermal reaction.

3.4. Evaluation of Photocatalytic Activity

The photocatalytic activity was measured through the phenol decomposition rate under visible-light irradiation. For this aim, we dispersed 0.125 g of the obtained photocatalyst in 25 mL of phenol aqueous solution (C_0 = 25 mg/L), inside a cylindrical reactor with a circular quartz window. In use, there was a reactor fitted with a cooling jacket; during the reaction, it was cooled by the constant flow of water at $\leq$10 °C, supplied in aeration at 5 dm^3/h. The reactor's quartz-windowed side was exposed to the illumination of intensity equaling to 3 mW/cm^2 (by 1000 W Xenon lamp, 6271H Oriel; optical filter >420 nm, GG 420).

In order to establish the absorption–desorption equilibrium between the phenol and photocatalyst prior to the reaction, the suspension was allocated to 30 min long stirring in the dark, preliminarily to the photo-initiation of the catalysis; then, about 1 mL of the suspension was sampled from the reactor. During the irradiation, the samples were taken in 3 and 20 min intervals. Each sample was filtered through syringe filters (Φ = 0.2 µm) for the removal of the photocatalyst micro-particles, anterior to the due evaluation. The concentration of the remaining phenol was measured colorimetrically (λmax = 480 nm), after the derivatization with diazo-p-nitroaniline with a UV-VIS spectrometer (Evolution 220, Thermo-Scientific, Waltham, MA, USA).

The controlled photoactivity experiments were carried out using different scavengers (ammonium oxalate as a scavenger of photogenerated holes, AgNO$_3$ for electrons, benzoquinone for superoxide radical species, and tert-butyl alcohol for hydroxyl radical species). The scavenger concentration was equal to the phenol content, and experiments were performed analogously to the photocatalytic degradation of the phenol described in the manuscript, except that the scavengers were added to the reaction system.

4. Conclusions

This study is the first step towards enhancing our understanding of the effect of ethylammonium nitrate ionic liquids on the surface properties of the TiO$_2$ spheres formed in the solvothermal synthesis. In summary, the TiO$_2$ microspheres with a superior visible-light photocatalytic activity were prepared in the presence of the ethylammonium nitrate ionic liquid, using a solvothermal method followed by a calcination process. It should be highlighted that the most active TiO$_2$ samples formed in the presence of [EAN][NO$_3$] possessed almost the same activity induced by visible light than P25 TiO$_2$ under UV radiation. The phenol degradation rate was equal to 3.12 µmol/dm^3/min for the TiO$_2$_EAN/Vis system, and 3.46 µmol/dm^3/min for the P25/UV system. In this paper, the kinetics of the highly active TiO$_2$ microsphere formation in the presence of ethylammonium nitrate ionic liquid was examined. The obtained results revealed that the micro-particles of TiO$_2$_EAN(1:1)_3h that formed during only 3 h of synthesis time revealed a really high photoactivity under visible irradiation at 75%. This value increased to 80% and 82% after 12 and 24 h, respectively. However, the reaction yield for the 3 h synthesis time was relatively low (only 19%) and significantly increased with the increasing reaction time (38% for 6 h, 65% for 12 h, and 93% for 24 h). The photoactivity increase was accompanied by an increase in the specific surface area, thus the pore sizes as well as the ability to absorb the VIS irradiation.

The effective interactions between the ionic liquid components, mainly carbon, nitrogen, and the micro-particles surface of TiO$_2$, were clearly demonstrated by the XPS analysis. This factor could result in excellent visible-light photocatalytic activity for the IL–TiO$_2$ samples prepared. The radical trapping experiments revealed that O$_2$•$^-$ and OH• were the main active species during the degradation process.

Author Contributions: Conceptualization, A.Z.-M. and J.Ł.; funding acquisition, J.Ł.; investigation, A.G., M.C.-S., M.P.-G., W.L., E.R., T.K., Ż.P., and J.Ł.; project administration, E.G.; supervision, A.Z.-M. and J.Ł.; writing (original draft), A.G., W.L., E.R., T.K., E.G., and J.Ł.; writing (review and editing), A.Z.-M. and J.Ł.

Funding: This research was funded by the National Science Center within the program SONATA 8, research grant: 'Influence of the ionic liquid structure on interactions with TiO$_2$ particles in ionic liquid assisted hydrothermal synthesis', contract No. 2014/15/D/ST5/02747.

Conflicts of Interest: The authors declare no conflict of interest.

References

1. Hoffmann, M.R.; Martin, S.T.; Choi, W.; Bahnemann, D.W. Environmental applications of semiconductor photocatalysis. *Chem. Rev.* **1995**, *95*, 69–96. [CrossRef]
2. Fujishima, A.; Rao, T.N.; Tryk, D.A. Titanium dioxide photocatalysis. *J. Photochem. Photobiol. C Photochem. Rev.* **2000**, *1*, 1–21. [CrossRef]
3. Pelaez, M.; Nolan, N.T.; Pillai, S.C.; Seery, M.K.; Falaras, P.; Kontos, A.G.; Dunlop, P.S.M.; Hamilton, J.W.J.; Byrne, J.A.; O'Shea, K.; et al. A review on the visible light active titanium dioxide photocatalysts for environmental applications. *Appl. Catal. B Environ.* **2012**, *125*, 331–349. [CrossRef]
4. Fox, M.A.; Dulay, M.T. Heterogeneous photocatalysis. *Chem. Rev.* **1993**, *93*, 341–357. [CrossRef]
5. Thompson, T.L.; Yates, J.T. Surface science studies of the photoactivation of TiO$_2$ new photochemical processes. *Chem. Rev.* **2006**, *106*, 4428–4453. [CrossRef] [PubMed]
6. Zhang, B.; Xue, Z.; Xue, Y.; Huang, Z.; Li, Z.; Hao, J. Ionic liquid-assisted synthesis of morphology-controlled TiO$_2$ particles with efficient photocatalytic activity. *RSC Adv.* **2015**, *5*, 81108–81114. [CrossRef]
7. Ramanathan, R.; Bansal, V. Ionic liquid mediated synthesis of nitrogen, carbon and fluorine-codoped rutile TiO$_2$ nanorods for improved UV and visible light photocatalysis. *RSC Adv.* **2015**, *5*, 1424–1429. [CrossRef]
8. Xu, H.; Ouyang, S.; Liu, L.; Reunchan, P.; Umezawa, N.; Ye, J. Recent advances in TiO$_2$-based photocatalysis. *J. Mater. Chem. A* **2014**, *2*, 12642–12661. [CrossRef]
9. Zhang, F.; Sun, D.; Yu, C.; Yin, Y.; Dai, H.; Shao, G. A sol–gel route to synthesize SiO$_2$/TiO$_2$ well-ordered nanocrystalline mesoporous photocatalysts through ionic liquid control. *New J. Chem.* **2015**, *39*, 3065–3070. [CrossRef]
10. Alammar, T.; Noei, H.; Wang, Y.; Mudring, A.-V. Mild yet phase-selective preparation of TiO$_2$ nanoparticles from ionic liquids—A critical study. *Nanoscale* **2013**, *5*, 8045–8055. [CrossRef] [PubMed]
11. Ahmed, E.; Breternitz, J.; Groh, M.F.; Ruck, M. Ionic liquids as crystallisation media for inorganic materials. *CrystEngComm* **2012**, *14*, 4874–4885. [CrossRef]
12. Bhattacharyya, K.; Majeed, J.; Dey, K.K.; Ayyub, P.; Tyagi, A.K.; Bharadwaj, S.R. Effect of Mo-Incorporation in the TiO$_2$ Lattice: A mechanistic basis for photocatalytic dye degradation. *J. Phys. Chem. C* **2014**, *118*, 15946–15962. [CrossRef]
13. Yu, S.; Liu, B.; Wang, Q.; Gao, Y.; Shi, Y.; Feng, X.; An, X.; Liu, L.; Zhang, J. Ionic Liquid Assisted Chemical Strategy to TiO$_2$ Hollow Nanocube Assemblies with Surface-Fluorination and Nitridation and High Energy Crystal Facet Exposure for Enhanced Photocatalysis. *ACS Appl. Mater. Interfaces* **2014**, *6*, 10283–10295. [CrossRef] [PubMed]
14. Gindri, I.M.; Frizzo, C.P.; Bender, C.R.; Tier, A.Z.; Martins, M.A.P.; Villetti, M.A.; Machado, G.; Rodriguez, L.C.; Rodrigues, D.C. Preparation of TiO$_2$ nanoparticles coated with ionic liquids: A supramolecular approach. *ACS Appl. Mater. Interfaces* **2014**, *6*, 11536–11543. [CrossRef] [PubMed]
15. Łuczak, J.; Paszkiewicz, M.; Krukowska, A.; Malankowska, A.; Zaleska-Medynska, A. Ionic liquids for nano-and microstructures preparation. Part 1: Properties and multifunctional role. *Adv. Colloid Interface Sci.* **2016**, *230*, 13–28. [CrossRef] [PubMed]
16. Paszkiewicz, M.; Łuczak, J.; Lisowski, W.; Patyk, P.; Zaleska-Medynska, A. The ILs-assisted solvothermal synthesis of TiO$_2$ spheres: The effect of ionic liquids on morphology and photoactivity of TiO$_2$. *Appl. Catal. B Environ.* **2016**, *184*, 223–237. [CrossRef]
17. Kaur, N.; Singh, V. Current status and future challenges in ionic liquids, functionalized ionic liquids and deep eutectic solvent-mediated synthesis of nanostructured TiO$_2$: A review. *New J. Chem.* **2017**, *41*, 2844–2868. [CrossRef]
18. Wender, H.; Feil, A.F.; Diaz, L.B.; Ribeiro, C.S.; Machado, G.J.; Migowski, P.; Weibel, D.E.; Dupont, J.; Teixeira, S.R. Self-organized TiO$_2$ nanotube arrays: Synthesis by anodization in an ionic liquid and assessment of photocatalytic properties. *ACS Appl. Mater. Interfaces* **2011**, *3*, 1359–1365. [CrossRef] [PubMed]
19. Qi, L.; Yu, J.; Jaroniec, M. Enhanced and suppressed effects of ionic liquid on the photocatalytic activity of TiO$_2$. *Adsorption* **2013**, *19*, 557–561. [CrossRef]
20. Marr, P.C.; Marr, A.C. Ionic liquid gel materials: Applications in green and sustainable chemistry. *Green Chem.* **2016**, *18*, 105–128. [CrossRef]

21. Chang, S.-M.; Lee, C.-Y. A salt-assisted approach for the pore-size-tailoring of the ionic-liquid-templated TiO$_2$ photocatalysts exhibiting high activity. *Appl. Catal. B Environ.* **2013**, *132*, 219–228. [CrossRef]
22. Lopes, C.W.; Finger, P.H.; Mignoni, M.L.; Emmerich, D.J.; Mendes, F.M.T.; Amorim, S.; Pergher, S.B.C. TiO$_2$-TON zeolite synthesis using an ionic liquid as a structure-directing agent. *Microporous Mesoporous Mater.* **2015**, *213*, 78–84. [CrossRef]
23. Yu, N.; Gong, L.; Song, H.; Liu, Y.; Yin, D. Ionic liquid of [Bmim]$^+$Cl$^-$ for the preparation of hierarchical nanostructured rutile titania. *J. Solid State Chem.* **2007**, *180*, 799–803. [CrossRef]
24. Gołąbiewska, A.; Paszkiewicz-Gawron, M.; Sadzińska, A.; Lisowski, W.; Grabowska, E.; Zaleska-Medynska, A.; Łuczak, J. Fabrication and photoactivity of ionic liquid–TiO$_2$ structures for efficient visible-light-induced photocatalytic decomposition of organic pollutants in aqueous phase. *Beilstein J. Nanotechnol.* **2018**, *9*, 580–590. [CrossRef] [PubMed]
25. Han, C.-C.; Ho, S.-Y.; Lin, Y.-P.; Lai, Y.-C.; Liang, W.-C.; Chen-Yang, Y.-W. Effect of π–π stacking of water miscible ionic liquid template with different cation chain length and content on morphology of mesoporous TiO$_2$ prepared via sol–gel method and the applications. *Microporous Mesoporous Mater.* **2010**, *131*, 217–223. [CrossRef]
26. Chen, Y.; Li, W.; Wang, J.; Gan, Y.; Liu, L.; Ju, M. Microwave-assisted ionic liquid synthesis of Ti^{3+} self-doped TiO$_2$ hollow nanocrystals with enhanced visible-light photoactivity. *Appl. Catal. B Environ.* **2016**, *191*, 94–105. [CrossRef]
27. Łuczak, J.; Paszkiewicz-Gawron, M.; Długokęcka, M.; Lisowski, W.; Grabowska, E.; Makurat, S.; Rak, J.; Zaleska-Medynska, A. Visible light photocatalytic activity of ionic liquid-TiO$_2$ spheres: Effect of the ionic liquid's anion structure. *ChemCatChem* **2017**, *9*, 4377–4388. [CrossRef]
28. Kim, S.; Ko, K.C.; Lee, J.Y.; Illas, F. Single oxygen vacancies of (TiO$_2$) 35 as a prototype reduced nanoparticle: Implication for photocatalytic activity. *Phys. Chem. Chem. Phys.* **2016**, *18*, 23755–23762. [CrossRef] [PubMed]
29. Jiang, Y.; Zhu, Y.J.; Cheng, G.F. Synthesis of Bi$_2$Se$_3$ Nanosheets by Microwave Heating Using an Ionic Liquid. *Cryst. Growth Des.* **2006**, *6*, 2174–2176. [CrossRef]
30. Kaper, H.; Sallard, S.B.; Djerdj, I.; Antonietti, M.; Smarsly, B.M. Toward a Low-Temperature Sol−Gel Synthesis of TiO$_2$(B) Using Mixtures of Surfactants and Ionic Liquids. *Chem. Mater.* **2010**, *22*, 3502–3510. [CrossRef]
31. Verma, Y.L.; Tripathi, A.K.; Singh, V.K.; Balo, L.; Gupta, H.; Singh, S.K.; Singh, R.K. Preparation and properties of titania based ionogels synthesized using ionic liquid 1-ethyl-3-methyl imidazolium thiocyanate. *Mater. Sci. Eng. B* **2017**, *220*, 37–43. [CrossRef]
32. Jing, L.; Wang, M.; Li, X.; Xiao, R.; Zhao, Y.; Zhang, Y.; Yan, Y.-M.; Wu, Q.; Sun, K. Covalently functionalized TiO$_2$ with ionic liquid: A high-performance catalyst for photoelectrochemical water oxidation. *Appl. Catal. B Environ.* **2015**, *166*, 270–276. [CrossRef]
33. Ravishankar, T.N.; Nagaraju, G.; Dupont, J. Photocatalytic activity of Li-doped TiO$_2$ nanoparticles: Synthesis via ionic liquid-assisted hydrothermal route. *Mater. Res. Bull.* **2016**, *78*, 103–111. [CrossRef]
34. Liu, H.; Liang, Y.; Hu, H.; Wang, M. Hydrothermal synthesis of mesostructured nanocrystalline TiO$_2$ in an ionic liquid–water mixture and its photocatalytic performance. *Solid State Sci.* **2009**, *11*, 1655–1660. [CrossRef]
35. Shahi, S.K.; Kaur, N.; Singh, V. Fabrication of phase and morphology controlled pure rutile and rutile/anatase TiO$_2$ nanostructures in functional ionic liquid/water. *Appl. Surf. Sci.* **2016**, *360*, 953–960. [CrossRef]
36. Yan, X.; Ohno, T.; Nishijima, K.; Abe, R.; Ohtani, B. Is methylene blue an appropriate substrate for a photocatalytic activity test? A study with visible-light responsive Titania. *Chem. Phys. Lett.* **2006**, *429*, 606–610. [CrossRef]
37. Ohtani, B. Photocatalysis A to Z—What we know and what we do not know in a scientific sense. *J. Photochem. Photobiol. C Photochem. Rev.* **2010**, *11*, 157–178. [CrossRef]
38. Li, F.-T.; Wang, X.-J.; Zhao, Y.; Liu, J.-X.; Hao, Y.-J.; Liu, R.-H.; Zhao, D.-S. Ionic-liquid-assisted synthesis of high-visible-light-activated N–B–F-tri-doped mesoporous TiO$_2$ via a microwave route. *Appl. Catal. B Environ.* **2014**, *144*, 442–453. [CrossRef]
39. Mirhoseini, F.; Salabat, A. Ionic liquid based microemulsion method for the fabrication of poly (methyl methacrylate)–TiO$_2$ nanocomposite as a highly efficient visible light photocatalyst. *RSC Adv.* **2015**, *5*, 12536–12545. [CrossRef]
40. Raj, K.J.A.; Viswanathan, B. Effect of surface area, pore volume and particle size of P25 titania of the phase transformation of anatase to rutile. *Indian J. Chem.* **2009**, *48A*, 1378–1382.

41. Djerdj, I.; Tonejc, A.M. Structural investigations of nanocrystalline TiO$_2$ samples. *J. Alloys Compd.* **2006**, *413*, 159–174. [CrossRef]
42. Naumkin, A.V.; Kraut-Vass, A.; Gaarenstroom, S.W.; Powell, C.J. NIST X-ray Photoelectron Spectroscopy Database, NIST Standard Reference Database 20, Version 4.1. 2012. Available online: http://srdata.nist.gov/xps/ (accessed on 25 March 2013).
43. Hu, X.; Zhang, T.; Jin, Z.; Zhang, J.; Xu, W.; Yan, J.; Zhang, J.; Zhang, L.; Wu, Y. Fabrication of carbon-modified TiO$_2$ nanotube arrays and their photocatalytic activity. *Mater. Lett.* **2008**, *62*, 4579–4581. [CrossRef]
44. Janus, M.; Inagaki, M.; Tryba, B.; Toyoda, M.; Morawski, A.W. Carbon-modified TiO$_2$ photocatalyst by ethanol carbonisation. *Appl. Catal. B Environ.* **2006**, *63*, 272–276. [CrossRef]
45. Kusiak-Nejman, E.; Janus, M.; Grzmil, B.; Morawski, A.W. Methylene Blue decomposition under visible light irradiation in the presence of carbon-modified TiO$_2$ photocatalysts. *J. Photochem. Photobiol. A Chem.* **2011**, *226*, 68–72. [CrossRef]
46. Li, Y.; Wang, Y.; Kong, J.; Jia, H.; Wang, Z. Synthesis and characterization of carbon modified TiO$_2$ nanotube and photocatalytic activity on methylene blue under sunlight. *Appl. Surf. Sci.* **2015**, *344*, 176–180. [CrossRef]
47. Wei, X.-N.; Wang, H.-L.; Wang, X.-K.; Jiang, W.-F. Facile synthesis of tunable carbon modified mesoporous TiO$_2$ for visible light photocatalytic application. *Appl. Surf. Sci.* **2017**, *412*, 357–365. [CrossRef]
48. Qi, H.-P.; Liu, Y.-Z.; Chang, L.; Wang, H.-L. In-situ one-pot hydrothermal synthesis of carbon-TiO$_2$ nanocomposites and their photocatalytic applications. *J. Environ. Chem. Eng.* **2017**, *5*, 6114–6121. [CrossRef]
49. Wu, D.; Long, M.; Cai, W.; Chen, C.; Wu, Y. Low temperature hydrothermal synthesis of N-doped TiO$_2$ photocatalyst with high visible-light activity. *J. Alloys Compd.* **2010**, *502*, 289–294. [CrossRef]
50. Dolat, D.; Quici, N.; Kusiak-Nejman, E.; Morawski, A.W.; Puma, G.L. One-step, hydrothermal synthesis of nitrogen, carbon co-doped titanium dioxide (N, CTiO$_2$) photocatalysts. Effect of alcohol degree and chain length as carbon dopant precursors on photocatalytic activity and catalyst deactivation. *Appl. Catal. B Environ.* **2012**, *115*, 81–89. [CrossRef]
51. Peng, F.; Cai, L.; Huang, L.; Yu, H.; Wang, H. Preparation of nitrogen-doped titanium dioxide with visible-light photocatalytic activity using a facile hydrothermal method. *J. Phys. Chem. Solids* **2008**, *69*, 1657–1664. [CrossRef]

 *catalysts*

Article

Electrochemically Obtained TiO$_2$/Cu$_x$O$_y$ Nanotube Arrays Presenting a Photocatalytic Response in Processes of Pollutants Degradation and Bacteria Inactivation in Aqueous Phase

Magda Kozak [1,*], Paweł Mazierski [1], Joanna Żebrowska [2], Marek Kobylański [1], Tomasz Klimczuk [3], Wojciech Lisowski [4], Grzegorz Trykowski [5], Grzegorz Nowaczyk [6] and Adriana Zaleska-Medynska [1]

[1] Department of Environmental Technology, Faculty of Chemistry, University of Gdansk, 80-308 Gdansk, Poland; pawel.mazierski@phdstud.ug.edu.pl (P.M.); marek.kobylanski@phdstud.ug.edu.pl (M.K.); adriana.zaleska@ug.edu.pl (A.Z.-M.)
[2] Department of Molecular Biotechnology, Faculty of Chemistry, University of Gdansk, 80-308 Gdansk, Poland; joanna.zebrowska@ug.edu.pl
[3] Faculty of Applied Physics and Mathematics, Gdansk University of Technology, 80-233 Gdansk, Poland; tomek@mif.pg.gda.pl
[4] Institute of Physical Chemistry, Polish Academy of Sciences, 01-224 Warsaw, Poland; wlisowski@ichf.edu.pl
[5] Faculty of Chemistry, Nicolaus Copernicus University, 87-100 Torun, Poland; tryki@umk.pl
[6] NanoBioMedical Centre, Adam Mickiewicz University, Umultowska 85, 61-614 Poznan, Poland; nowag@amu.edu.pl
* Correspondence: magda.kozak@ug.edu.pl; Tel.: +48-58-523-5159

Received: 15 May 2018; Accepted: 2 June 2018; Published: 5 June 2018

Abstract: TiO$_2$/Cu$_x$O$_y$ nanotube (NT) arrays were synthesized using the anodization method in the presence of ethylene glycol and different parameters applied. The presence, morphology, and chemical character of the obtained structures was characterized using a variety of methods—SEM (scanning electron microscopy), XPS (X-ray photoelectron spectroscopy), XRD (X-ray crystallography), PL (photoluminescence), and EDX (energy-dispersive X-ray spectroscopy). A p-n mixed oxide heterojunction of Ti-Cu was created with a proved response to the visible light range and the stable form that were in contact with Ti. TiO$_2$/Cu$_x$O$_y$ NTs presented the appearance of both Cu$_2$O (mainly) and CuO components influencing the dimensions of the NTs (1.1–1.3 µm). Additionally, changes in voltage have been proven to affect the NTs' length, which reached a value of 3.5 µm for Ti$_{90}$Cu$_{10}$_50V. Degradation of phenol in the aqueous phase was observed in 16% of Ti$_{85}$Cu$_{15}$_30V after 1 h of visible light irradiation (λ > 420 nm). Scavenger tests for phenol degradation process in presence of NT samples exposed the responsibility of superoxide radicals for degradation of organic compounds in Vis light region. Inactivation of bacteria strains *Escherichia coli* (*E. coli*), *Bacillus subtilis* (*B. subtilis*), and *Clostridium* sp. in presence of obtained TiO$_2$/Cu$_x$O$_y$ NT photocatalysts, and Vis light has been studied showing a great improvement in inactivation efficiency with a response rate of 97% inactivation for *E. coli* and 98% for *Clostridium* sp. in 60 min. Evidently, TEM (transmission electron microscopy) images confirmed the bacteria cells' damage.

Keywords: heterogeneous photocatalysis; TiO$_2$/Cu$_2$O nanotubes; anodization; nanomaterials fabrication; removal of microbiological pollutants

---

1. Introduction

Processes connected with the photocatalysis phenomenon are in area that is receiving great attention nowadays. Once we add nanomaterials to this combination, we will produce an interesting mix that not only scientists but also industry have been paying attention to lately. The main reason for this interest is the scope of application—i.e., the healthy nature of the environmental engineering and car industries.

Intensive studies on nanomaterials—research on properties, attempts to reduce their size or modifications—lead to the extension of the application possibilities. In recent years, more attention has been directed towards transformation processes that involve a light source—i.e., solve energy and pollution problems in the presence of semiconductors (photocatalysts). One of the semiconductors, considered as the most attractive photocatalyst in previous research, is TiO_2, which is characterized by a wide range of advantages, i.e., high chemical stability and relatively low price, and it is not toxic. It has been confirmed that it is applicable to the degradation of pollutants in both the gaseous and liquid phases [1,2], as well as in many different areas such as CO_2 reduction [3], water splitting [4], or antibacterial activity [5]. There is also one main disadvantage of TiO_2 photocatalysts, which is their minimal energy value (Eg, c.a 3.2 eV), which is necessary for the electron excitation that needed to generate holes in the valence band and carry on photocatalytic reactions. This value corresponds to 388 nm, so it can be activated in the UV irradiation range (300–380 nm), which really limits its applications [6].

The limitations have led to worldwide research focusing on TiO_2 modifications to extend its activity to the range of light irradiation of the Vis, such as sensitization with a semiconductor with a narrow band gap, metal ion doping, or nonmetal doping or dye sensitization [7]. These processes not only increase the activity but also modify the active surface area. There are two approaches in terms of the nanofabrication process: "top-down" and "bottom-up." The first represents the idea of using larger structures, which can be further controlled in the nanoscale, whereas the second includes the miniaturization of components with the self-assembly process. There are a variety of forms of modified nanostructures—nanoparticles, nanotubes (NTs), nanosheets, or nanocubes obtained using many methods such as electrodeposition [6], self-assembly examples, atomic layer deposition, or anodization [8–10].

Anodization of titanium and its alloys was performed in many environments [3,11], but modification of Ti alloys with this method was not so common. During the oxidation process, a big influence on the results of the experiment had some parameters such as applied voltage, the composition of the electrolyte, or the time of the process [12]. The anodic growth of compact oxides in metal surfaces and the formation of tubes are governed by a competition between anodic oxide formation and chemical dissolution of the oxide as a soluble fluoride complex. In 2001, Grimes et al. first reported the influence of hydrofluoric (HF) acid in the anodic oxidation process on the porosity of the titanium NT arrays [13]. It is proved that longer NTs are formed in electrolytes based on organic compounds like ethylene glycol or glycerol. Studies on literature show that usage of viscous solvents for anodization results in smooth wall structures of the NTs [14,15].

One promising aspect of TiO_2 modification is its ability to combine with Cu, including Cu_2O and CuO species, which can function as an electron mediator to widen the wavelength region for absorption [16]; however, Cu is not stable in terms of nano-scale size [17]. Furthermore, Cu_2O is also promising with regard to the formation of p-n heterojunctions with TiO_2 [18], which can lead to the improvement of the modified photocatalysts features. It was already reported that photocatalysts containing titanium and copper have the ability to cause pollutant degradation [19,20]; what is more, there is great potential in these materials in terms of inactivation of bacteria that has studied since 1985. Metsunga et al. were among the pioneers who studied the application of titanium dioxide (TiO_2) as a promising photocatalyst in terms of antimicrobial reactions [21]. Copper possesses high antimicrobial features [22] with the potential for drug degradation [23] and, due to lower toxicity, price, and increased cytocompatibility [24], it is more favorable than silver or gold. This is the reason why it has been more intensively studied among other catalysts nowadays. Cu and TiO_2 combinations were studied in different modifications including photodeposition of copper [25], microwave co-precipitation technique [26], layer films [27], alloy rods [28], nanocrystals [19], composite coating on wooden substrate [29], copper decorated TiO_2 nanorods [30], combination of magnetron sputtering and annealing treatment [31], doped nanoparticles [32] or radiolytic deposition of copper species at the surface of TiO_2 nanotubes [33]. However, properties of nanostructures containing copper obtained *via* anodic oxidation using alloys are still rather unknown, and very little research has been undertaken to understand the quantity influence of copper in the Cu-TiO_2 nanostructured composite regarding the photocatalytic activities of Cu-doped TiO_2. Furthermore, the influence of different parameters on oxidation processes such as voltage, time, and electrolyte composition has not been fully studied yet. In view of this, in this work, it was decided to extend knowledge about the anodized Ti-Cu alloys while considering different parameters of the process. Moreover, the impact of anodized Ti-Cu alloys on bacteria inactivation (*E. coli, B. subtilis, Clostridium* sp.) in Vis light was investigated for the first time. The expected outcomes are as follows: (a) the anodic oxidation of Ti-Cu alloys will form TiO_2 NT arrays in various production conditions, (b) TiO_2/Cu_xO_y will show increased activity under Vis light, and (c) TiO_2/Cu_xO_y will present intense bacteria inactivation features.

2. Results and Discussion

2.1. Morphology and Formation of TiO$_2$/Cu$_x$O$_y$ NTs

The surface morphology of obtained TiO_2/Cu_xO_y NTs was determined by SEM (scanning electron microscopy) and is presented in Figure 1a–e, while SEM images of references samples, namely, pristine TiO_2, are displayed in Figure S1 (Supplementary Materials). The upper layer of NTs remains unveiled, and only a small amount of initial barrier layer is visible, which suggests that cleaning the surface of samples after anodization in an ultrasonic bath was effective. Characteristic dimensions of NTs such as length, diameter, and wall thickness were calculated based on SEM images and are gathered in Table 1. Among the TiO_2/Cu_xO_y NTs, the length of tubes and external diameter varied from 1.1 to 3.5 µm and 83–98 nm, respectively, indicating that the shape of NTs strictly depends on applied voltage during anodization process and amount of Cu in the Ti-Cu alloys. In general, the length of all TiO_2/Cu_xO_y NTs is shorter than pristine TiO_2 NTs prepared from Ti foil under the same anodization conditions. Additionally, the length TiO_2/Cu_xO_y NTs decreased with the increase of Cu content in the Ti-Cu alloy. These phenomena have already been observed for the anodization of titanium alloys [34,35] and can be ascribed to the accelerated dissolution of TiO_2/Cu_xO_y NTs [28,34].

As will be described later (in XPS (X-ray photoelectron spectroscopy) part) by anodizing Ti-Cu alloys, we obtained p-n heterojunction consisting of TiO_2 NTs and Cu_xO_y species, in which Cu species appear in the form of Cu_2O (mostly) and CuO. Based on TEM image depicted in Figure 1f, we can conclude that Cu_xO_y species are evenly spaced along the NTs, and no nanoparticles formation was observed. On the other hand, the formation of copper oxides inside TiO_2 NTs at various oxidation states (Cu^{1+} and Cu^{2+}) can be represented by the following reactions [36,37]:

$$Cu - e^- \rightarrow Cu^+, \ 2Cu^+ + 2OH^- \rightarrow Cu_2O + H_2O \tag{1}$$

$$Cu - 2e^- \rightarrow Cu^{2+}, \ Cu^{2+} + 2OH^- \rightarrow CuO + H_2O \tag{2}$$

These reactions occur simultaneously with the formation of TiO_2, which can be represented as

$$Ti + 2H_2O \rightarrow TiO_2 + 4H^+ + 4e^- \tag{3}$$

$$TiO_2 + 4H^+ + 6F^- \rightarrow TiF_6{}^{2-} + 2H_2O \tag{4}$$

Reaction (4) affects the anodization process and is responsible for the growth of TiO_2 in the form of NTs [38,39].

Table 1. Sample labels, preparation conditions, NTs dimensions (based on SEM measurements), and Cu content (based on EDX (Energy-dispersive X-ray spectroscopy)).

Sample Label	Material of Working Electrode	Anodization Voltage (V)	External Diameter (nm)	Tubes Length (µm)	Wall Thickness (nm)	Cu Content (wt.%)
Ti_30V	Ti foil	30	80	1.5	10	-
Ti_40V		40	100	3.0	13	-
Ti_50V		50	120	6.0	18	-
Ti$_{95}$Cu$_5$_30V	Ti(95%)/Cu(5%) alloy	30	85	1.3	12	3.57
Ti$_{90}$Cu$_{10}$_30V	Ti(90%)/Cu(10%) alloy	30	88	1.2	11	6.20
Ti$_{85}$Cu$_{15}$_30V	Ti(85%)/Cu(15%) alloy	30	83	1.1	15	9.45
Ti$_{90}$Cu$_{10}$_40V	Ti(90%)/Cu(10%) alloy	40	98	2.5	14	6.25
Ti$_{90}$Cu$_{10}$_50V	Ti(90%)/Cu(10%) alloy	50	97	3.5	14	3.44

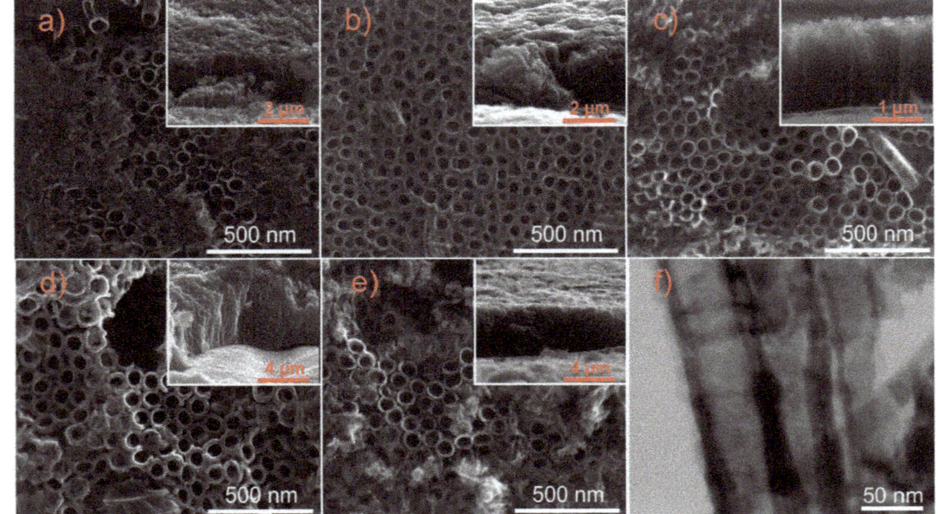

Figure 1. Top-view and cross-sectional SEM images of Ti$_{95}$Cu$_5$_30V (**a**), Ti$_{90}$Cu$_{10}$_30V (**b**), Ti$_{85}$Cu$_{15}$_30V (**c**), Ti$_{90}$Cu$_{10}$_40V (**d**), Ti$_{90}$Cu$_{10}$_50V (**e**) samples, and TEM image of Ti$_{95}$Cu$_5$_30V (**f**).

2.2. XRD Analysis

Phase composition of the tested samples was checked by using powder X-ray diffractometer (X'Pert Pro MPD, PANalytical, Almelo, The Netherlands). Figure 2 details diffraction patterns for Ti-Cu alloys after the anodic oxidation process. The patterns for Ti anodized samples are in Table S1 and Figure S2 (supplementary materials). The experimental data and refined model (LeBail) are represented by circles and a solid red line, respectively. The sets of vertical bars show expected positions for TiO_2—anatase (black), Ti (red), and $CuTi_2$ alloy (olive). The Miller indices for the anatase are shown in panel (d). The strongest reflection (101) is observed at around 25.3 deg. Ti foil is always observed with the strongest reflection at 40.3 deg. Foils with Cu concentration 10% and 15% reveal presence of $CuTi_2$ alloy—clearly seen by growing reflection near 16.6 deg. (Figure 2a,b). Interestingly, a relative signal of $CuTi_2$ is weaker for $Ti_{90}Cu_{10}$ samples anodized in 40V and 50V, compared to the one anodized in 30 V ($Ti_{90}Cu_{10}$_30V). It is likely caused by thicker film of anatase on the surface for $Ti_{90}Cu_{10}$_40V and $Ti_{90}Cu_{10}$_50V. We do not observe neither Cu metal, nor Cu oxides.

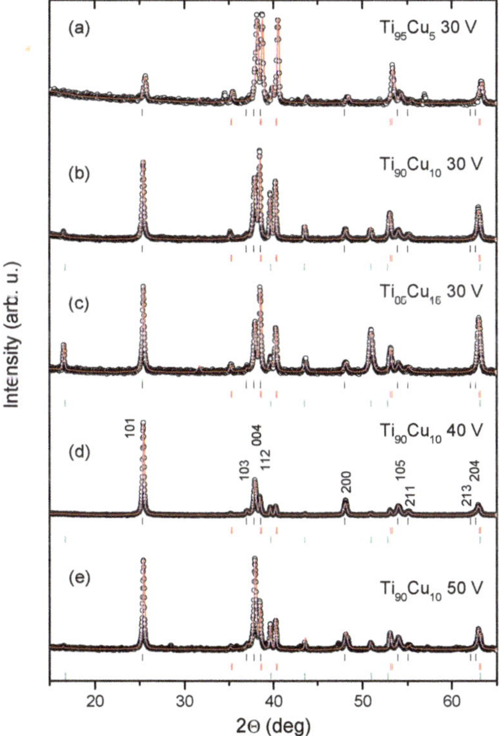

Figure 2. X-ray diffraction patterns for Ti-Cu alloys. A red solid line is a LeBail fit to the experimental data (open circles). Vertical bars represent positions of expected Bragg peaks for (from top): TiO_2—anatase (black), Ti (red), and $CuTi_2$ alloy (olive), respectively.

More qualitative results were obtained by using LeBail refinement. Table S1 contains obtained lattice parameters for TiO_2—anatase (tetragonal I41/amd, ICSD code: 063711), Ti—metal (hexagonal P63/mmc, ICSD code: 076265), and $CuTi_2$ (tetragonal I4/mmm, ICSD code: 015807) compounds. A corrected full width at half maximum (FWHM) of the strongest (101) anatase reflection was used for calculations of the average crystallite size using the Scherrer equation. The estimated size is between 250 Å and 460 Å for $Ti_{95}Cu_5$_30V and Ti_{90}Cu10_50V, respectively. However, because of complexity of studied photocatalysts, presence of three oxides (TiO_2, Cu_2O, CuO), and different wall thickness of NTs, it is hard to obtain more information about the crystallite size. There is no obvious change of the anatase lattice parameters with the increasing voltage. The a and c parameters for Ti_30V, Ti_40V, and Ti_50V, as well as for $Ti_{90}Cu_{10}$_30V, $Ti_{90}Cu_{10}$_40V, and $Ti_{90}Cu_{10}$_50V series, remain almost the same. However, for the samples anodized in 30V, with different Cu content, there is increase in both a and c values. It is reflected also by increase in unit cell volume from 136.0 $Å^3$ ($Ti_{95}Cu_5$_30V) to 136.6 $Å^3$ ($Ti_{90}Cu_{10}$_30V) and 136.0 $Å^3$ ($Ti_{85}Cu_{15}$_30V). Higher Cu concentration is likely responsible for larger crystallite size: 250 Å, 360 Å, and 390 Å for 5%, 10%, and 15% Cu, respectively.

2.3. XPS Analysis

Elemental composition (in at.%) in the surface layer of Cu-modified TiO_2 NTs was evaluated by XPS analysis and reported in Table 2 and Table S2. Detection of Cu in addition to Ti and O confirm the effective modification of this element in all TiO_2 NTs. Carbon, nitrogen, and fluorine species were also detected. Results obtained in different studies by our group and also knowledge gathered from literature confirm that source of fluoride, carbon, and nitrogen species lies in electrolyte used in NTs preparation. Level of contamination depends on porosity and thickness of the films used in experiment, as well as composition of electrolyte and time of the anodization. However, it was confirmed by D. Regonini et al. that these species do not influence the degradation process [40,41]. They are common contaminants of TiO_2 nanotubes obtained by anodic oxidation, and all derive from the electrolyte [40,42].

The chemical character of Ti originating from pristine TiO_2 and Cu from the Cu-modified TiO_2 NTs were identified from the Ti 2p and Cu 2p HR XPS spectra, respectively. The selected spectra for 30 V are summarized in Table 2 and shown in Figure 3. The Ti 2p spectrum is resolved into two doublet-components at BE of Ti $2p_{3/2}$ signal at 458.7 and 457.3 eV and are assigned to Ti^{4+} and Ti^{3+}, respectively (Figure 3a). The Ti^{4+} is the dominant surface state and relative contribution of the Ti^{3+} species is similar for all NTs (Table 2). Following the curve-fitting procedure of Cu $2p_{3/2}$ spectra (Figure 3b), we separated the XPS peaks at the BE of 932.1 and 933.8 eV, which are characteristic of Cu(I) and Cu(II) oxide species, respectively [27,40,43–45]. The Cu(I) composites are the dominant Cu fraction for all samples, and their relative contribution depends on applied voltage during anodization process and amount of Cu in the Ti-Cu alloys (Table 2).

Inspection of the data presented in Table 2 reveal the relative contribution of Cu(I) fraction to be systematically increased as the Cu amount in the Ti-Cu alloys increased (see the XPS data for TiCu alloys anodized at 30 V).

Table 2. Surface properties and photoactivity of TiO_2/Cu_xO_y NTs and reference samples (pristine TiO_2 NTs).

Sample Label	Average Crystallite Size (nm)	XPS Analysis						Photocatalytic Reaction Rate, r ($\mu mol \cdot dm^{-3} \cdot min^{-1}$)	
		$\sum$ Ti (at.%)	Ti^{4+} 458.7 eV (%)	Ti^{3+} 457.3 eV (%)	Cu (at.%)	Cu^{1+} 932.2 eV (%)	Cu^{2+} 933.8 eV (%)	UV-Vis Light (λ > 350 nm)	Vis Light (λ > 420 nm)
Ti_30V	33	16.20	98.42	1.58	0	-	-	1.25	0.04
Ti_40V	27	24.79	97.38	2.62	0	-	-	1.35	0.13
Ti_50V	36	26.29	97.38	2.62	0	-	-	1.44	0.15
$Ti_{95}Cu_5$_30V	25	21.79	95.89	4.11	0.13	81.27	18.73	1.02	0.41
$Ti_{90}Cu_{10}$_30V	36	25.32	97.49	2.51	0.11	88.38	11.62	1.16	0.51
$Ti_{85}Cu_{15}$_30V	39	23.99	97.30	2.70	0.14	93.28	6.72	0.81	0.55
$Ti_{90}Cu_{10}$_40V	46	25.41	98.07	1.93	0.11	76.61	23.39	1.62	0.37
$Ti_{90}Cu_{10}$_50V	46	25.06	97.16	2.84	0.08	95.78	4.22	3.31	0.32

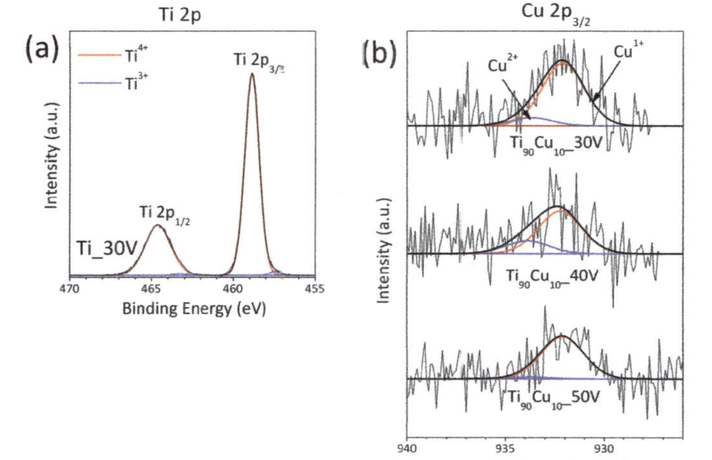

Figure 3. (**a**) Ti 2p XPS spectrum from the surface of pristine TiO_2 NTs (Ti_30V) and (**b**) Cu 2p$_{3/2}$ XPS spectra of selected TiO_2/Cu_xO_y NTs ($Ti_{90}Cu_{10}$ alloys anodized at 30–50 V).

2.4. UV-Vis Spectra and Photoluminescence Properties

Figure 4a demonstrates UV-Vis spectra that were prepared for all investigated samples, not only alloys but also pristine NTs. The analysis results indicate clearly UV signals at the region 300–390 nm, which are strictly connected with the excitation state of the electrons and the movement from valence to conduction band [17]. $Ti_{95}Cu_5$_30V reflects high absorbance of UV light ($\lambda < 380$ nm) in comparison to pristine NTs. All modified NTs generally present more intensive absorbance in Vis irradiation range than pristine NTs, which can lead to intensified response during experiments in visible light. Maximum absorbance was observed at 600 nm with $Ti_{90}Cu_{10}$_30V, $Ti_{95}Cu_5$_30V, and $Ti_{85}Cu_{15}$_30V, and stands for red shift compared to pristine NTs. For $Ti_{90}Cu_{10}$_40V alloy, the absorbance maximum was not registered. Increased absorption values in Vis range in comparison to pristine NTs is a result of Cu_xO_y appearance in the system and their narrower band gap ($Cu_2O = 2.1$ eV; $CuO = 1.7$ eV), which work as a photo sensitizer that broadens the photo response of TiO_2 NTs to the visible region. The wide absorption band could have occurred because of inter-band transition in the Ti_xCu_y alloys. Figure 4a reveals a slight shift in band-gap transition of modified NTs to longer wavelengths. This effect can be assigned to stronger stabilization of the conduction band of TiO_2, Cu_2O, and CuO than their valence band [17].

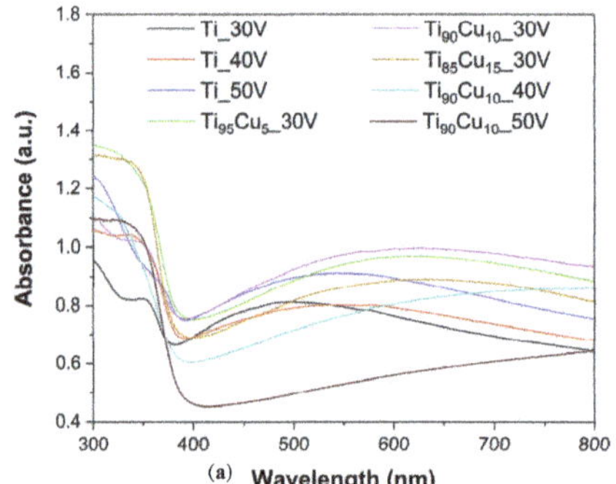
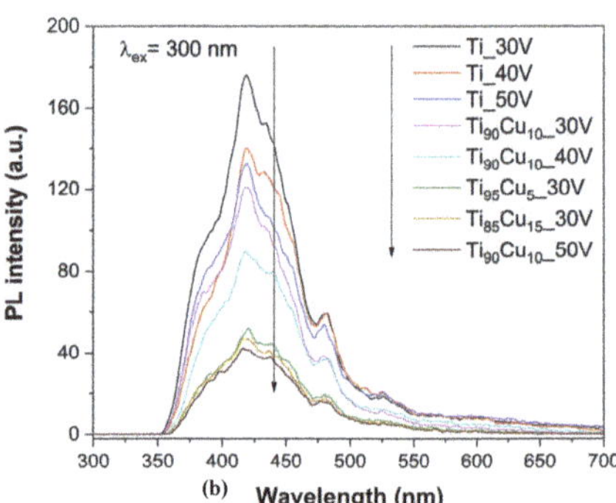

Figure 4. Absorbance (**a**) and photoluminescence (**b**) spectra of pristine TiO_2 and TiO_2/Cu_xO_y NTs.

The photoluminescence (PL) is a nondestructive spectroscopic technique often applied in study of intrinsic and extrinsic properties of both bulk semiconductors and nanostructures [46]. Mainly, the analysis of spectra of nanostructures can help in the characterization of the structure, providing information on the interface morphology and the quality of the materials. Photoluminescence (PL) signals and their intensity are closely related to photocatalytic activity. Possibly, the lower the PL intensity, the higher the separation rate of photo-induced charges, and, possibly, the higher the photocatalytic activity [47]. There are three intensive bands presented in Figure 4b, which can be distinguished. Two of them lie in the range between 400 and 440 nm and originate from charge recombination in the surface state defects [48]. The bands at 445–500 nm are attributed to different intrinsic defects in the TiO_2 lattice such as oxygen vacancies, titanium vacancies, and interstitial defects [49]. Because of different thicknesses of obtained NTs, the recombination rate intensity may be highly influenced. Figure 4b reveals TiO_2-Cu_xO_y NTs and pristine TiO_2 NTs response on PL, showing that TiO_2-Cu_xO_y NTs were more efficient than pristine TiO_2 NTs because of better charge carriers separation ability, which leads to improved photocatalytic properties.

2.5. Photodegradation Ability in Aqueous Phase

Photodegradation ability of obtained NTs was studied in phenol degradation process proceed in two different light sources: UV-Vis ($\lambda > 350$ nm) and Vis ($\lambda > 420$ nm) conditions. At the beginning, a direct photolysis was performed to confirm the necessity of photocatalyst appearance. In case of UV-Vis irradiation, 3% of phenol was degraded, and in Vis light no action was observed. The significant results for UV-Vis were presented by samples with applied voltage above 30 V. $Ti_{90}Cu_{10}\_50V$ and $Ti_{90}Cu_{10}\_40V$ have degradation efficiencies close to 85% and 45%, respectively, while pristine NTs showed efficiency between 35–40% (Figure 5a). NTs obtained with voltage equal to 30 V revealed the efficiency lower than pristine NTs under UV-Vis light. The reason of obtained results is strictly connected with length of the NTs, as pristine NTs are longer (1.5–6 μm) than NTs from copper alloys anodized in 30 V, while in terms of $Ti_{90}Cu_{10}\_40V$ and $Ti_{90}Cu_{10}\_50V$, NTs are just a little shorter (2.5–3.5 μm) than pristine NTs but sufficient for conducting the degradation process. Combination of length and Cu_2O and CuO species in total gives higher activity compering to pristine NTs. In terms of Vis light irradiation (Figure 5b), whose results were the main area of interest, the outcome appears to be different. The efficiency increased in the following order: $Ti\_30V < Ti\_40V < Ti\_50V < Ti_{90}Cu_{10}\_50V < Ti_{90}Cu_{10}\_40V < Ti_{95}Cu_5\_30V < Ti_{90}Cu_{10}\_30V < Ti_{85}Cu_{15}\_30V$ showing that NTs obtained with 30 V presents the best results, reaching a value of 16% for the $Ti_{85}Cu_{15}\_30V$ sample. Obtained result can relate to the correlance of penetration depth of visible light and thickness of the NTs, in which only small amounts of photons can reach lower parts of the NTs. The level of influence on phenol degradation in terms of Vis light can lay also in amount of Cu in the sample. EDX results presented in Table 1 show that in $Ti_{85}Cu_{15}\_30V$ the highest amount of Cu was measured with the value of 9.45 (wt.%). The same results were obtained by XPS measurements (Table 2). Similar effect was observed during our previous work with the NTs obtained by anodic oxidation of Ti-Ag alloys [50].

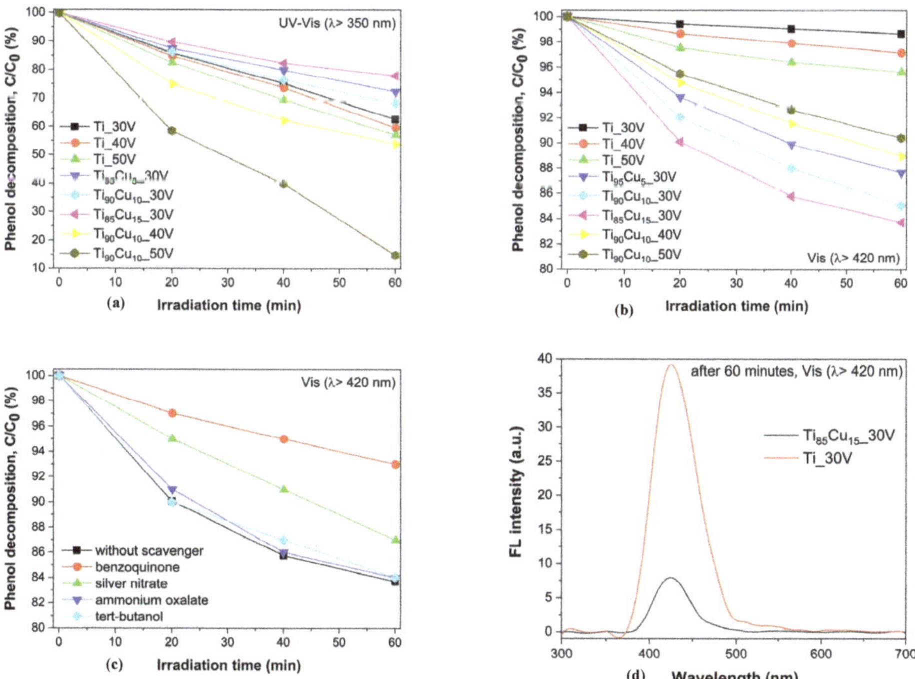

Figure 5. Kinetics of photocatalytic degradation of phenol under UV-Vis (**a**) and Vis (**b**) irradiation, photocatalytic decomposition of phenol under visible light irradiation in the presence of scavengers, and $Ti_{95}Cu_{15}\_30V$ sample (**c**) and •OH radical generation efficiency under Vis irradiation (**d**) in the presence of selected samples.

2.6. The Excitation Mechanism of TiO$_2$-Cu$_x$O$_y$ NTs

For better understanding of the visible light excitation mechanism of TiO$_2$, NTs modified with Cu species additional experiments were performed. First of all, to confirm phenol degradation mechanism, reactions with different scavengers-benzoquinone (for O$_2\bullet^-$ radicals), silver nitrate (for e$^-$), ammonium oxalate (for h$^+$), and tert-butanol (for hydroxyl radicals) were performed for the most active sample Ti$_{85}$Cu$_{15}$_30V, as shown in Figure 5c. The photocatalytic efficiency of phenol degradation for benzoquinone was 13% and for silver nitrate 6%. In terms of other scavengers, the degradation was negligible in comparison with the reaction without scavengers what can lead to conclusions that O$_2\bullet^-$ radicals are the main initiator of photocatalytic degradation under visible light irradiation. Secondly, to confirm above results, the $\bullet$OH radical generation tests were taken (see Figure 5d), confirming that larger amounts of $\bullet$OH radicals were produced in case of pristine NTs than in presence of TiO$_2$-Cu$_x$O$_y$ NTs. It can incline that other forms of reactive oxygen species are the source of phenol degradation under Vis light (as demonstrated in Figure 6a).

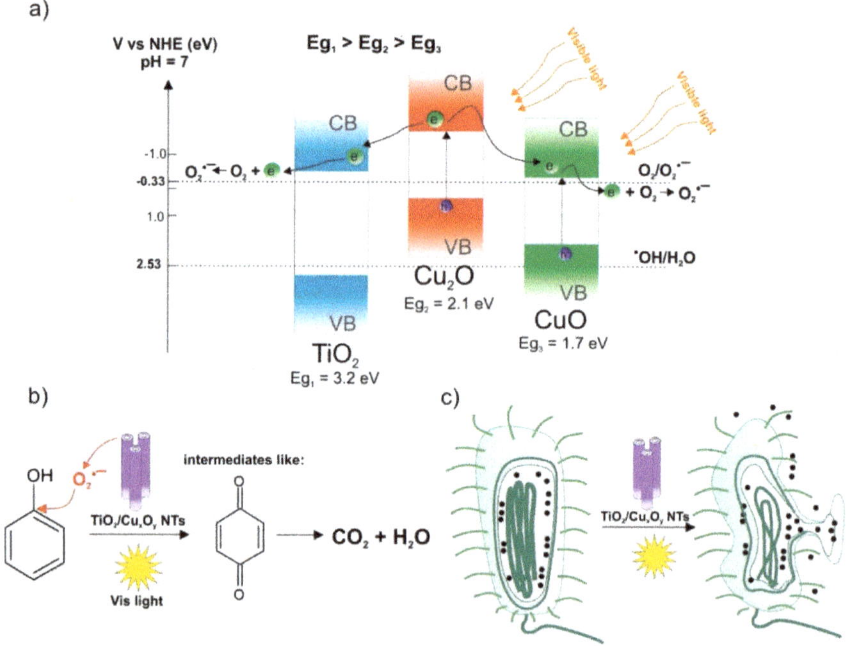

Figure 6. A suggested scheme of (**a**) visible light excitation mechanism in presence of TiO$_2$-Cu$_2$O and CuO NTs, (**b**) phenol degradation pathway under Vis light, and (**c**) inactivation of bacterial cell by O$_2\bullet^-$ radicals.

As the results of XPS showed division of the TiO$_2$-Cu$_x$O$_y$ NTs in terms of Cu type, the Vis light photocatalytic mechanism (Figure 6a) and the stability of heterojunctions should be considered for both types Cu$^+$ and Cu^{2+}. As the TiO$_2$ NTs band gap (3.2 eV) is too wide to absorb Vis light, TiO$_2$-Cu$_2$O were investigated for band gaps much narrower 2.1 eV and 1.7 eV, respectively, to confirm the application of materials in Vis light. TiO$_2$-Cu$_2$O and TiO$_2$-CuO NTs with the Vis light conditions can activate electrons and create the pair electron-hole [51]. In terms of TiO$_2$-Cu$_2$O NTs, electron from CB Cu$_2$O is transferred to the conduction band of TiO$_2$ NTs, whereas holes from valence band of Cu$_2$O remain immobile and are unable to generate $\bullet$OH radicals, because the band edge potential is lower than edge for $\bullet$OH radical generation with potential of 2.53 eV. The electron path from CB of TiO$_2$ NTs transfers further to the environment reacting with the oxygen and creating superoxide anion radicals $O_2\bullet^-$ and then H_2O_2 and $HO_2\bullet$. When analyzing the reaction with TiO_2-CuO NTs, the situation is very much different. As the band gap of CuO (1.7 eV) enables electrons to move conduction band of CuO, the electrons react with oxygen forming superoxide anion radicals $O_2\bullet^-$ and then H_2O_2 and $HO_2\bullet$. Furthermore, the reaction with pollutant (phenol) after several processes led to formation of intermediates and finally H_2O and CO_2 (as shown in Figure 6b).

Moreover, it is confirmed that redox potentials for the reduction of Cu_2O to Cu and for oxidation Cu_2O to CuO occur and should be taken into consideration. However, Weng et al reported that in presence of even small amount of CuO, which works as a protection shield, the possibility of Cu_2O photocorrosion is less probable [52]. XPS analysis showed (Table 2) that there are trace amounts of CuO in tested NTs alloys that can lead to the conclusion that Cu_2O is resistant to photocorrosion process. All results above and analysis presented by Luna et al. [53] can lead to proposed Cu_2O-CuO-TiO_2 complex mechanism (Figure 6a).

2.7. Assessment of Antibacterial Properties of TiO_2/Cu_xO_y NTs

Microbial contamination of the environment can be a critical issue for many aspects of our lives. That is why the modification of surface with different additives, which can foster the process of bacterial inactivation, is important [54]. The assessment of the antibacterial properties of photocatalytic layer—TiO_2/Cu_xO_y NTs obtained from $Ti_{95}Cu_5$ alloy in Vis light—was investigated in three different configurations in presence of microorganisms, Ti-Cu NTs, Vis light, microorganisms and Vis light only, and microorganisms and Ti-Cu NTs in the dark. Three different bacterial strain were used: *E. coli*, *B. subtilis*, and *Clostridium* sp. The amount of bacteria after the process was measured quantitatively (CFU/mL (colony-forming unit)) and with usage of TEM (Table 3 and Figure 7, respectively).

Table 3. Efficiency of bacteria inactivation in aqueous phase in the presence of the $Ti_{95}Cu_5\_30V$ sample and visible light ($\lambda > 420$ nm).

Bacterial Strain	Experimental Conditions	Efficiency after 60 min
	Light source: switched on Bacteria: present Photocatalytic layer: present	97%
E. coli — OD = 0.09 **STARTING CFU/mL: 3.3 $\times$ 10^2**	Light source: switched off Bacteria: present Photocatalytic layer: present	12%
	Light source: switched on Bacteria: present Photocatalytic layer: absent	3%
	Light source: switched on Bacteria: present	Did not grow
B. subtilis — OD = 0.09 **STARTING CFU/mL: 2.5 $\times$ 10^2**	Photocatalytic layer: present Light source: switched off Bacteria: present Photocatalytic layer: present	Did not grow
	Light source: switched on Bacteria: present Photocatalytic layer: absent	16%
	Light source: switched on Bacteria: present Photocatalytic layer: present	98%
Clostridium sp. — OD = 0.1 **STARTING CFU/mL: 3.8 $\times$ 10^2**	Light source: switched off Bacteria: present Photocatalytic layer: present	0%
	Light source: switched on Bacteria: present Photocatalytic layer: absent	5%

The quantitative analysis shows that correlation between efficiency of the process and experimental conditions are clearly noticeable. The significant results were obtained for measurements in presence of Vis light source and photocatalytic layers in case of two strains: *E. coli* (97% damage in 60 min) and *Clostridium* sp. (98% damage in 60 min). What is more, in example of *B. subtilis* there was no response of bacteria and it never grew (all results presented in Table S3). The lack of *B. subtilis* growth can be related to conditions of the experiment or the influence of Cu^{2+} ions leached from surface of nanotubes.

B. subtilis can form endospores which are small, metabolically dormant cells remarkably resistant to heat, desiccation, radiation, and chemical insult. Sporulation is the last response to nutrient starvation or other stressful conditions. Alternative condition to forming the spores is high cell density favors sporulation. The process of endospore formation has profound morphological and physiological consequences [55,56]. Endospores are produced in a process called sporulation, which is reversible, because the vegetative cell can be regenerated again when only in the environment there will be favorable conditions. This reverse process to sporulation is called germination of spores. This process requires a number of spore specific proteins. Most proteins are associated with the inner spore membrane [57]. Germination of spores depends on many conditions present in bacterial growth and formulation way of spores. Stress conditions for bacteria (irradiation of the catalytic layer connected radiation) led to the formation of spores, which under normal conditions of Bacillus sp. growth were unable to undergo germination.

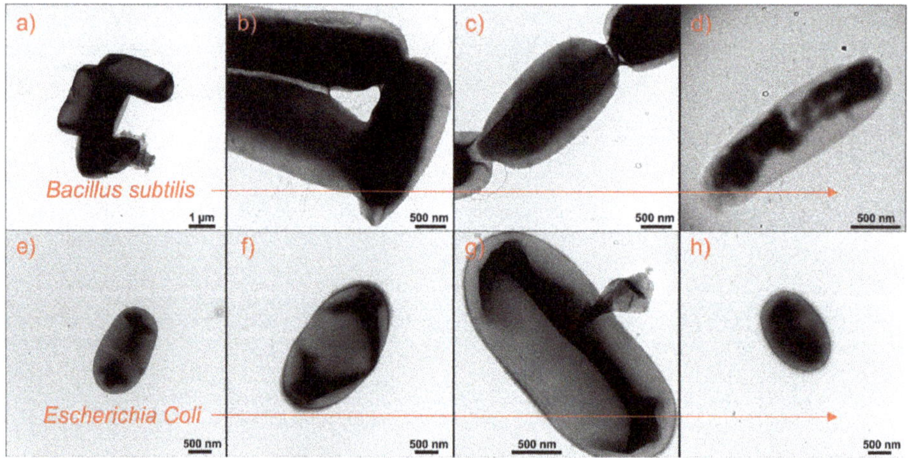

Figure 7. TEM images of bacteria after 60 min of various processes: (**a**,**e**) reference bacteria (**b**,**f**) light switched off, bacteria present, photocatalyst present, (**c**,**g**) light switched on, bacteria present, photocatalyst absent and (**d**,**h**) light switched on, bacteria present, photocatalyst present.

However, there are also differences between germination of *Bacillus spores* and *Clostridium spores*. First, while germinant receptor (GR) function in germination *Bacillus* sp. requires all three GR subunits. It appears that only a GRc subunit alone can facilitate *Clostridia spore* germination. Second, spores of some *Clostridia* lack GRs with any similarity to those in spores of bacilli, likely a reflection of the great diversity in the *Clostridia*. However, *C. difficile spores* do germinate well with specific bile salts and also respond to various amino acids [57]. In the case of *E. coli*, these types of processes are not observed because they do not sporulate.

For confirmation of previous analysis, the effect of Cu^{2+} on the growth of *E. coli* and *B. subtilis* was investigated (Figure S3 and Table S4). The growth of bacterial cultures (with the addition of Cu^{2+} ions) was controlled by measuring the optical density at 600 nm. Growth inhibition was found for both strains in the first two dilutions. The minimal inhibitory concentration (MIC) is 0.1 mM of Cu^{2+} ions. A similar effect was observed by Zong et al. [34]. After obtaining above results, only *E. coli* and *B. subtilis* were sent for TEM analysis to confirm the transformation of bacteria. Images revealed changes in bacteria shape in every tested configuration with the biggest influence of photocatalytic layer on *B. subtilis* (clear damage of the inside structure) and light source on *E. coli* (whole deformation of bacteria). Deformed bacterial cell cannot grow and is no longer a serious environmental problem. Finally, it is likely that the cell membrane will undergo complete mineralization [58]. In summary, the as-prepared photocatalyst irradiated with Vis light has very high, indisputable bactericidal effects (in both cases, efficiency of bacteria inactivation reached 97%). *E. coli*, *Clostridium* sp., and *B. subtilis* are removed from water and the surface of the photocatalysts within 60 min. of irradiation.

2.8. Suggested Mechanism of Bacteria Inactivation

A full knowledge about the mechanism of bacteria inactivation is essential for further development of nanocomposites involved in disinfection processes. Matsunga et al. were the pioneers with hypothesis in which Coenzyme A is degraded by ROS in the presence of light [21]. The enzyme cut off can cause respiration problems, possibly leading to death. However further research performed by Saito et al. confirmed that main reason for cell death is the burst of the cell wall membrane leading to leakages [59]. Obtained TEM images (Figure 7) for experiment with $Ti_{95}Cu_5\_30V$ confirmed the destruction of bacteria cell structure and wall membrane in presence of photocatalyst and light, which led to conclusion that copper as a component photocatalyst influenced the process. Moreover, based on analysis made by Kikuchi et al., who confirmed the role of the ROS by addition of hydroxyl scavengers to the reaction system [60], it can be concluded that for TiO_2-Cu_XO_y, another source of bacteria inactivation lies in oxygen radicals. The suggested scheme presenting damage of bacteria cell during the inactivation process is presented in Figure 6c.

3. Materials and Methods

3.1. Materials

The titanium foils and alloys were purchased from HMW Hauner (Röttenbach, Germany). Isopropanol, acetone, and methanol (p.a., POCh S.A., Gliwice, Poland) were used for cleaning Ti foil and alloys surface. NH_4F (p.a.) and ethylene glycol (99.0%, p.a.) purchased from POCh S.A. were the components of the electrolyte, which were used for preparation of the TiO_2 nanotubes. Deionized water used during experiments had conductivity of 0.05 μS.

The following bacterial strains were used in this work: *E. coli* DSMZ collection no 1116, *B. subtilis* DSMZ collection no 347, and *Clostridium* sp. DSMZ collection no 2634. *E. coli* is a Gram-negative, facultatively anaerobic, rod-shaped bacteria. *E. coli* is found in the gut of animals, including human gut, as well as commonly in soil and water. Bacteria does not create endospores. *B. subtilis* is Gram-positive, facultative anaerobe, and rod-shaped found in soil and the gastrointestinal tract of ruminants and humans. It can form a tough, protective endospore, allowing it to tolerate extreme environmental conditions. *Clostridium* sp. is Gram-positive, obligate anaerobes, rod-shaped bacteria. They are producing endospores commonly found mainly in the soil and digestive tract of animals, including humans, female reproductive organs, as well as in water and sewage. These bacteria are characterized by the possibility of binding atmospheric nitrogen and reduction of sulphites [61].

To measure the copper influence on growth of bacteria cultures, the bacterial cultures mentioned above were conducted in LB medium (1% tryptone, 0.5% yeast extract, 1% NaCl, pH 7.0), supplemented with 98% copper acetate ($C_4H_6CuO_4 \times H_2O$) from Avantor Performance Materials S.A. (Gliwice, Poland).

3.2. Preparation of NTs

Ti and Ti-Cu alloys containing different amounts of copper (5, 10, 15 wt.%) in the form of sheets were cut into pieces of size 2 × 3 cm. In the first step, experiment samples were cleaned with usage of acetone, isopropanol, methanol, and deionized water separately, one by one, in ultrasonic bath for 10 min. Cleaning process was finalized with drying alloys in the air steam. In the second step, the set up for anodization process was established with two electrodes-platinum mesh as the cathode and Ti/Ti-Cu alloy as a working electrode. Moreover, the Ag/AgCl reference electrode was incorporated into the system to gather information about the definite potential of the electrode. The anodic oxidation process was performed in the presence of electrolyte solution (98 vol % ethylene glycol, 2 vol % water and 0.09 M NH_4F) for 60 min with applied voltage in a range between 30 and 50 V. The anodization was monitored with DC power supply (MANSON SDP 2603, Hong Kong, China). After all, in third step, samples were sonicated in deionized water for 5 min, dried in air stream at 80 °C for 24 h, and calcinated at 450 °C (heating rate of 2 °C/min) for 1 h.

3.3. Characterization Systems

To understand the morphology of NTs obtained by electrochemical method, scanning electron microscopy was performed (SEM-FEI Quanta 250 FEG FEI Company, Brno, Czech Republic). To locate the anomalies, structure high-resolution transmission electron microscopy (HRTEM Jeol ARM 200F, Akishama, Tokio, Japan)) was carried out. The X-Ray photoelectron spectroscopy (XPS) measurements were completed at PHI 5000 VersaProbeTM (ULVAC-PHI, Chigasaki, Japan) spectrometer with monochromatic Al $K\lambda$ radiation (hv = 1486.6 eV). Phase composition on the surface was checked using a room temperature powder X-ray X'Pert Pro MPD diffractometer (PANalytical, Almelo, The Netherlands)(CuK$\alpha\lambda$ 1.5406 Å). A LeBail refinement was performed using the HighScore software (Ver. 3.0d, PANanalytical, Almelo, The Netherlands). The binding energy (BE) scale of all detected high resolution (HR) spectra was referenced by setting the BE of the aliphatic carbon peak (C-C) signal to 284.6 eV. The photoluminescence (PL) measurements were taken at room temperature with LS-50B Luminescence Spectrophotometer with Xenon discharge lamp as an excitation source and special detector—a R928 photomultiplier (HAMAMATSU, Hamamatsu, Japan). The excitation radiation (360 nm) was directed on the sample's surface at an angle of 90°. The UV-Vis reflectance and absorbance spectra of pure and copper doped NTs were obtained with usage of Shimadzu UV-Vis. Spectrophotometer (UV 2600) (SHIMADZU, Kioto, Japan), with reference samples of barium sulphate. The range for the spectra registration was between 300 and 800 nm in room temperature and set scanning speed of 250 nm/min.

3.4. Photocatalytic Activity

The equipment used for all photocatalytic activity measurements was as follows: a quartz reactor with the capacity of 10 mL, 1000 W Xenon Lamp (Oriel 66021 Stratford, CT, USA), and cut off filters. Light intensity was measured for both filters with value of 40 mW/cm^2 for UV-Vis and 2 mW/cm^2 for Vis range.

3.4.1. Phenol Degradation Process

The process of photocatalytic degradation in model reaction with usage of phenol for two light sources was performed. The irradiation was controlled by two cut off filters GG350 (UV-Vis λ > 350 nm) and GG 420 (Vis λ > 420 nm). To perform the experiment, prepared phenol solution (20 mg/L) in amount of 8 mL was applied to the reactor, followed by immersing the examined alloy. Prepared sample was placed on a stirrer (500 rpm) and irradiated with preferable light for 60 min. Not only reference phenol solution (0.5 mL) was taken before the beginning of procedure, but also samples (0.5 mL) after each 20 min of the irradiated process were collected. We chose colorimetric method in the presence of p-nitroaniline and with usage of UV-Vis spectrophotometer (λ_{max} = 480 nm) to investigate the phenol concentration.

3.4.2. Microorganisms Inactivation Process

The experiment connected with bacteria inactivation process was performed with application of three microorganisms species—*E. coli*, *B. subtilis*, and *Clostridium* sp. All vessels and media used in bacterial experiments were pre-sterilized by steam. The preparation of bacteria were harvested in LB medium by shaken in air shaker at 37 °C for 16 h. The bacteria pellet was isolated from medium by centrifugation at 2739 $\times$ *g* for 10 min. The resulting pellet was resuspended in sterile water to final concentration *E. coli* 3.3 $\times$ 10^2, *B. subtilis* 2.5 $\times$ 10^2, *Clostridium* sp. 3.8 $\times$ 10^2 (OD_{600nm} = 0.1). As it is proven that microorganisms die in UV irradiation range, the Vis range with the cut off filter GG420 was used. As a photocatalyst, $Ti_{95}Cu_5$ alloy was applied.

The idea of the experiment was to check the inactivation process in three control tests—in presence of microorganisms, Ti-Cu NTs, and Vis light; in presence of microorganisms and Vis light only; and in the presence of microorganisms and Ti-Cu NTs in the dark. To perform the experiment, the suspension of bacteria (8 mL) was applied in the same reactor as in phenol degradation process. Depending on the configuration, Ti-Cu NTs were immersed in bacteria suspension or not. Prepared set up (with or without NTs) was placed on a stirrer (500 rpm) and irradiated with preferable light or kept in the dark for 60 min. Reference samples (1 mL) were collected just before each experiment, and consecutive samples (1 mL) were taken every 20 min. After the complete process there were two procedures of measurements. First, the collected samples were prepared by serial dilution (100 µL of sample in 900 µL of sterile 1 $\times$ PBS (phosphate-buffered saline)) and subsequent 10 µL of each were seeded on agar plate PCA (Plate Count Agar). The plates were incubated at 37 °C in an incubator for 16 h. Grown bacteria were counted, along with the amount of microorganisms in 1 mL of the dilution with the formula CFU = number of colonies $\times$ dilution/volume of inoculum. Second time, more diluted consecutive samples (1 mL taken every 20 min) were put in the ice and sent to resolution transmission electron microscopy.

3.4.3. Measurement of Copper Cu^{2+} Influence on Growth of Bacterial Cultures

The aim of the measurement was to check the influence of Cu^{2+} on the growth of bacterial cultures. This method was chosen to identify if the copper alone influenced the bacteria. The diluted solution of copper actetite was used to perform this measurement. The bacteria were harvested in LB medium by shaken in air shaker at 37 °C for 16 h. *B. subtilis* was added to LB medium to OD_{600nm} = 0.13 and *E. coli* to OD_{600nm} = 0.1 and were harvested in air shaker at 37 °C for 30 min. In the test tube, serial dilution of Cu^{2+} ions were prepared from 1–1 $\times$ 10^{-9} mM. Then, OD_{600nm} was measured; the negative control was LB medium, and the positive control was LB medium with bacteria with appropriate OD_{600nm}.

To determine the influence of Cu^{2+} ions on *B. subtilis* culture, the 2-fold serial dilution test of ions was tested. The *B. subtilis* was culture in LB medium by shaken in air shaker at 37 °C for 16 h. *B. subtilis* was added to LB medium to OD_{600nm} = 0.1 and harvested in air shaker at 37 °C for 1.5 h. In the test tube 2-fold serial dilution of Cu^{2+}, ions were prepared from 10–1.9 $\times$ 10^{-2}. The OD_{600nm} was measured every 30 min; the negative control was LB medium, and the positive control was LB medium with bacteria with appropriate OD_{600nm}.

3.4.4. Measurement of Hydroxyl Radicals

As •OH radicals are considered the most powerful in the oxidation process of many organic compounds [62], the role and amount was determined to understand the photocatalytic properties. A terephthalic acid, as a substance which effectively captures •OH radicals and generates highly fluorescent product, was used and investigated for photoluminescence intensity. The tests were performed in the same laboratorium set up as in the phenol degradation process with the initial concentration of terephthalic acid equal Co. = 0.5 mM. The reactor was irradiated for 60 min, with 20 min intervals for sample collection. The photoluminescence spectra of all collected samples were measured at LS–50 B luminescence spectrophotometer (HAMAMATSU, Hamamatsu, Japan) with lamp with excitation wavelength at 315 nm and photomultiplier detector.

3.4.5. Investigation of Photodegradation Mechanism

In order to understand the mechanism of photodegradation of phenol solution silver nitrate, ammonium oxalate, tert-buthyl alcohol, and benzoquinone were applied, which are scavengers for e^-, h^+, $\bullet OH$, and $O_2\bullet^-$ radicals, respectively. The photodegradation experiment conditions were the same as in phenol degradation process.

4. Conclusions

The analysis of new spectra of nanostructures developed by electrochemical method TiO_2/Cu_xO_y NT arrays due to their unique features can be used not only in the photodegradation process but also in bacteria inactivation. In this study, the influence of copper amount and applied voltage in NTs formation was investigated, as well as the correlation of NTs composition with the efficiency of the photodegradation of pollutants in the aqueous phase. All prepared Ti_xCu_y foils after the anodization process have proved to be self-organized nanotubes arrays with external diameter of 85–97 nm and length 1.1–3.5 μm. The dimensions were directly induced by applied voltage, as the NTs were usually shorter and thicker than pristine TiO_2 NTs. Furthermore, photodegradation of phenol in the aqueous phase with the presence of NT photocatalysts was performed under Vis light irradiation ($\lambda > 420$ nm). The best performance was marked in the case of TiO_2/Cu_xO_y NTs arrays with a Cu content of 15 wt.%. Additionally, PL analysis and absorbance measurements were performed that revealed promising results in terms of NTs application in Vis light region. It was proved that oxygen vacancies activated by Cu dopant in the NTs could be the main factor in the phase stabilization of the anataze phase.

Moreover, the assessment of the antibacterial properties of new NTs was performed with very promising results regarding the efficiency of bacteria degradation in terms of *E. coli* (97% degradation in 60 min) and *Clostridium* sp. (98% degradation in 60 min) under visible light irradiation (using low CFU/mL of bacteria).

The anodization method with Cu-Ti alloys developed in this study is a simple and effective method that can be implemented on a larger scale, not only in environmental applications but also in medical industry. Our results are encouraging and should be continued in the study of complexes with more than two components during anodization. Further work needs to be carried out in direction of creating nanomaterials with cascade heterojunctions possessing accurate photoelectrochemical characterization, which can be cheaper and more controlled in comparison to two composite photocatalysts. Research in this area was already undertaken by analyzing Cu-Ag-Ti alloys. It is also a challenge to consider the antibiotic resistant strains of mutant microbs, which are a growing problem nowadays. Finally, what seems to be an important matter is the variety of standards used by research group to measure the efficiency of the photocatalysts and their influence on microorganisms.

Supplementary Materials: The following are available online at http://www.mdpi.com/2073-4344/8/6/237/s1, Figure S1: Top-view and cross-sectional SEM images of pristine TiO_2 NTs, Figure S2: X-ray diffraction patterns for pristine TiO_2 NTs, Table S1: Refined lattice parameters for TiO_2—anatase, Ti—metal and $CuTiO_2$—alloy. The crystallite size was calculated for the anatase only, Table S2: Elemental composition (in at.%) in the surface layer of TiO_2 and Cu-modified TiO_2 NTs, evaluated by XPS analysis, Table S3: Efficiency of bacteria inactivation after 20, 40 and 60 min of various processes, Figure S3: Image of influence of Cu^{2+} ions on the growth of *B.Subtilis*, Table S4: The influence of Cu^{2+} ions on the growth of *E. coli* and *B. subtilis*. (a) OD measurements at 600 nm of *E. coli* and *B. subtilis* cultures with the addition of Cu^{2+} ions from $1–1 \times 10^{-9}$ mM (serial dilution), (b) OD measurements at 600 nm of *B. subtilis* culture with addition of Cu^{2+} ions from $10–1.9 \times 10^{-2}$ from $1–10^{-9}$ mM. To, T1, T2—the subsequent measurement points; Kp—positive control, culture of strain without Cu^{2+} ions; Kn—negative control, the medium.

Author Contributions: Conceptualization: A.Z.-M. and P.M.; methodology: A.Z.-M. and P.M.; investigation: M.K. (Magda Kozak), P.M., J.Ż., M.K. (Marek Kobylański), T.K., W.L., G.T., and G.N.; visualization: M.K. (Magda Kozak) and P.M.; writing-original draft preparation: M.K. (Magda Kozak) ; writing review & reediting: P.M and A.Z.-M.; supervision: A.Z.-M.; funding acquisition: A.Z.-M.

Funding: This research was financially supported by the Polish National Science Center (grant No. NCN 2014/15/B/ST5/00098, ordered TiO_2/M_xO_y nanostructures obtained by electrochemical method).

Conflicts of Interest: The authors declare no conflict of interest.

References

1. Hoffmann, M.R.; Martin, S.T.; Choi, W.; Bahnemann, D.W. Environmental Applications of Semiconductor Photocatalysis. *Chem. Rev.* **1995**, *95*, 69–96. [CrossRef]
2. Fujishima, A.; Rao, T.N.; Tryk, D.A. Titanium dioxide photocatalysis. *J. Photochem. Photobiol. C Photochem. Rev.* **2000**, *1*, 1–21. [CrossRef]
3. Low, J.; Cheng, B.; Yu, J. Surface modification and enhanced photocatalytic CO_2 reduction performance of TiO_2: A review. *Appl. Surf. Sci.* **2017**, *392*, 658–686. [CrossRef]
4. Ni, M.; Leung, M.K.H.; Leung, D.Y.C.; Sumathy, K. A review and recent developments in photocatalytic water-splitting using TiO_2 for hydrogen production. *Renew. Sustain. Energy Rev.* **2007**, *11*, 401–425. [CrossRef]
5. Yadav, H.M.; Kim, J.-S.; Pawar, S.H. Developments in photocatalytic antibacterial activity of nano TiO_2: A review. *Korean J. Chem. Eng.* **2016**, *33*, 1989–1998. [CrossRef]
6. Zhang, S.; Zhang, S.; Peng, F.; Zhang, H.; Liu, H.; Zhao, H. Electrodeposition of polyhedral Cu_2O on TiO_2 nanotube arrays for enhancing visible light photocatalytic performance. *Electrochem. Commun.* **2011**, *13*, 861–864. [CrossRef]
7. Pelaez, M.; Nolan, N.T.; Pillai, S.C.; Seery, M.K.; Falaras, P.; Kontos, A.G.; Dunlop, P.S.M.; Hamilton, J.W.J.; Byrne, J.A.; O'Shea, K.; et al. A review on the visible light active titanium dioxide photocatalysts for environmental applications. *Appl. Catal. B Environ.* **2012**, *125*, 331–349. [CrossRef]
8. Nevárez-Martínez, M.; Kobylański, M.; Mazierski, P.; Wółkiewicz, J.; Trykowski, G.; Malankowska, A.; Kozak, M.; Espinoza-Montero, P.; Zaleska-Medynska, A. Self-Organized TiO_2–MnO_2 Nanotube Arrays for Efficient Photocatalytic Degradation of Toluene. *Molecules* **2017**, *22*, 564. [CrossRef]
9. Daraio, C.; Jin, S. Synthesis and Patterning Methods for Nanostructures Useful for Biological Applications. In *Nanotechnology for Biology and Medicine*; Silva, G.A., Parpura, V., Eds.; Springer: New York, NY, USA, 2012; pp. 27–44. ISBN 978-0-387-31282-8.
10. Nevárez-Martínez, M.; Mazierski, P.; Kobylański, M.; Szczepańska, G.; Trykowski, G.; Malankowska, A.; Kozak, M.; Espinoza-Montero, P.; Zaleska-Medynska, A. Growth, Structure, and Photocatalytic Properties of Hierarchical V_2O_5–TiO_2 Nanotube Arrays Obtained from the One-step Anodic Oxidation of Ti–V Alloys. *Molecules* **2017**, *22*, 580. [CrossRef]
11. Mohammed, M.T.; Khan, Z.A.; Siddiquee, A.N. Surface Modifications of Titanium Materials for developing Corrosion Behavior in Human Body Environment: A Review. *Procedia Mater. Sci.* **2014**, *6*, 1610–1618. [CrossRef]
12. Jung, M.; Scott, J.; Ng, Y.H.; Jiang, Y.; Amal, R. CuO x dispersion and reducibility on TiO_2 and its impact on photocatalytic hydrogen evolution. *Int. J. Hydrogen Energy* **2014**, *39*, 12499–12506. [CrossRef]
13. Gong, D.; Grimes, C.A.; Varghese, O.K.; Hu, W.; Singh, R.S.; Chen, Z.; Dickey, E.C. Titanium oxide nanotube arrays prepared by anodic oxidation. *J. Mater. Res.* **2001**, *16*, 3331–3334. [CrossRef]
14. Macak, J.M.; Schmuki, P. Anodic growth of self-organized anodic TiO_2 nanotubes in viscous electrolytes. *Electrochim. Acta* **2006**, *52*, 1258–1264. [CrossRef]
15. Xie, Z.B.; Blackwood, D.J. Effects of anodization parameters on the formation of titania nanotubes in ethylene glycol. *Electrochim. Acta* **2010**, *56*, 905–912. [CrossRef]
16. Erjavec, B.; Tišler, T.; Tchernychova, E.; Plahuta, M.; Pintar, A. Self-Doped Cu-Deposited Titania Nanotubes as Efficient Visible Light Photocatalyst. *Catal. Lett.* **2017**, *147*, 1686–1695. [CrossRef]
17. Hai, Z.; EL Kolli, N.; Chen, J.; Remita, H. Radiolytic synthesis of Au–Cu bimetallic nanoparticles supported on TiO_2: Application in photocatalysis. *New J. Chem.* **2014**, *38*, 5279–5286. [CrossRef]
18. Lalitha, K.; Sadanandam, G.; Kumari, V.D.; Subrahmanyam, M.; Sreedhar, B.; Hebalkar, N.Y. Highly Stabilized and Finely Dispersed CuO/TiO: A Promising Visible Sensitive Photocatalyst for Continuous Production of Hydrogen from Glycerol:Water Mixtures. *J. Phys. Chem. C* **2010**, *114*, 22181–22189. [CrossRef]

19. Geng, Z.; Zhang, Y.; Yuan, X.; Huo, M.; Zhao, Y.; Lu, Y.; Qiu, Y. Incorporation of Cu_2O nanocrystals into TiO_2 photonic crystal for enhanced UV–visible light driven photocatalysis. *J. Alloys Compd.* **2015**, *644*, 734–741. [CrossRef]

20. Colón, G.; Maicu, M.; Hidalgo, M.C.; Navío, J.A. Cu-doped TiO_2 systems with improved photocatalytic activity. *Appl. Catal. B Environ.* **2006**, *67*, 41–51. [CrossRef]

21. Matsunaga, T.; Tomoda, R.; Nakajima, T.; Wake, H. Photoelectrochemical sterilization of microbial cells by semiconductor powders. *FEMS Microbiol. Lett.* **1985**, *29*, 211–214. [CrossRef]

22. *Polymeric Materials with Antimicrobial Activity: From Synthesis to Applications*; Muñoz-Bonilla, A.; Cerrada, M.; Fernández-García, M. (Eds.) Polymer Chemistry Series; Royal Society of Chemistry: Cambridge, UK, 2013; ISBN 978-1-84973-807-1.

23. Chen, L.; Luo, T.; Yang, S.; Xu, J.; Liu, Z.; Wu, F. Efficient metoprolol degradation by heterogeneous copper ferrite/sulfite reaction. *Environ. Chem. Lett.* **2018**, *16*, 599–603. [CrossRef]

24. Wojcieszak, D.; Mazur, M.; Kaczmarek, D.; Poniedziałek, A.; Osękowska, M. An impact of the copper additive on photocatalytic and bactericidal properties of TiO_2 thin films. *Mater. Sci. Pol.* **2017**, *35*. [CrossRef]

25. Reilche, H.; Dunn, W.W.; Bard, J.A. Allen Heterogeneous photocatalytic and photosynthetic deposition of copper on TiO_2 and WO_3 powders. *J. Phys. Chem.* **1979**, *83*, 2248–2251.

26. Moniz, S.J.A.; Tang, J. Charge Transfer and Photocatalytic Activity in CuO/TiO_2 Nanoparticle Heterojunctions Synthesised through a Rapid, One-Pot, Microwave Solvothermal Route. *ChemCatChem* **2015**, *7*, 1659–1667. [CrossRef]

27. Hu, Q.; Huang, J.; Li, G.; Chen, J.; Zhang, Z.; Deng, Z.; Jiang, Y.; Guo, W.; Cao, Y. Effective water splitting using CuO x/TiO_2 composite films: Role of Cu species and content in hydrogen generation. *Appl. Surf. Sci.* **2016**, *369*, 201–206. [CrossRef]

28. Ma, Q.; Liu, S.J.; Weng, L.Q.; Liu, Y.; Liu, B. Growth, structure and photocatalytic properties of hierarchical Cu–Ti–O nanotube arrays by anodization. *J. Alloys Compd.* **2010**, *501*, 333–338. [CrossRef]

29. Gao, L.; Qiu, Z.; Gan, W.; Zhan, X.; Li, J.; Qiang, T. Negative Oxygen Ions Production by Superamphiphobic and Antibacterial TiO_2/Cu_2O Composite Film Anchored on Wooden Substrates. *Sci. Rep.* **2016**, *6*. [CrossRef]

30. Cheng, M.; Yang, S.; Chen, R.; Zhu, X.; Liao, Q.; Huang, Y. Copper-decorated TiO_2 nanorod thin films in optofluidic planar reactors for efficient photocatalytic reduction of CO_2. *Int. J. Hydrogen Energy* **2017**, *42*, 9722–9732. [CrossRef]

31. He, X.; Zhang, G.; Wang, X.; Hang, R.; Huang, X.; Qin, L.; Tang, B.; Zhang, X. Biocompatibility, corrosion resistance and antibacterial activity of TiO_2/CuO coating on titanium. *Ceram. Int.* **2017**, *43*, 16185–16195. [CrossRef]

32. Yadav, H.M.; Otari, S.V.; Koli, V.B.; Mali, S.S.; Hong, C.K.; Pawar, S.H.; Delekar, S.D. Preparation and characterization of copper-doped anatase TiO_2 nanoparticles with visible light photocatalytic antibacterial activity. *J. Photochem. Photobiol. A Chem.* **2014**, *280*, 32–38. [CrossRef]

33. Nischk, M.; Mazierski, P.; Wei, Z.; Siuzdak, K.; Kouame, N.A.; Kowalska, E.; Remita, H.; Zaleska-Medynska, A. Enhanced photocatalytic, electrochemical and photoelectrochemical properties of TiO_2 nanotubes arrays modified with Cu, AgCu and Bi nanoparticles obtained via radiolytic reduction. *Appl. Surf. Sci.* **2016**, *387*, 89–102. [CrossRef]

34. Zong, M.; Bai, L.; Liu, Y.; Wang, X.; Zhang, X.; Huang, X.; Hang, R.; Tang, B. Antibacterial ability and angiogenic activity of Cu-Ti-O nanotube arrays. *Mater. Sci. Eng. C* **2017**, *71*, 93–99. [CrossRef]

35. Kim, D.; Fujimoto, S.; Schmuki, P.; Tsuchiya, H. Nitrogen doped anodic TiO_2 nanotubes grown from nitrogen-containing Ti alloys. *Electrochem. Commun.* **2008**, *10*, 910–913. [CrossRef]

36. He, J.-B.; Lu, D.-Y.; Jin, G.-P. Potential dependence of cuprous/cupric duplex film growth on copper electrode in alkaline media. *Appl. Surf. Sci.* **2006**, *253*, 689–697. [CrossRef]

37. Jiang, X.; Herricks, T.; Xia, Y. CuO Nanowires Can Be Synthesized by Heating Copper Substrates in Air. *Nano Lett.* **2002**, *2*, 1333–1338. [CrossRef]

38. Mazierski, P.; Nadolna, J.; Lisowski, W.; Winiarski, M.J.; Gazda, M.; Nischk, M.; Klimczuk, T.; Zaleska-Medynska, A. Effect of irradiation intensity and initial pollutant concentration on gas phase photocatalytic activity of TiO_2 nanotube arrays. *Catal. Today* **2017**, *284*, 19–26. [CrossRef]

39. Roy, P.; Berger, S.; Schmuki, P. TiO_2 Nanotubes: Synthesis and Applications. *Angew. Chem. Int. Ed.* **2011**, *50*, 2904–2939. [CrossRef]

40. Regonini, D.; Bowen, C.R.; Jaroenworaluck, A.; Stevens, R. A review of growth mechanism, structure and crystallinity of anodized TiO_2 nanotubes. *Mater. Sci. Eng. R Rep.* **2013**, *74*, 377–406. [CrossRef]

41. Grimes, C.A.; Mor, G.K. *TiO_2 Nanotube Arrays: Synthesis, Properties, and Applications*; Springer: Dordrecht, The Netherlands; New York, NY, USA, 2009; ISBN 978-1-4419-0067-8.

42. Valota, A.; LeClere, D.J.; Hashimoto, T.; Skeldon, P.; Thompson, G.E.; Berger, S.; Kunze, J.; Schmuki, P. The efficiency of nanotube formation on titanium anodized under voltage and current control in fluoride/glycerol electrolyte. *Nanotechnology* **2008**, *19*, 355701. [CrossRef]

43. Biesinger, M.C.; Lau, L.W.M.; Gerson, A.R.; Smart, R.S.C. Resolving surface chemical states in XPS analysis of first row transition metals, oxides and hydroxides: Sc, Ti, V, Cu and Zn. *Appl. Surf. Sci.* **2010**, *257*, 887–898. [CrossRef]

44. Naumkin, A.V.; Kraut-Vass, A.; Gaarenstroom, S.W.; Powell, C.J. NIST X-ray Photoelectron Spectroscopy Database. 2012. Available online: https://srdata.nist.gov/xps/ (accessed on 4 June 2018).

45. Suzuki, S.; Hirabayashi, K.; Mimura, K.; Okabe, T.; Isshiki, M.; Waseda, Y. Surface Layer Formed by Selective Oxidation in High-Purity Copper-Titanium Binary Alloys. *Mater. Trans.* **2002**, *43*, 2303–2308. [CrossRef]

46. Sanguinetti, S.; Guzzi, M.; Gurioli, M. Accessing structural and electronic properties of semiconductor nanostructures via photoluminescence. In *Characterization of Semiconductor Heterostructures and Nanostructures*; Elsevier: New York, NY, USA, 2008; pp. 175–208. ISBN 978-0-444-53099-8.

47. Liqiang, J.; Yichun, Q.; Baiqi, W.; Shudan, L.; Baojiang, J.; Libin, Y.; Wei, F.; Honggang, F.; Jiazhong, S. Review of photoluminescence performance of nano-sized semiconductor materials and its relationships with photocatalytic activity. *Sol. Energy Mater. Sol. Cells* **2006**, *90*, 1773–1787. [CrossRef]

48. Garlisi, C.; Szlachetko, J.; Aubry, C.; Fernandes, D.L.A.; Hattori, Y.; Paun, C.; Pavliuk, M.V.; Rajput, N.S.; Lewin, E.; Sá, J.; Palmisano, G. N-TiO_2/Cu-TiO_2 double-layer films: Impact of stacking order on photocatalytic properties. *J. Catal.* **2017**, *353*, 116–122. [CrossRef]

49. Beltran-Huarac, J.; Guinel, M.J.-F.; Weiner, B.R.; Morell, G. Bifunctional Fe_3O_4/ZnS:Mn composite nanoparticles. *Mater. Lett.* **2013**, *98*, 108–111. [CrossRef]

50. Mazierski, P.; Malankowska, A.; Kobylański, M.; Diak, M.; Kozak, M.; Winiarski, M.J.; Klimczuk, T.; Lisowski, W.; Nowaczyk, G.; Zaleska-Medynska, A. Photocatalytically Active TiO_2/Ag_2O Nanotube Arrays Interlaced with Silver Nanoparticles Obtained from the One-Step Anodic Oxidation of Ti–Ag Alloys. *ACS Catal.* **2017**, *7*, 2753–2764. [CrossRef]

51. Park, S.-M.; Razzaq, A.; Park, Y.H.; Sorcar, S.; Park, Y.; Grimes, C.A.; In, S.-I. Hybrid Cu_xO–TiO_2 Heterostructured Composites for Photocatalytic CO_2 Reduction into Methane Using Solar Irradiation: Sunlight into Fuel. *ACS Omega* **2016**, *1*, 868–875. [CrossRef]

52. Zhang, Z.; Wang, P. Highly stable copper oxide composite as an effective photocathode for water splitting via a facile electrochemical synthesis strategy. *J. Mater. Chem.* **2012**, *22*, 2456–2464. [CrossRef]

53. Luna, A.L.; Valenzuela, M.A.; Colbeau-Justin, C.; Vázquez, P.; Rodriguez, J.L.; Avendaño, J.R.; Alfaro, S.; Tirado, S.; Garduño, A.; De la Rosa, J.M. Photocatalytic degradation of gallic acid over CuO–TiO_2 composites under UV/Vis LEDs irradiation. *Appl. Catal. A Gen.* **2016**, *521*, 140–148. [CrossRef]

54. Chen, L.; Tang, M.; Chen, C.; Chen, M.; Luo, K.; Xu, J.; Zhou, D.; Wu, F. Efficient Bacterial Inactivation by Transition Metal Catalyzed Auto-Oxidation of Sulfite. *Environ. Sci. Technol.* **2017**, *51*, 12663–12671. [CrossRef]

55. Stephens, C. Bacterial sporulation: A question of commitment? *Curr. Biol.* **1998**, *8*, R45–R48. [CrossRef]

56. Grossman, A.D.; Losick, R. Extracellular control of spore formation in *Bacillus subtilis*. *Proc. Natl. Acad. Sci. USA* **1988**, *85*, 4369–4373.

57. Setlow, P. Germination of Spores of Bacillus Species: What We Know and Do Not Know. *J. Bacteriol.* **2014**, *196*, 1297–1305. [CrossRef]

58. Jacoby, W.A.; Maness, P.C.; Wolfrum, E.J.; Blake, D.M.; Fennell, J.A. Mineralization of Bacterial Cell Mass on a Photocatalytic Surface in Air. *Environ. Sci. Technol.* **1998**, *32*, 2650–2653. [CrossRef]

59. Saito, T.; Iwase, T.; Horie, J.; Morioka, T. Mode of photocatalytic bactericidal action of powdered semiconductor TiO_2 on mutans streptococci. *J. Photochem. Photobiol. B Biol.* **1992**, *14*, 369–379. [CrossRef]

60. Kikuchi, Y.; Sunada, K.; Iyoda, T.; Hashimoto, K.; Fujishima, A. Photocatalytic bactericidal effect of TiO_2 thin films: Dynamic view of the active oxygen species responsible for the effect. *J. Photochem. Photobiol. A Chem.* **1997**, *106*, 51–56. [CrossRef]

61. Kunicki-Goldfinger, W.; Baj, J.; Markiewicz, Z.; Kobyliński, S. *Życie Bakterii*; Wydawnictwo Naukowe PWN: Warszawa, Poland, 2008; ISBN 978-83-01-14378-7.
62. Bubacz, K.; Kusiak-Nejman, E.; Tryba, B.; Morawski, A.W. Investigation of OH radicals formation on the surface of TiO_2/N photocatalyst at the presence of terephthalic acid solution. Estimation of optimal conditions. *J. Photochem. Photobiol. A Chem.* **2013**, *261*, 7–11. [CrossRef]

Article

Synthesis of Rectorite/Fe₃O₄/ZnO Composites and Their Application for the Removal of Methylene Blue Dye

Huanhuan Wang [1], Peijiang Zhou [1,*], Rui Guo [1], Yifei Wang [1], Hongju Zhan [2] and Yunfei Yuan [1]

[1] School of Resources and Environmental Science, Hubei Biomass-Resource Chemistry and Environmental Biotechnology Key Laboratory, Eco-Environment Technology R&D and Service Center, Wuhan University, Wuhan 430079, China; doublewhh@163.com (H.W.); gregoryguo@163.com (R.G.); yifeiwangzyh@gmail.com (Y.W.); yuanyunfei0523@126.com (Y.Y.)

[2] Chemical Engineering and Pharmaceutical Department, Jingchu University of Technology, Jingmen 448000, China; blackcat927@tom.com

* Correspondence: zhoupj@whu.edu.cn; Tel.: +86-027-6876-6930

Received: 10 February 2018; Accepted: 6 March 2018; Published: 9 March 2018

Abstract: A novel series of rectorite-based magnetic zinc oxide (ZnO) photocatalysts (REC/Fe₃O₄/ZnO) was synthesized and characterized in the present work. The fabricated REC/Fe₃O₄/ZnO composite possessed a high specific surface area and high capacity of adsorption and photocatalysis toward methylene blue (MB) dye. The adsorption isotherm of the dye on the composite fitted well to the Langmuir model, with a maximum adsorption of 35.1 mg/g. The high adsorption capacity increased the interactions between the dye and the REC/Fe₃O₄/ZnO, which enabled efficient decomposition of the dye under simulated solar radiation using REC/Fe₃O₄/ZnO as the photocatalyst. The degradation kinetics of MB dye followed the Langmuir–Hinshelwood model. More importantly, the degradation of MB dye and the mass loss of REC/Fe₃O₄/ZnO after three repetitive experiments were quite small. This suggests that the magnetic composite has great potential as an effective, stable, and easily recovered catalyst. Four major intermediates were detected during the degradation of MB dye and the degradation pathway was proposed.

Keywords: adsorption; magnetic ZnO; methylene blue; photodegradation; rectorite

---

1. Introduction

Synthetic dyes are used in a wide range of industries, including textiles, cosmetics, printing, pharmaceuticals, and food. Of these dyes, approximately 1–2% and 1–10%, are discharged into the environment by manufacturing processes and end-users, respectively [1–5], which poses a serious threat to human health and the environment. Several physical, chemical and biological techniques have been used to treat dyestuff waste, but the high cost, formation of hazardous coproducts, and intensive energy requirements have limited their extensive application [6]. Additionally, traditional wastewater treatment processes are inefficient in handling dye pollutants because of their biological resistance and chemical stability [7–9]. The removal of dyestuff waste from aqueous solutions is technically very challenging.

Photocatalysis could be an effective wastewater treatment technology for dye pollutants because of its potential high activity, low energy consumption, mild treatment conditions, and ease of handling [10–16]. The irradiation of wide bandgap semiconductors by ultraviolet light produces various reactive oxygen species (ROS) including hydroxyl radicals (•OH) and singlet oxygen ($^1O_2$), which can efficiently oxidize or mineralize organic compounds. Moreover, direct reaction between valence-band holes and organic pollutants could induce oxidation or decomposition of these target pollutants. The contributions of ROS and holes toward the degradation of pollutants depend on the electronic properties of the target substance and the photocatalyst [17,18]. Photocatalysis can degrade organic dyes into water, carbon dioxide, and other non-toxic inorganic compounds without causing secondary pollution.

Complementary metal-oxide semiconductors have received significant attention because of their efficient application in photocatalysis such as titanium dioxide (TiO_2), cuprous oxide (Cu_2O), and so on [19,20]. Zinc oxide (ZnO) is a well-known photocatalyst with a bandgap of 3.37 eV that permits the absorption of ultraviolet (UV)-visible light. It also has the advantages of high photoactivity, non-toxicity, and low manufactured cost [21–26]. Photocatalysis using ZnO has been used to remove many pollutants from aqueous solutions because ZnO is effective in producing •OH [27–29]. However, the application of such photocatalysis to wastewater treatment is limited because of the difficulty in separating and recovering ZnO powder from the treated solutions. Using magnetic heterogeneous catalysts could facilitate efficient, rapid, and economical separation of the photocatalyst [30]. Water treatment agents modified with magnetic nanoparticles are particularly interesting; examples of these materials include magnetic iron oxide/clay composite materials such as magnetite (Fe_3O_4) [31–38] and maghemite (γ-Fe_2O_3) [35,39].

One strategy to improve ZnO performance has focused on the development of ZnO composites that enhance the photocatalytic efficiency via improved adsorption [40–45]. Rectorite (REC), a silicate clay mineral, is composed of alternating pairs of nonexpandable dioctahedral mica-like and expandable dioctahedral smectite-like layers [46]. REC efficiently adsorbs organic compounds both on its external surfaces and within its interlaminar spaces because of its high specific surface area and ion exchange properties [47]. This mineral is also of considerable interest as a catalyst support because of its low cost, small size, and unusual intercalated structure. Consequently, considerable research has been devoted to exploring the use of REC-based materials to adsorb or catalytically decompose environmental pollutants [48–51].

In this study, REC/Fe_3O_4 was prepared by mixing REC with Fe_3O_4. This was further mixed with ZnO to obtain a REC/Fe_3O_4/ZnO composite via an improved hydrothermal process. The composite was characterized by transmission electron microscopy (TEM), scanning electron microscopy (SEM), X-ray powder diffraction (XRD) and Fourier transform-infrared (FT-IR). The absorption and photocatalytic properties of REC/Fe_3O_4/ZnO were evaluated by the decomposition of methylene blue (MB) dye under simulated sunlight irradiation. Such a composite material has potential as a new low-cost and recyclable agent for the efficient eliminate treatment of dyestuff wastewater.

2. Results and Discussion

2.1. Morphology and BET Surface Area of the REC/Fe₃O₄/ZnO Composites

The morphologies of the various REC/Fe_3O_4/ZnO composites were investigated by SEM and TEM (Figure 1). REC is an interstratified clay mineral (Figure 1a,d). Breakage of the structure (Figure 1b) formed particles having diameters of ca. 6–35 nm (Figure 1e), in which Fe_3O_4 was integrated with the mineral. The structure became further disrupted with the deposition of ZnO (Figure 1c) when many more particles were found with the REC (Figure 1f).

BET surface area of REC applied in this study was 11.7 m^2 g^{-1} (Table 1), which is 25% smaller than that of REC/Fe_3O_4. The increase of BET after magnetization was mainly because of the breakage of REC as shown in Figure 1e. BET surface area was approximately 16.8 and 16.0 for REC/ZnO and REC/Fe_3O_4/ZnO composites, respectively.

Table 1. Specific surface area of ZnO, REC, REC/Fe$_3$O$_4$ (1:0.3), REC/ZnO (1:0.5), REC/Fe$_3$O$_4$/ZnO (1:0.3:0.5) powder.

BET Surface Area (m^2·g^{-1})	ZnO	REC	REC/Fe$_3$O$_4$	REC/ZnO	ZnO/REC/Fe$_3$O$_4$
	5.3	11.7	15.6	16.8	16.0

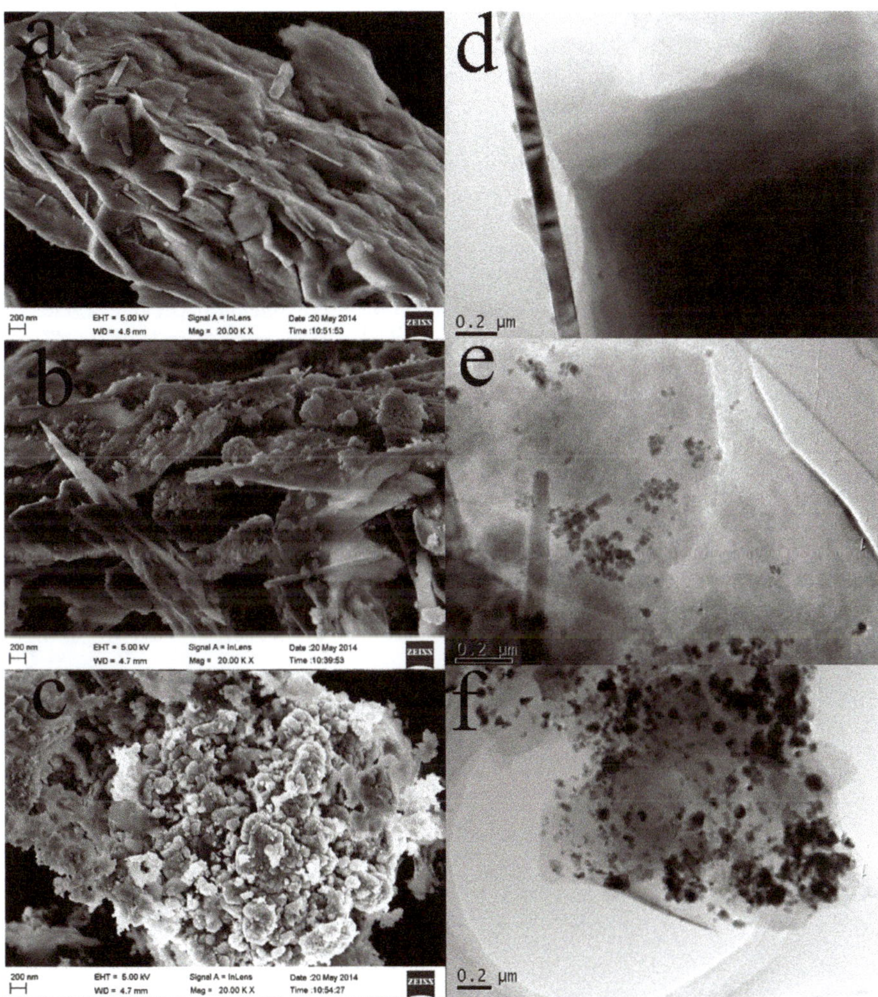

Figure 1. Scanning electron microscopy (SEM) images of (**a**) rectorite (REC), (**b**) REC/magnetite (Fe$_3$O$_4$) (1:0.3), and (**c**) REC/Fe$_3$O$_4$/zinc oxide (ZnO) (1:0.3:0.5) and transmission electron microscopy (TEM) images of (**d**) REC, (**e**) REC/Fe$_3$O$_4$ (1:0.3), and (**f**) REC/Fe$_3$O$_4$/ZnO (1:0.3:0.5).

2.2. Structural Characterization of the REC/Fe$_3$O$_4$/ZnO Composites

Figure 2 shows the XRD patterns of ZnO, REC and their composite materials for 2θ ranging from 10 to 70°. The diffraction peak at 2θ = 19.8° was a characterized peak of REC [52], which was also observed in REC/ZnO, REC/Fe$_3$O$_4$ and REC/Fe$_3$O$_4$/ZnO. The relative intensities of diffraction peak at 2θ = 19.8° was changed in composites, compared with that in pure REC. The finding indicates that the structure of rectorite distorted to some extents during the synthesis of composites. Similar results were also reported during the synthesis of other REC-based photocatalyst [52]. Peaks observed at 31.7°, 34.4°, 36.2°, 47.5°, 56.5°, 62.8°, 66.3°, 67.9°, and 69.0° were indexed to hexagonal wurtzite (ZnO; JCPDS Data Card no. 36-1451). Diffraction peaks at 33.1°, 35.7°, 40.8°, 54.1°, and 64.0°, characteristic of Fe_3O_4 were observed for the REC/Fe_3O_4 and $REC/Fe_3O_4/ZnO$ composites [48]. The diameters of ZnO and Fe_3O_4 were approximately 45.7 and 22.4 nm in $REC/Fe_3O_4/ZnO$, putting diffraction peaks at $2\theta = 36.3°$ and 19.8° and the corresponding FWHM (full width at half maxima) into Scherrer equation. The relative size between ZnO and Fe_3O_4 was correlated with that shown in Figure 1e,f.

Figure 3 compares the Fourier transform-infrared (FT-IR) spectra of REC and its composites. The spectra of REC, ZnO/REC, and $REC/Fe_3O_4/ZnO$ were quite similar, with only small differences in band locations and intensities. The bands at 3645–3647 and 3429–3440 cm^{-1} were attributed to the hydrogen-bonding bending vibration of water and hydroxyl stretching vibration in all three samples [53]. Bands observed at 1635 and 1386 cm^{-1} in ZnO were attributed to the asymmetrical and symmetrical stretching of the zinc carboxylate [54], respectively. The band at 1635 cm^{-1} overlapped with the band at 1640 cm^{-1}, which was assigned to a water bending vibration in REC [55,56]. Bands at 1021 and 1053 cm^{-1} were assigned as In-plane Si–O–Si stretching [56], which was found to be 1012 and 1047 cm^{-1} in REC/ZnO and $REC/Fe_3O_4/ZnO$ in this study. Bands at 478 and 545 cm^{-1} were assigned as Si–O bending and Si–O–Al bending [56], which was 486 and 548 cm^{-1} in REC/ZnO, and 490 and 552 cm^{-1} in $REC/Fe_3O_4/ZnO$, respectively. Notably, band at approximately 530–580 cm^{-1} was also reported to Fe–O bond [57] in Fe_3O_4, which could be emerged in Si–O–Al bending band.

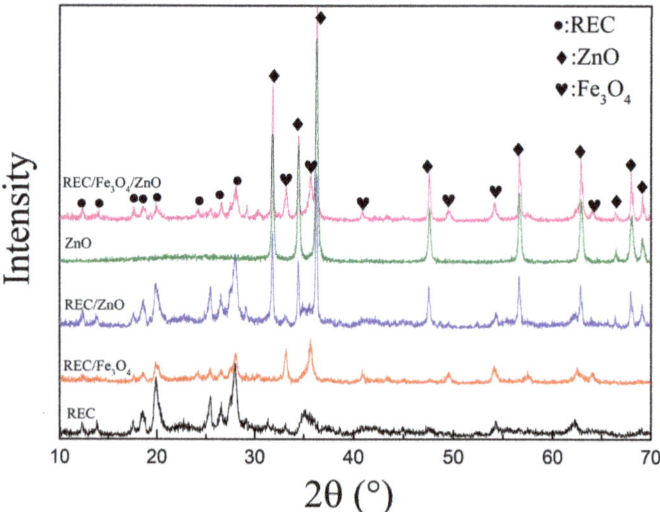

Figure 2. X-ray diffraction (XRD) patterns of REC, ZnO, REC/ZnO (1:0.5), REC/Fe_3O_4 (1:0.3) and $REC/Fe_3O_4/ZnO$ (1:0.3:0.5).

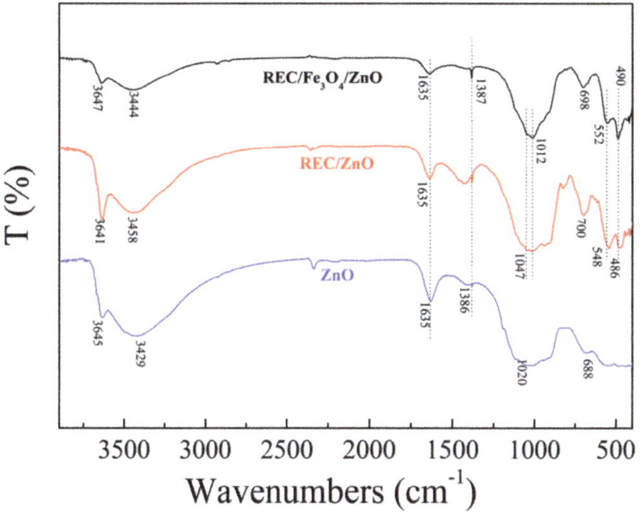

Figure 3. Fourier transform-infrared (FT-IR) spectra of ZnO, REC/ZnO (1:0.5), and ZnO/REC/Fe$_3$O$_4$ (1:0.3:0.5).

2.3. Thermogravimetric Analysis (TGA) Analysis REC and REC/Fe$_3$O$_4$/ZnO

TGA analysis was investigated to investigate the stability of ternary composite. As shown in Figure 4, REC barely lose its weight up to 800 °C mainly because of the loss of binding water. As for REC/Fe$_3$O$_4$/ZnO, the loss of weight could be divided into two regions. Loss of water could result in the loss weight below 200 °C. Decomposition of ZnCO$_3$·2Zn(OH)$_2$ impurity could lead to the loss of weight at approximately 600 °C.

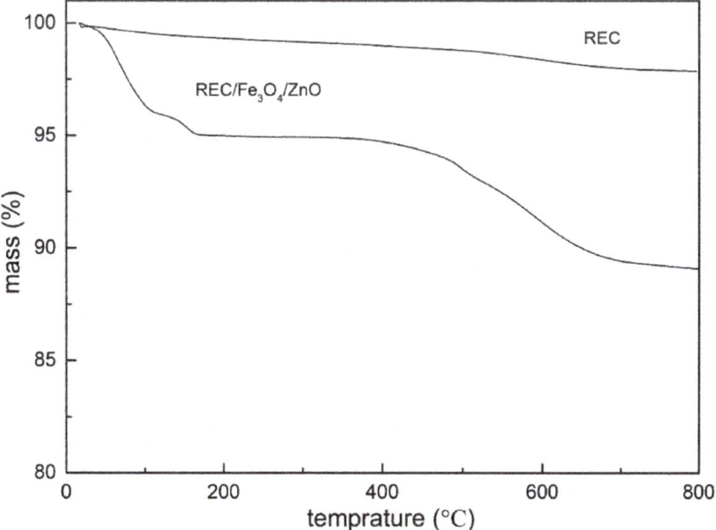

Figure 4. TGA of REC and REC/Fe$_3$O$_4$/ZnO (1:0.3:0.5).

2.4. Adsorption Equilibrium and Isotherm of MB Dye on REC/Fe$_3$O$_4$/ZnO

Figure 5a illustrates the results of equilibrium adsorption studies. The adsorption of MB dye increased with increasing time to 50 min, reaching equilibrium within 60 min for REC/Fe$_3$O$_4$/ZnO. The adsorption kinetics were fitted using different equations including pseudo-first-order, pseudo-second-order and Elovich models; the adjusted R-squared (adj. r^2) values were 0.98, 0.98, and 0.93, respectively. The calculated equilibrium adsorption was 1.5 and 2.2 mg/g from the pseudo-first-order and pseudo second-order models, respectively. The experimental equilibrium adsorption of 1.2 mg/g of MB dye on REC/Fe$_3$O$_4$/ZnO was thus consistent with pseudo-first-order kinetics.

The Freundlich and Langmuir isotherms are commonly used to describe the adsorption properties of pollutants. Figure 5b shows that the adsorption data fitted the Langmuir (adj. r^2 = 0.999) better than the Freundlich isotherm model (adj. r^2 = 0.974). The Langmuir isotherm is typically used to model monolayer adsorption on adsorbents having homogeneous and energetically uniform surfaces. Thus, our MB dye adsorption data are consistent with adsorption on the outer layer of the composites. The estimated recovered maximum adsorption capacity of MB dye on REC/Fe$_3$O$_4$/ZnO was 35.1 mg/g, which was 13% and 72% larger than that of REC/ZnO (31.1 mg/g) and ZnO (20.4 mg/g). These data clearly showed REC component contributed significantly to the adsorption of MB to REC/Fe$_3$O$_4$/ZnO.

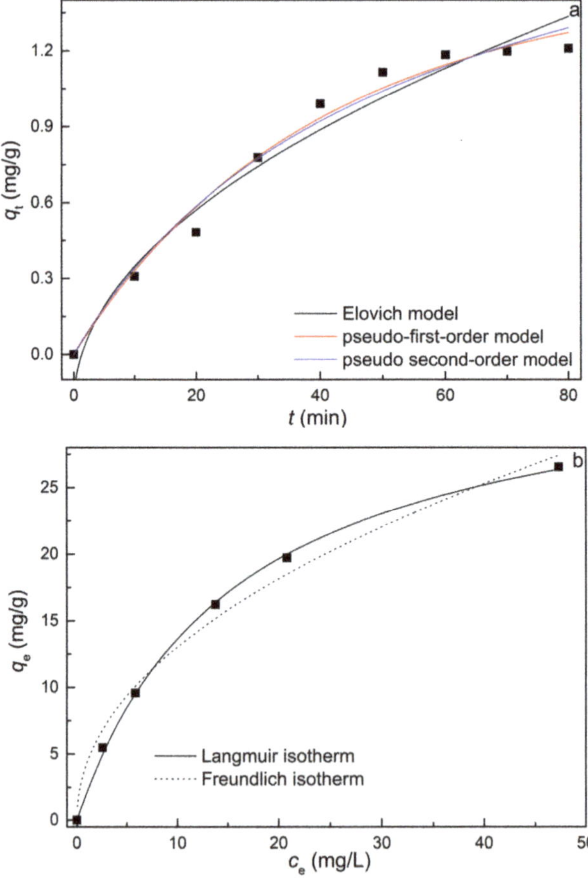

Figure 5. (**a**) Adsorption kinetics and (**b**) isotherm of methylene blue (MB) dye on REC/Fe$_3$O$_4$/ZnO (1:0.3:0.5).

2.5. Effect of Component Mass Ratio on the Degradation of MB Dye

The fabricated composites contained REC, Fe_3O_4, and ZnO. The impact of the component mass ratio on the removal of MB dye was investigated to obtain the optimal composition. Figure 6a shows that the dark adsorption of MB dye on $REC/Fe_3O_4/ZnO$ was only slightly influenced by the ZnO content. After a 60-min adsorption period, the MB dye concentration in the bulk phase decreased to 4.2, 4.0, and 4.0 mg/L for ZnO contents of 1.5, 1.0, and 0.5, respectively. Control experiment showed that degradation of MB was negligible by direct photolysis in the absence of catalysts because of the low irradiance (1.9 mW/cm^2). However, photodegradation of the MB dye was greatly influenced by the ZnO content, with the degradation clearly decreasing with increasing ZnO content. The observed degradation kinetic constant (k_{obs}) for the degradation of MB dye was 0.0056, 0.0086, and 0.012 min^{-1} for ZnO contents of 1.5, 1.0, and 0.5, respectively. A plausible explanation for the decreased photoactivity with increasing ZnO content is that larger ZnO particles were formed, which may render the $REC/Fe_3O_4/ZnO$ less photoactive.

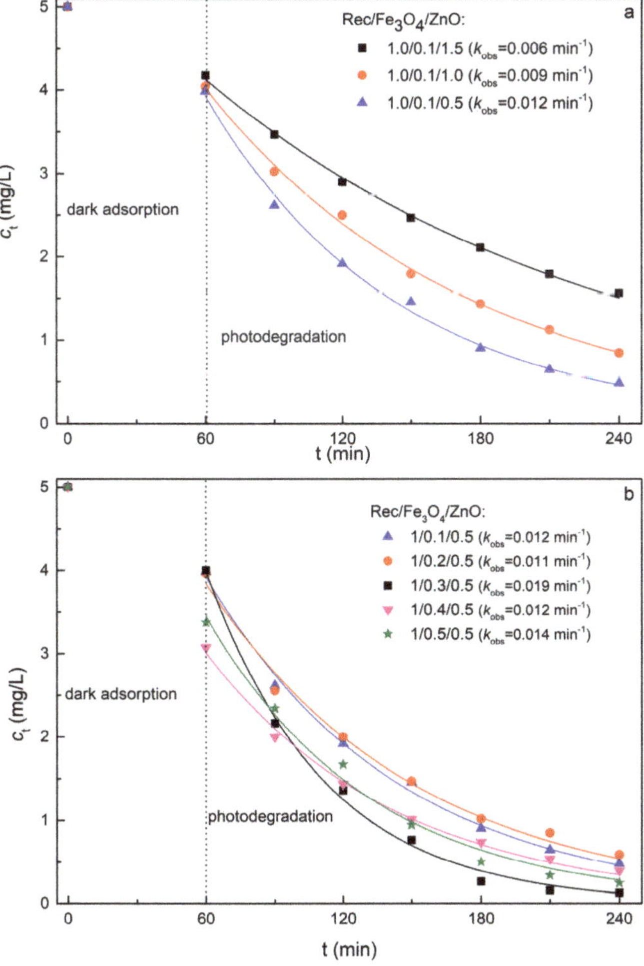

Figure 6. Effects of (**a**) ZnO and (**b**) Fe_3O_4 mass contents of $REC/Fe_3O_4/ZnO$ on the adsorption and photodegradation of MB dye.

Figure 6b shows that the dark adsorption of MB dye was highly influenced by the Fe_3O_4 content of $REC/Fe_3O_4/ZnO$. The adsorption ratio first increased from 20 to 45% as the Fe_3O_4 content increased from 0.1 to 0.3; it then decreased to 33% as the Fe_3O_4 content increased to 0.5. However, k_{obs} for the MB photodegradation only slightly increased, from 0.012 to 0.015 min^{-1}, as the Fe_3O_4 content increased from 0.1 to 0.3, which indicated that the photodegradation was not significantly affected by the Fe_3O_4 content. Figure 1 shows that the layered structure of REC was damaged by the introduction of Fe_3O_4, which was detrimental to MB dye adsorption. However, Fe_3O_4 was not an effective catalyst even under simulated solar radiation, particularly at neutral pH values. Therefore, the photoactivity of $REC/Fe_3O_4/ZnO$ (as indicated by the degradation of adsorbed MB dye) was likely not greatly influenced by the Fe_3O_4 content. These results established that REC and ZnO acted mainly as adsorbent and photocatalyst, respectively, for the removal of MB dye by $REC/Fe_3O_4/ZnO$.

2.6. Effect of REC/Fe₃O₄/ZnO Dosage on the Degradation of MB Dye

Figure 7 shows that the dark adsorption of MB dye on $REC/Fe_3O_4/ZnO$ slightly increased as the composite dosage increased from 0.3 to 1.1 g/L; this was attributed to increasing availability of adsorption sites. The photodegradation of MB dye also greatly increased with increasing dosage up to 0.9 g/L. The k_{obs} for the photodegradation of the MB dye increased from 0.0084 to 0.019 min^{-1} as the $REC/Fe_3O_4/ZnO$ dosage increased from 0.3 to 0.9 g/L. These results are consistent with many studies where the reaction accelerates with increasing catalyst dosage when the amount of reactive species, such as •OH, is determined by the concentration of the catalyst in the bulk solution. However, a dosage of 1.1 g/L had an inhibitory effect on the degradation of the dye. This was attributed to agglomeration of $REC/Fe_3O_4/ZnO$ particles at the high concentration, which reduced light transmission. The optimal amount of added $REC/Fe_3O_4/ZnO$ is the least amount required for complete photon absorption for the photodegradation reaction; in our experiments, we fixed the $REC/Fe_3O_4/ZnO$ concentration at 0.9 g/L.

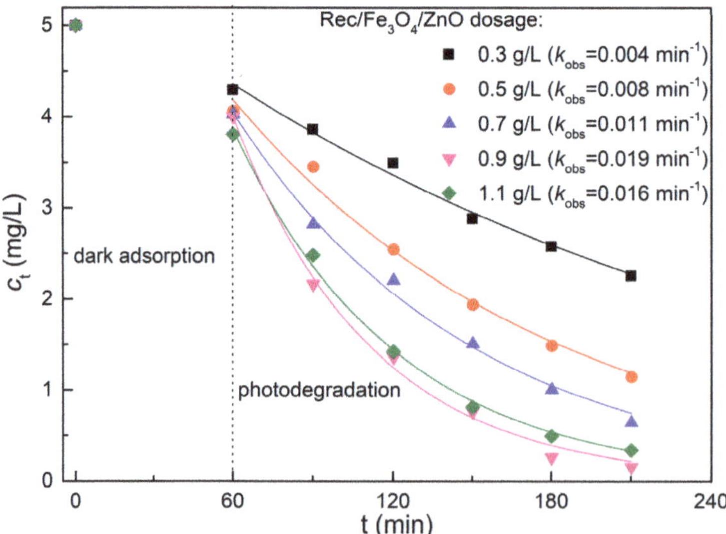

Figure 7. Effect of $REC/Fe_3O_4/ZnO$ (1:0.3:0.5) dosage on the adsorption and photodegradation of MB dye.

2.7. Effect of Solution pH on the Degradation of MB Dye

The impact of solution pH on the adsorption and photodegradation of MB dye on $REC/Fe_3O_4/ZnO$ is shown in Figure 8. It clearly shows that acidic solution favored the adsorption of the dye and that the adsorption ratio at pH 5.0 was about twice that at other pH values. Conversely, degradation of the dye in acidic solution was much slower below pH 6.0. The rate constant k_{obs} increased from 0.011 to 0.019 min^{-1} as the solution pH increased from 5.0 to 6.0, and then decreased to 0.0076 min^{-1} at pH 8.0. These findings demonstrated that the solution pH had variable effect on the adsorption and degradation process of MB dye on REC/Fe$_3$O$_4$/ZnO. This variable effect of pH is related to its multiple roles in electrostatic interactions with the catalyst surface and substrate, and to the formation of charged radicals during the reaction process. This makes the interpretation of pH effects on the photodegradation of organic pollutants very difficult. The REC/Fe$_3$O$_4$/ZnO surface was negatively charged and the MB dye was present in its cationic form at pH 5.0. Therefore, the electrostatic attraction generated between cationic dye molecules and the negative surface charge contributed to the high adsorption ratio measured in acidic solutions. Typically, ZnO exhibits higher photoactivity at neutral pH. The degradation of MB dye on REC/Fe$_3$O$_4$/ZnO was faster at pH 6.0 because ZnO was the most important photocatalyst present in the composite at that pH.

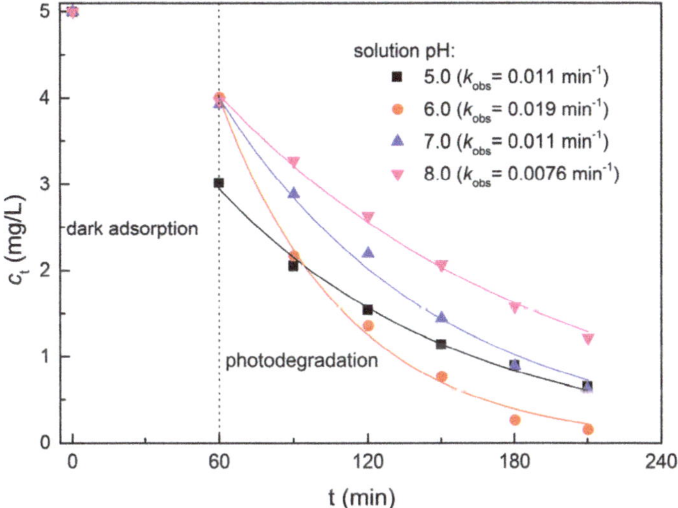

Figure 8. Effect of solution pH on the adsorption and photodegradation of MB dye on REC/Fe$_3$O$_4$/ZnO (1:0.3:0.5).

2.8. Kinetics for the Degradation of MB Dye on REC/Fe$_3$O$_4$/ZnO

The effect of initial MB dye concentration on its degradation was investigated for initial concentrations ranging from 5.0 to 30.0 mg/L (Figure 9a). After reaching adsorption equilibrium, the dye gradually decomposed with increasing irradiation time. The initial rate of photodegradation increased and reached a plateau with increasing initial dye concentration (Figure 9b). The data were fitted to the Langmuir–Hinshelwood kinetic model, which is frequently used to model the initial photocatalytic degradation rates of organic compounds. The rate law is given by Equation (1):

$$r_0 = -\frac{dc}{dt} = k_{re}K_s c_0/(1 + K_s c_0), \tag{1}$$

where r_0 is the initial rate of disappearance of MB dye, c_0 is the initial concentration of the dye, k_{re} is the reaction rate constant, and K_s is the Langmuir adsorption constant. The calculated values for k_{re} and K_s were 0.122 mg/(L·min) and 0.069 L/mg, respectively.

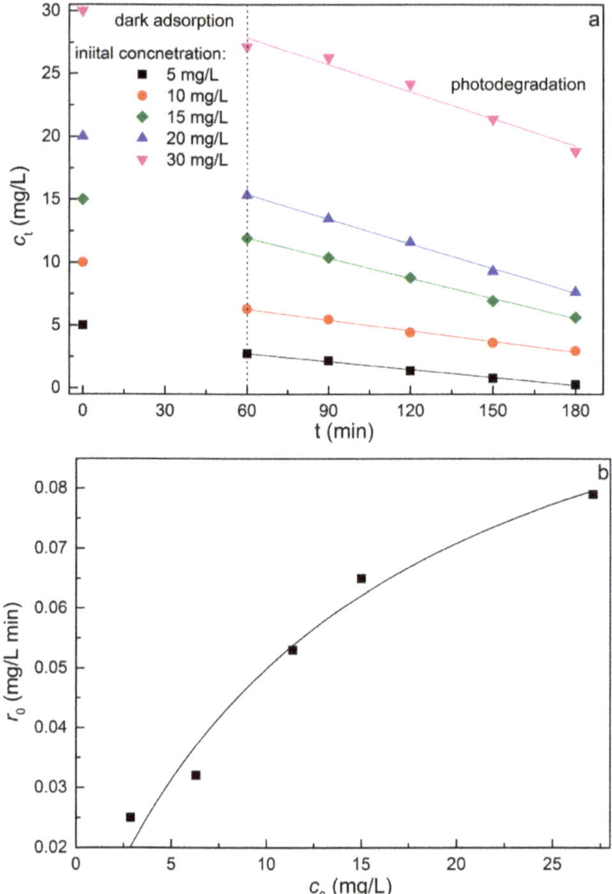

Figure 9. (a) Correlation between the initial rate of loss of MB dye on REC/Fe$_3$O$_4$/ZnO (1:0.3:0.5) and the initial dye concentration. (b) The solid line represents fitting of the data to the Langmuir–Hinshelwood kinetic model.

2.9. Recovery and Stability of REC/Fe$_3$O$_4$/ZnO

The performance of a single catalyst sample for the removal of MB dye over several cycles was determined to assess the photostability and possible reuse of REC/Fe$_3$O$_4$/ZnO. After 1-h adsorption and 5-h irradiation sequences, the composite was recovered from the reaction solution using an external magnet and redispersed in fresh 5-mg/L MB dye solution. Figure 10 shows that a noticeable decrease in the removal ratio of the dye occurred during repeated use of the REC/Fe$_3$O$_4$/ZnO catalyst. Only 80% remained after three cycles, corresponding to a mass loss of 14%. The results demonstrated satisfactory photostability of the REC/Fe$_3$O$_4$/ZnO catalyst.

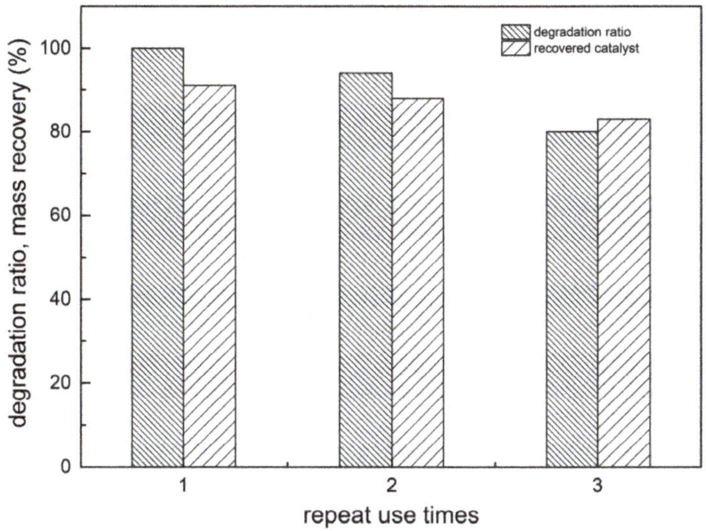

Figure 10. Performance and recovery of REC/Fe$_3$O$_4$/ZnO (1:0.3:0.5) during repeated use.

2.10. Mechanism for the Degradation of MB Dye on REC/Fe$_3$O$_4$/ZnO

The degradation intermediates were identified by liquid chromatography-mass spectrometry (LC-MS). The signal at $m/z = 284$ present before degradation was assigned to the MB dye molecule (Scheme 1). Oxidation of MB dye could follow different paths. Addition of •OH to MB dye molecules would lead to the formation of an intermediate with $m/z = 300$ [58]. Oxidation of sulfur atoms and bond cleavage at nitrogen-bridged sites would lead to the formation of a sulfoxide intermediate having $m/z = 303$ [59]. However, neither $m/z = 300$ nor 303 was detected in the present study. Instead, signals were found at $m/z = 317, 274, 138$, and 345. An intermediate with $m/z = 317$ could be formed through the addition of two •OH radicals per MB dye molecule, or through oxidation of methyl (–CH$_3$) groups of the sulfoxide intermediate, and the oxidation of more methyl groups would lead to the formation of a species having $m/z = 345$. Addition of •OH to an intermediate of $m/z = 317$ and cleavage at a sulfur-bridged site might lead to the formation of an intermediate with $m/z = 138$. Additionally, loss of a methyl group could produce an intermediate with $m/z = 274$. Contribution of •OH to the degradation of MB was proved by a much smaller degradation kinetic constant in the presence of isopropanol acting as •OH scavenger. Oxidation of these intermediates could form ring cleavage products, and even mineralization to CO$_2$ and H$_2$O is plausible. The chemical structures of any intermediates were not identified because of the complexity of the degradation process.

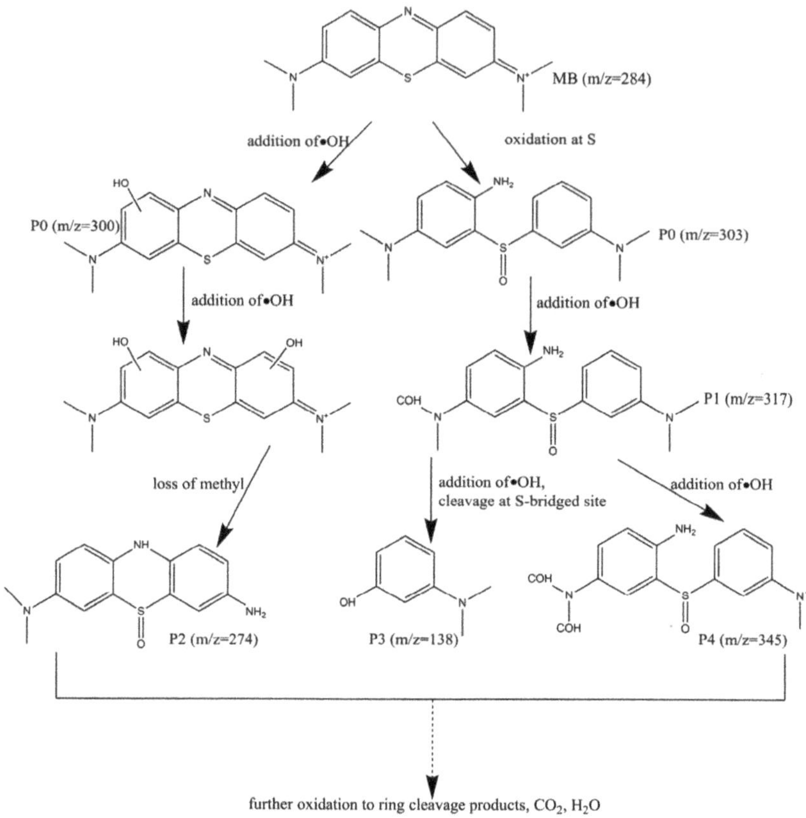

Scheme 1. Proposed pathway for the degradation of MB dye on REC/Fe$_3$O$_4$/ZnO (1:0.3:0.5).

3. Materials and Methods

3.1. Chemicals

Ferric chloride, ferric sulfate, MB, zinc acetate dihydrate (Zn(CH$_3$COO)$_2$·2H$_2$O) (purity: 99.9%), hydrazine hydrate, dimethyl benzene, ethylene glycol, ethyl alcohol, and sodium hydroxide were purchased from Alpha-Aesar (Shanghai, China) and used as-received. They were of analytical grade. Refined REC was provided by Hubei Mingliu Inc. Co. (Wuhan, China). The water used in the experiments had been pretreated with an ultrapure water system (Liyuan Electric Instrument Co., Beijing, China).

3.2. Synthesis of REC/Fe₃O₄/ZnO

3.2.1. Synthesis of Magnetic REC

Fe$_3$O$_4$ was prepared by a co-precipitation method as follows. A solution of FeCl$_3$ (1.625 g) and FeSO$_4$ (1.219 g) was prepared at a molar ratio of 4:3, and 1 mol/L of NaOH solution (250 mL) was quickly added to the mixture while mixing at high speed with a magnetic mixer to adjust the pH to 11. Mixing was continued at 60 °C for 1.5 h, and then the solution was placed in a thermostated water bath at 80 °C for 1.5 h to crystallize the product. The Fe$_3$O$_4$ product was isolated by filtration, rinsed with pure water until the pH of the filtrate was neutral, dried at 105 °C for 6 h, and finally ground to a particle size of 74 μm.

Magnetic REC was prepared as follows. A mixture of REC in water was sonicated to provide a uniform suspension. A ferrofluid containing 50 wt. % of the Fe_3O_4 described above was slowly dropped into REC suspensions to provide REC/Fe_3O_4 mixtures having weight ratios of 1:0.1, 1:0.2, 1:0.3, 1:0.4, and 1:0.5. Each mixture was ultrasonically dispersed for 60 min. The product was recovered by filtration, dried at 105 °C for 6 h, and ground to a particle size of 74 μm.

3.2.2. Synthesis of $REC/Fe_3O_4/ZnO$

The $REC/Fe_3O_4/ZnO$ nanocomposites were prepared via a mild liquid-phase synthesis method. The reaction of zinc acetate dihydrate ($Zn(CH_3COO)_2 \cdot 2H_2O$; 2.1951 g (10 mmol)) with hydrazine hydrate ($N_2H_4 \cdot H_2O$; 80%; 0.726 mL (15 mmol)) provided a theoretical yield of 0.8137 g of ZnO. Based on this calculation, appropriate amounts of the magnetic REC were added to form $REC/Fe_3O_4/ZnO$ suspensions having weight ratios of 1:0.1:0.5, 1:0.2:0.5, 1:0.3:0.5, 1:0.4:0.5, 1:0.5:0.5, 1:0.1:0.5, 1:0.1:1.0 and 1:0.1:1.5. In the experiments, 2.1951 g (10 mmol) of $Zn(CH_3COO)_2 \cdot 2H_2O$ and the appropriate calculated weights of magnetic REC were dissolved in 200 mL of a mixed solvent of dimethyl benzene and ethylene glycol under vigorous stirring for 30 min. Then, a solution of hydrazine hydrate (0.726 mL, 15 mmol) in anhydrous ethanol (30 mL) was added dropwise to the suspension. The resulting dispersion was vigorously stirred for 5 h at room temperature, and then transferred to a separating funnel and allowed to stand for 1.5 h. Centrifugation provided the $REC/Fe_3O_4/ZnO$ nanocomposites as gray solids. These were isolated by filtration, rinsed three times with anhydrous ethanol, and calcined for 6 h in a muffle furnace at temperatures of 200, 300, 400, 500, and 600 °C.

3.3. Characterization of the Synthesized Magnetic Materials

The Brunauer–Emmett–Teller (BET) surface area was determined using a Micromeritics model ASAP 2020 Instrument (Micromeritics, Norcross, GA, USA). The XRD patterns of the products were determined using a Dmax-rA powder diffractometer (Rigaku, Akishima, Japan), which used Cu Kα radiation source at a scanning rate of $2°$ min^{-1}. SEM images were acquired using a QUANTA 200 instrument (FEI, Hillsboro, OR, USA). Transmission electron microscopy (TEM) images were obtained with a JEM 2010HT instrument (JEOL, Akishima, Japan) at an accelerating voltage of 200 kV. Thermogravimetric analysis (TGA) was conducted on a TGA 2050 thermogravimetric analyzer with a heating rate of 10 °C/min from 50 to 800 °C under a nitrogen atmosphere (TA Instruments, NewCastle, DE, USA). The Brunauer–Emmett–Teller (BET) surface areas of the two TiO_2 were determined using a Micromeritics ASAP 2020 setup (Micromeritics, Norcross, GA, USA).

3.4. Adsorption of MB Dye on REC/Fe₃O₄/ZnO

Adsorption kinetics: A dispersion of $REC/Fe_3O_4/ZnO$ (0.9 g/L) was prepared by adding $REC/Fe_3O_4/ZnO$ to 100 mL of an aqueous MB solution (concentration: 5 mg/L) and the dispersion was shaken at 25 °C. Samples (2 mL) were withdrawn from the flask at different time intervals. The adsorbent and MB dye solution were quickly separated by a magnet, and the concentration of dye in the supernatant was analyzed by UV-visible spectroscopy.

Adsorption isotherms: Batch adsorption studies were performed using aqueous suspensions containing MB dye at different initial concentrations; the dosage of $REC/Fe_3O_4/ZnO$, REC/Fe_3O_4 or ZnO as adsorbent was held at 0.9 g/L. The suspension was continuously stirred at constant temperature using a mechanical stirrer for 2 h. After reaching equilibrium, a 2-mL aliquot of the suspension was withdrawn to determine the equilibrium concentration, c_t. Adsorption isotherm experiments were carried out at pH 6.0 in the absence of electrolytes.

3.5. Photocatalytic Degradation of MB Dye under Simulated Solar Radiation

Photodegradation of MB dye was carried out in a home-made photoreactor. The radiation source was an incandescent light bulb lamp that provided radiation at ≥350 nm with an irradiance of 1900 μW/cm^2 (Figure 11). Aqueous solutions of MB dye (200 mL; initial concentration: 5 to 20 mg/L) were mixed magnetically with the various catalysts in a 250 mL Pyrex beaker. After equilibrating in the dark for 1 h, aliquots (2 mL) of those suspensions were withdrawn to determine the initial MB concentration, c_0. Aliquots (2 mL) were also collected at selected time intervals as the MB degraded; these were magnetically separated and used to determine c_t. The degradation of the MB dye was monitored using a 2550 UV-visible spectrophotometer (Shimadzu, Kyoto, Japan) with a 10-mm cuvette. The slope of a linear fit of the data provided the initial photodegradation rate, R_0.

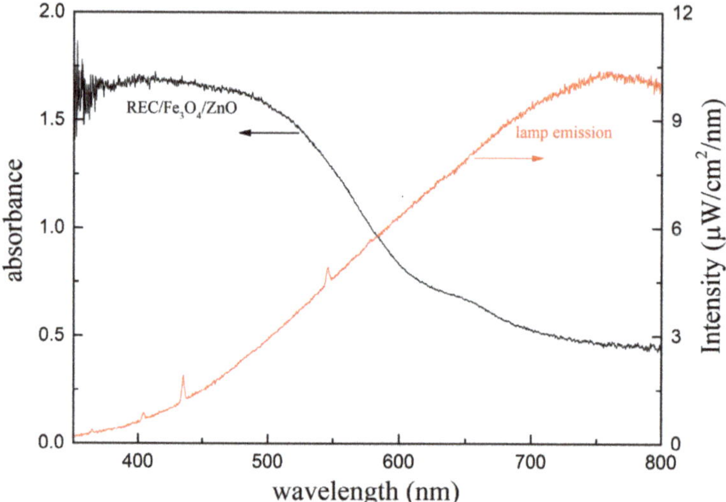

Figure 11. Absorption spectrum of REC/Fe$_3$O$_4$/ZnO (1:0.3:0.5) and the irradiance of lamp using in this study.

3.6. Sample and Data Analyses

Sample analysis: The MB dye concentration was determined according to its absorbance at 665 nm.

Data analysis: The mass of MB dye adsorbed per gram of adsorbent at different times (q_t, mg/g) and at equilibrium (q_e, mg/g) were calculated using Equations (2) and (3), respectively:

$$q_t = \frac{(c_0 - c_t) \times V}{m}, \tag{2}$$

$$q_e = \frac{(c_0 - c_e) \times V}{m}, \tag{3}$$

where c_0, c_t, and c_e are the initial concentration, concentration at time t, and equilibrium concentration of the MB dye (all in mg/g), respectively.

The adsorption ratio (%) of the MB dye was then calculated using Equation (4):

$$R = \left(1 - \frac{c_e}{c_0}\right) \times 100\%, \tag{4}$$

The Freundlich (Equation (5)) and Langmuir (Equation (6)) isotherms were applied to describe the adsorption properties of the MB dye on REC/Fe$_3$O$_4$/ZnO, REC/Fe$_3$O$_4$ and ZnO, as follows [60]:

$$q_e = K_F \times c_e^{1/n}, \tag{5}$$

where K_F ($(mg/g) \times (L/g)^{1/n}$) is the Freundlich affinity coefficient and $1/n$ is the Freundlich exponential coefficient. Additionally:

$$q_e = \frac{q_{max} \times K_L \times c_e}{1 + K_L \times c_e},\tag{6}$$

where q_{max} (mg/g) is the maximum adsorption of MB dye on the adsorbents and K_L (L/mg) is the Langmuir adsorption constant.

4. Conclusions

A series of REC/Fe$_3$O$_4$/ZnO composites was synthesized and characterized. The Fe$_3$O$_4$ phase destroyed the layered structure of REC and increased the adsorption of MB dye. The ZnO component greatly assisted the degradation of the dye, with the activity of REC/Fe$_3$O$_4$/ZnO decreasing with increasing ZnO content. REC/Fe$_3$O$_4$/ZnO exhibited the highest photoactivity for the removal of MB dye at pH 6.0. The adsorption isotherm and degradation kinetics followed the Langmuir and Langmuir–Hinshelwood models, respectively. The mass loss and photoactivity of REC/Fe$_3$O$_4$/ZnO only slightly decreased after three cycles. The primary degradation mechanism was also proposed based on the detected intermediates. Our study demonstrated that REC/Fe$_3$O$_4$/ZnO composites have great potential as catalysts for the treatment of dye pollutants in aqueous solutions.

Acknowledgments: This research is supported by the National High Technology Research and Development Program of China (No. 2007AA06Z418), the National Natural Science Foundation of China (Nos. 20577036, 20777058, 20977070), the National Natural Science Foundation of Hubei province in China (No. 2015CFA137), and the Open Fund of Hubei Biomass-Resource Chemistry and Environmental Biotechnology Key Laboratory, and the Fund of Eco-environment Technology R&D and Service Center (Wuhan University).

Author Contributions: Peijiang Zhou and Huanhuan Wang conceived and designed the experiments; Rui Guo, Yifei Wang and Yunfei Yuan performed the experiments; Huanhuan Wang and Hongju Zhan analyzed the data; Peijiang Zhou contributed reagents/materials/analysis tools; Huanhuan Wang wrote the paper.

References

1. Forgacs, E.; Cserhati, T.; Oros, G. Removal of synthetic dyes from wastewaters: A review. *Environ. Int.* **2004**, *30*, 953–971. [CrossRef] [PubMed]
2. An, A.K.; Guo, J.; Lee, E.J.; Jeong, S.; Zhao, Y.; Wang, Z.; Leiknes, T. PDMS/PVDF hybrid electrospun membrane with superhydrophobic property and drop impact dynamics for dyeing wastewater treatment using membrane distillation. *J. Membr. Sci.* **2017**, *525*, 57–67. [CrossRef]
3. Zinatloo-Ajabshir, S.; Salavati-Niasari, M.; Zinatloo-Ajabshir, Z. Facile size-controlled preparation of highly photocatalytically active praseodymium zirconate nanostructures for degradation and removal of organic pollutants. *Sep. Purif. Technol.* **2017**, *177*, 110–120. [CrossRef]
4. Giannakoudakis, D.A.; Kyzas, G.Z.; Avranas, A.; Lazaridis, N.K. Multi-parametric adsorption effects of the reactive dye removal with commercial activated carbons. *J. Mol. Liq.* **2016**, *213*, 381–389. [CrossRef]
5. Spagnoli, A.A.; Giannakoudakis, D.A.; Bashkova, S. Adsorption of methylene blue on cashew nut shell based carbons activated with zinc chloride: The role of surface and structural parameters. *J. Mol. Liq.* **2017**, *229*, 465–471. [CrossRef]
6. Chang, S.H.; Wang, K.S.; Chao, S.J.; Peng, T.H.; Huang, L.C. Degradation of azo and anthraquinone dyes by a low-cost Fe^0/air process. *J. Hazard. Mater.* **2009**, *166*, 1127–1133. [CrossRef] [PubMed]
7. Ganesh, R.; Boardman, G.D.; Michelsen, D. Fate of azo dyes in sludges. *Water Res.* **1994**, *28*, 1367–1376. [CrossRef]
8. Razo-Flores, E.; Luijten, M.; Donlon, B.; Lettinga, G.; Field, J. Biodegradation of selected azo dyes under methanogenic conditions. *Water Sci. Technol.* **1997**, *36*, 65–72.
9. Lucas, M.; Peres, J. Decolorization of the azo dye reactive black 5 by Fenton and photo-Fenton oxidation. *Dyes Pigment.* **2006**, *71*, 236–244. [CrossRef]

10. Zhang, X.; Wang, Y.; Liu, B.; Sang, Y.; Liu, H. Heterostructures construction on TiO$_2$ nanobelts: A powerful tool for building high-performance photocatalysts. *Appl. Catal. B Environ.* **2017**, *202*, 620–641. [CrossRef]

11. Tan, C.; Cao, X.; Wu, X.J.; He, Q.; Yang, J.; Zhang, X.; Chen, J.; Zhao, W.; Han, S.; Nam, G.H.; et al. Recent advances in ultrathin two-dimensional nanomaterials. *Chem. Rev.* **2017**, *117*, 6225–6331. [CrossRef] [PubMed]

12. Pirhashemi, M.; Habibi-Yangjeh, A. Ultrasonic-assisted preparation of plasmonic ZnO/Ag/Ag$_2$WO$_4$ nanocomposites with high visible-light photocatalytic performance for degradation of organic pollutants. *J. Colloid Interface Sci.* **2017**, *491*, 216–229. [CrossRef] [PubMed]

13. Karunakaran, C.; Dhanalakshmi, R. Photocatalytic performance of particulate semiconductors under natural sunshine—Oxidation of carboxylic acids. *Sol. Energy Mater. Sol. Cells* **2008**, *92*, 588–593. [CrossRef]

14. Madhavan, J.; Muthuraaman, B.; Murugesan, S.; Anandan, S.; Maruthamuthu, P. Peroxomonosulphate, an efficient oxidant for the photocatalysed degradation of a textile dye, acid red 88. *Sol. Energy Mater. Sol. Cells* **2006**, *90*, 1875–1887. [CrossRef]

15. Kumar, S.; Karthikeyan, S.; Lee, A. G-C$_3$N$_4$-based nanomaterials for visible light-driven photocatalysis. *Catalysts* **2018**, *8*, 74. [CrossRef]

16. Jelle, A.A.; Hmadeh, M.; O'Brien, P.G.; Perovic, D.D.; Ozin, G.A. Photocatalytic properties of all four polymorphs of nanostructured iron oxyhydroxides. *ChemNanoMat* **2016**, *2*, 1047–1054. [CrossRef]

17. Zhang, G.; Lan, Z.A.; Wang, X. Conjugated polymers: Catalysts for photocatalytic hydrogen evolution. *Angew. Chem.-Int. Ed.* **2016**, *55*, 15712–15727. [CrossRef] [PubMed]

18. Zhang, L.; Li, J.; Chen, Z.; Tang, Y.; Yu, Y. Preparation of fenton reagent with H$_2$O$_2$ generated by solar light-illuminated nano-Cu$_2$O/MWNTs composites. *Appl. Catal. A Gen.* **2006**, *299*, 292–297. [CrossRef]

19. Ali, I.; Kim, J.O. Continuous-flow photocatalytic degradation of organics using modified TiO$_2$ nanocomposites. *Catalysts* **2018**, *8*, 43. [CrossRef]

20. Ho, W.; Tay, Q.; Qi, H.; Huang, Z.; Li, J.; Chen, Z. Photocatalytic and adsorption performances of faceted cuprous oxide (Cu$_2$O) particles for the removal of methyl orange (MO) from aqueous media. *Molecules* **2017**, *22*, 677. [CrossRef] [PubMed]

21. Mohd Hir, Z.; Abdullah, A.; Zainal, Z.; Lim, H. Photoactive hybrid film photocatalyst of polyethersulfone-ZnO for the degradation of methyl orange dye: Kinetic study and operational parameters. *Catalysts* **2017**, *7*, 313. [CrossRef]

22. Shu, H.Y.; Chang, M.C.; Tseng, T.H. Solar and visible light illumination on immobilized nano zinc oxide for the degradation and mineralization of orange G in wastewater. *Catalysts* **2017**, *7*, 164.

23. Sharma, D.; Sharma, S.; Kaith, B.S.; Rajput, J.; Kaur, M. Synthesis of zno nanoparticles using surfactant free in-air and microwave method. *Appl. Surf. Sci.* **2011**, *257*, 9661–9672. [CrossRef]

24. Giannakoudakis, D.A.; Arcibar-Orozco, J.A.; Bandosz, T.J. Key role of terminal hydroxyl groups and visible light in the reactive adsorption/catalytic conversion of mustard gas surrogate on zinc (hydr)oxides. *Appl. Catal. B Environ.* **2015**, *174–175*, 96–104. [CrossRef]

25. Ullah, R.; Dutta, J. Photocatalytic degradation of organic dyes with manganese-doped zno nanoparticles. *J. Hazard. Mater.* **2008**, *156*, 194–200. [CrossRef] [PubMed]

26. Chava, R.K.; Im, Y.; Kang, M. Nitrogen doped carbon quantum dots as a green luminescent sensitizer to functionalize zno nanoparticles for enhanced photovoltaic conversion devices. *Mater. Res. Bull.* **2017**, *94*, 399–407. [CrossRef]

27. He, W.; Jia, H.; Cai, J.; Han, X.; Zheng, Z.; Wamer, W.G.; Yin, J.-J. Production of reactive oxygen species and electrons from photoexcited zno and zns nanoparticles: A comparative study for unraveling their distinct photocatalytic activities. *J. Phys. Chem. C* **2016**, *120*, 3187–3195. [CrossRef]

28. Lakshmi Prasanna, V.; Vijayaraghavan, R. Insight into the mechanism of antibacterial activity of zno: Surface defects mediated reactive oxygen species even in the dark. *Langmuir ACS J. Surf. Colloids* **2015**, *31*, 9155–9162. [CrossRef] [PubMed]

29. Sun, L.; Shao, R.; Chen, Z.; Tang, L.; Dai, Y.; Ding, J. Alkali-dependent synthesis of flower-like zno structures with enhanced photocatalytic activity via a facile hydrothermal method. *Appl. Surf. Sci.* **2012**, *258*, 5455–5461. [CrossRef]

30. Mahmoodi, V.; Bastami, T.R.; Ahmadpour, A. Solar energy harvesting by magnetic-semiconductor nanoheterostructure in water treatment technology. *Environ. Sci. Pollut. Res. Int.* **2018**. [CrossRef] [PubMed]

31. Riahi-Madvaar, R.; Taher, M.A.; Fazelirad, H. Synthesis and characterization of magnetic halloysite-iron oxide nanocomposite and its application for naphthol green B removal. *Appl. Clay Sci.* **2017**, *137*, 101–106. [CrossRef]

32. Middea, A.; Spinelli, L.S.; Souza, F.G., Jr.; Neumann, R.; Fernandes, T.L.A.P.; Gomes, O.D.F.M. Preparation and characterization of an organo-palygorskite-Fe$_3$O$_4$ nanomaterial for removal of anionic dyes from wastewater. *Appl. Clay Sci.* **2017**, *139*, 45–53. [CrossRef]

33. Mu, B.; Tang, J.; Zhang, L.; Wang, A. Preparation, characterization and application on dye adsorption of a well-defined two-dimensional superparamagnetic clay/polyaniline/Fe$_3$O$_4$ nanocomposite. *Appl. Clay Sci.* **2016**, *132–133*, 7–16. [CrossRef]

34. Chang, J.; Ma, J.; Ma, Q.; Zhang, D.; Qiao, N.; Hu, M.; Ma, H. Adsorption of methylene blue onto Fe$_3$O$_4$/activated montmorillonite nanocomposite. *Appl. Clay Sci.* **2016**, *119*, 132–140. [CrossRef]

35. Hu, L.; Tang, X.; Wu, Z.; Lin, L.; Xu, J.; Xu, N.; Dai, B. Magnetic lignin-derived carbonaceous catalyst for the dehydration of fructose into 5-hydroxymethylfurfural in dimethylsulfoxide. *Chem. Eng. J.* **2015**, *263*, 299–308. [CrossRef]

36. Nur'aeni; Chae, A.; Jo, S.; Choi, Y.; Park, B.; Park, S.Y.; In, I. Synthesis of β-FeOOH/Fe$_3$O$_4$ hybrid photocatalyst using catechol-quaternized poly(*N*-vinyl pyrrolidone) as a double-sided molecular tape. *J. Mater. Sci.* **2017**, *52*, 8493–8501.

37. Shi, Z.; Yang, X.; Yao, S. Photocatalytic activity of cerium-doped mesoporous TiO$_2$ coated Fe$_3$O$_4$ magnetic composite under uv and visible light. *J. Rare Earths* **2012**, *30*, 355–360. [CrossRef]

38. Shekofteh-Gohari, M.; Habibi-Yangjeh, A. Novel magnetically separable Fe$_3$O$_4$@ZnO/AgCl nanocomposites with highly enhanced photocatalytic activities under visible-light irradiation. *Sep. Purif. Technol.* **2015**, *147*, 194–202. [CrossRef]

39. Majidnia, Z.; Idris, A. Combination of maghemite and titanium oxide nanoparticles in polyvinyl alcohol-alginate encapsulated beads for cadmium ions removal. *Korean J. Chem. Eng.* **2015**, *32*, 1094–1100. [CrossRef]

40. Jo, W.K.; Clament Sagaya Selvam, N. Enhanced visible light-driven photocatalytic performance of ZnO-g-C$_3$N$_4$ coupled with graphene oxide as a novel ternary nanocomposite. *J. Hazard. Mater.* **2015**, *299*, 462–470. [CrossRef] [PubMed]

41. Gu, N.; Gao, J.; Wang, K.; Yang, X.; Dong, W. ZnO–montmorillonite as photocatalyst and flocculant for inhibition of cyanobacterial bloom. *Water Air Soil Pollut.* **2015**, *226*, 136. [CrossRef]

42. Kolodziejczak-Radzimska, A.; Jesionowski, T. Zinc oxide-from synthesis to application: A review. *Materials* **2014**, *7*, 2833–2881. [CrossRef] [PubMed]

43. Feng, X.; Guo, H.; Patel, K.; Zhou, H.; Lou, X. High performance, recoverable Fe$_3$O$_4$/ZnO nanoparticles for enhanced photocatalytic degradation of phenol. *Chem. Eng. J.* **2014**, *244*, 327–334. [CrossRef]

44. Ökte, A.N.; Karamanis, D. A novel photoresponsive ZnO-flyash nanocomposite for environmental and energy applications. *Appl. Catal. B Environ.* **2013**, *142–143*, 538–552. [CrossRef]

45. Ahmad, M.; Ahmed, E.; Hong, Z.L.; Xu, J.F.; Khalid, N.R.; Elhissi, A.; Ahmed, W. A facile one-step approach to synthesizing ZnO/graphene composites for enhanced degradation of methylene blue under visible light. *Appl. Surf. Sci.* **2013**, *274*, 273–281. [CrossRef]

46. Bailey, S.W.; Brindley, G.W.; Kodama, H.; Martin, R.T. Report of the clay-minerals-society nomenclature committee for 1980–1981—Nomenclature for regular interstratifications. *Clays Clay Miner.* **1982**, *30*, 76–78. [CrossRef]

47. Guo, Y.; Zhang, G.; Gan, H. Synthesis, characterization and visible light photocatalytic properties of Bi$_2$WO$_6$/rectorite composites. *J. Colloid Interface Sci.* **2012**, *369*, 323–329. [CrossRef] [PubMed]

48. Lu, Y.; Chang, P.R.; Zheng, P.; Ma, X. Rectorite–TiO$_2$–Fe$_3$O$_4$ composites: Assembly, characterization, adsorption and photodegradation. *Chem. Eng. J.* **2014**, *255*, 49–54. [CrossRef]

49. Li, S.Q.; Zhou, P.J.; Zhang, W.S.; Chen, S.; Peng, H. Effective photocatalytic decolorization of methylene blue utilizing ZnO/rectorite nanocomposite under simulated solar irradiation. *J. Alloys Compd.* **2014**, *616*, 227–234. [CrossRef]

50. Wu, S.; Fang, J.; Xu, W.; Cen, C. Bismuth-modified rectorite with high visible light photocatalytic activity. *J. Mol. Catal. A Chem.* **2013**, *373*, 114–120. [CrossRef]

51. Zhang, Y.; Guo, Y.; Zhang, G.; Gao, Y. Stable TiO$_2$/rectorite: Preparation, characterization and photocatalytic activity. *Appl. Clay Sci.* **2011**, *51*, 335–340. [CrossRef]

52. Bu, X.Z.; Zhang, G.K.; Gao, Y.Y.; Yang, Y.Q. Preparation and photocatalytic properties of visible light responsive n-doped TiO_2/rectorite composites. *Microporous Mesoporous Mater.* **2010**, *136*, 132–137. [CrossRef]

53. Yang, L.; Liang, G.; Zhang, Z.; He, S.; Wang, J. Sodium alginate/Na^+-rectorite composite films: Preparation, characterization, and properties. *J. Appl. Polym. Sci.* **2009**, *114*, 1235–1240. [CrossRef]

54. Xiong, G.; Pal, U.; Serrano, J.G.; Ucer, K.B.; Williams, R.T. Photoluminesence and ftir study of ZnO nanoparticles: The impurity and defect perspective. *Phys. Status Solid* **2006**, *3*, 3577–3581. [CrossRef]

55. Wang, X.; Liu, B.; Ren, J.; Liu, C.; Wang, X.; Wu, J.; Sun, R. Preparation and characterization of new quaternized carboxymethyl chitosan/rectorite nanocomposite. *Compos. Sci. Technol.* **2010**, *70*, 1161–1167. [CrossRef]

56. Li, Z.; Jiang, W.T.; Hong, H. An ftir investigation of hexadecyltrimethylammonium intercalation into rectorite. *Spectrochim. Acta Part A* **2008**, *71*, 1525–1534. [CrossRef] [PubMed]

57. Nguyen, H.D.; Nguyen, T.D.; Nguyen, D.H.; Nguyen, P.T. Magnetic properties of Cr doped Fe_3O_4 porous nanoparticles prepared through a co-precipitation method using surfactant. *Adv. Nat. Sci. Nanosci. Nanotechnol.* **2014**, *5*, 035017. [CrossRef]

58. Oliveira, L.C.A.; Gonçalves, M.; Guerreiro, M.C.; Ramalho, T.C.; Fabris, J.D.; Pereira, M.C.; Sapag, K. A new catalyst material based on niobia/iron oxide composite on the oxidation of organic contaminants in water via heterogeneous fenton mechanisms. *Appl. Catal. A Gen.* **2007**, *316*, 117–124. [CrossRef]

59. Shirafuji, T.; Nomura, A.; Hayashi, Y.; Tanaka, K.; Goto, M. Matrix-assisted laser desorption ionization time-of-flight mass spectrometric analysis of degradation products after treatment of methylene blue aqueous solution with three-dimensionally integrated microsolution plasma. *Jpn. J. Appl. Phys.* **2016**, *55*. [CrossRef]

60. Yang, K.; Zhu, L.; Xing, B. Adsorption of polycyclic aromatic hydrocarbons by carbon nanomaterials. *Environ. Sci. Technol.* **2006**, *40*, 1855–1861. [CrossRef] [PubMed]

 *catalysts* 

Article

Cu Nanoparticles/Fluorine-Doped Tin Oxide (FTO) Nanocomposites for Photocatalytic H$_2$ Evolution under Visible Light Irradiation

Hui Liu, An Wang, Quan Sun, Tingting Wang * and Heping Zeng *

Key Laboratory of Functional Molecular Engineering of Guangdong Province, School of Chemistry & Chemical Engineering, South China University of Technology, Guangzhou 510640, China; cehuiliu@mail.scut.edu.cn (H.L.); ceakwong@mail.scut.edu.cn (A.W.); boldymansun@163.com (Q.S.)
* Correspondence: ttwang@scut.edu.cn (T.W.); hpzeng@scut.edu.cn (H.Z.);
  Tel.: +86-208-711-2631 (T.W.); +86-208-711-2631 (H.Z.)

Received: 12 August 2017; Accepted: 29 September 2017; Published: 12 December 2017

Abstract: Copper nanoparticles/fluorine-doped tin oxide (FTO) nanocomposites were successfully prepared by a simple hydrothermal method. The synthesized nanocomposites were characterized by X-ray diffraction (XRD), UV-visible diffuse-reflectance spectrum (UV-VIS DRS), energy dispersive X-ray (EDX), transmission electron microscopy (TEM), Raman spectra, and X-ray photoelectron spectroscopy (XPS). The obtained Cu/FTO nanocomposites exhibit high photocatalytic activity for H$_2$ evolution under visible light (λ > 420 nm) irradiation. When the content of Cu is 19.2 wt % for FTO, the Cu/FTO photocatalyst shows the highest photocatalytic activity and the photocatalytic H$_2$ evolution rate is up to 11.22 μmol·h^{-1}. Meanwhile, the photocatalyst exhibits excellent stability and repeatability. It is revealed that the transfer efficiency of the photogenerated electrons is improved greatly because of the intense interaction between Cu NPs and FTO. Furthermore, a possible mechanism is proposed for enhanced photocatalytic H$_2$ evolution of Cu/FTO photocatalysts under visible light irradiation.

Keywords: Cu nanoparticles; Cu/FTO nanocomposites; H$_2$ evolution; visible light; transfer efficiency

---

1. Introduction

Hydrogen (H$_2$) is regarded as one of the most promising energy sources because of its high energy content per mass. Photocatalytic H$_2$ evolution using solar energy has attracted much attention, which is a feasible method to solve energy shortages and environmental crises [1–3]. Since Fujishima and Honda firstly reported the photolysis of water based on TiO$_2$ electrodes in 1972 [4], semiconductor materials as photocatalysts have been extensively studied for highly-efficient solar water splitting. Subsequently, many semiconductor materials have been developed as efficient photocatalystys to make the utmost of solar energy for photocatalytic H$_2$ evolution [5,6].

It is well known that tin oxide (SnO$_2$) is one of the most promising semiconductors for photocatalytic H$_2$ evolution due to its excellent chemical and physical properties [7]. In the past few decades, SnO$_2$ has been extensively used in gas sensing, photoelectric conversion and photcatalysis, etc. However, because of the intrinsic band gap (ca. 3.6 eV), SnO$_2$ cannot fully use solar radiation in visible spectrum [8]. Many researchers have been trying to modify SnO$_2$ to increase its absorption of visible light, which accounts for 45% of sunlight [9]. Many different approaches can be used to promote visible light absorption, such as metal doping, ion doping, carbon materials doping, etc. [10–12]. Especially, fluorine (F) has been considered as one of the most effective dopants for enhancing light absorption. Fluorine-doped tin oxide (FTO) is widely used as conducting electrodes for photoelectrochemical reactions [13]. However, owing to the rapid recombination that photoinduction carries, the photocatalytic activity of fluorine-doped tin oxide (FTO) for H$_2$ evolution is usually very low, or non-existent, in the presence of a sacrificial agent [14].

As it is well known, loading metal nanoparticles (NPs) as co-catalysts on the surface of catalysts is one of the most effective ways to increase the separation efficiency of photoinduced electron-hole pairs [15]. Many researchers reported that metals, such as Au, Ag, Pt, Cu, and Pd, could be used to promote the migration efficiency of electrons [16–20]. Among of them, Cu has attracted much more attention because of its low cost and abundance [21]. Additionally, Cu NPs can generate the localized surface plasmon resonance (LSPR) effect and it is beneficial for the separation of photoinduced electron-hole pairs, which can greatly enhance the photocatalytic activity [22]. Therefore, Cu NPs as co-catalyst-modified semiconductor photocatalysts may be a promising system to promote the migration of photoinduced electrons for enhancing photocatalytic H_2 evolution. Though there are many reports about growth of copper or copper oxides on FTO for photoelectrochemical hydrogen evolution reaction (HER) and oxygen evolution reaction (OER) [23,24]. To the best of our knowledge, there is no report about Cu NP-modified FTO (Cu/FTO) nanocomposites for photocatalytic H_2 evolution under visible light irradiation.

Here, based on our previous research [15,22,25,26], we reported that FTO nanopowders were fabricated by a typical sol-gel method and then modified with Cu further by a facile hydrothermal method. The Cu/FTO nanocomposites were characterized by X-ray diffraction (XRD), UV-VIS, energy dispersive X-ray (EDX), transmission electron microscopy (TEM), X-ray photoelectron spectroscopy (XPS) and so forth. The photacatalytic H_2 evolution of Cu/FTO nanocomposites was evaluated under visible light illumination. The F dopant can greatly enhance the absorption ability of the visible light of SnO_2 and the Cu/FTO nanocomposites exhibit high photocatalytic activity for H_2 evolution. Furthermore, the possible factors in the improvement of photocatalytic activity were also investigated and discussed in this paper.

2. Results

2.1. Crystal Structure and Composition

Figure 1 shows the XRD patterns of as-prepared samples. For pure SnO_2, the diffraction peaks are indexed to the crystalline tetragonal structure of SnO_2 (JCPDS No. 41-1445) [27]. The FTO sample exhibits similar diffraction peaks with pure SnO_2, which indicates FTO has the similar crystal structure with pure SnO_2. For Cu/FTO samples, the diffraction peaks at 43.3°, 50.5°, and 74.1° are assigned to the (111), (200), and (220) planes of face-centered cubic Cu (JCPDS No. 04-0836) [26]. As the content of Cu increases, the intensity of characteristic peaks becomes stronger. No diffraction peak of CuO or Cu_2O is observed, which is in agreement with the result of the XPS analysis below. Additionally, the XRD patterns of FTO after the introduction of Cu show no significant change indicating that the introduction of Cu does not influence the crystal structure of FTO.

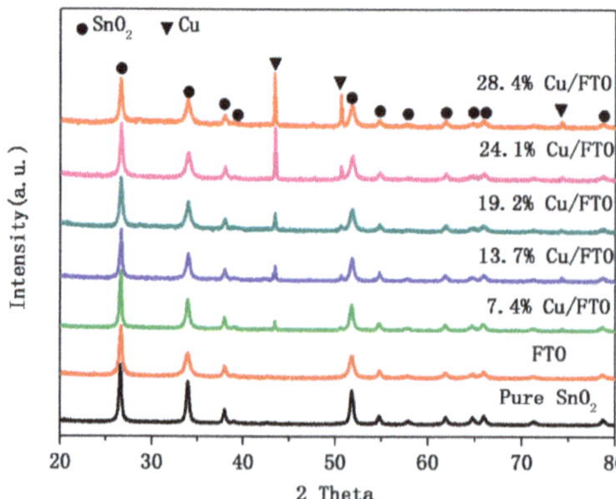

Figure 1. XRD (X-ray diffraction) patterns of pure SnO_2, FTO (Fluorine-doped tin oxide), and Cu/FTO samples with different contents of Cu.

To elaborate on the information of the as-prepared samples, Fourier transform infrared spectroscopy (FTIR) was carried out. Figure 2 depicts the FTIR spectra of pure SnO_2, FTO, and 19.2% Cu/FTO. The peaks at 924 cm^{-1} and 1530 cm^{-1} can be ascribed to the N–H out-of-plane bending vibration and in-plane bending vibration. The peak at 1678 cm^{-1} is attributed to the C=O stretching vibration [28]. The peaks around 3400–3800 cm^{-1} correspond to the N–H stretching vibration of amino groups and the O–H stretching vibrations of absorbed molecular water [6]. The characteristic peaks at 495 cm^{-1} and 642 cm^{-1} are ascribed to the O–Sn–O stretching vibration and the Sn–O stretching vibration, respectively [29]. The main characteristic peaks of SnO_2 are apparent in Cu/FTO nanocomposites, indicating the structure of SnO_2 remains relatively intact. Moreover, no peak of Cu is observed because Cu is not active for infrared spectra.

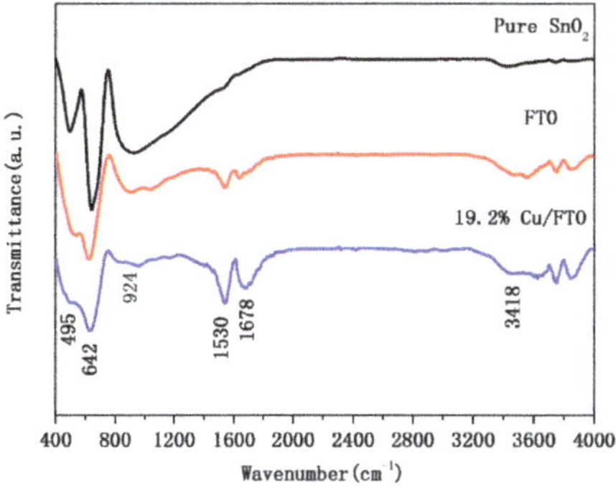

Figure 2. FTIR (Fourier transform infrared spectroscopy) spectra of pure SnO_2, FTO, and Cu/FTO samples.

Figure 3 displays the Raman spectra of pure SnO_2, FTO, and Cu/FTO. There are three fundamental Raman scattering peaks in good agreement with those of rutile SnO_2 single crystal [7]. For pure SnO_2, the characteristic and strong band at 625 cm^{-1} is attributed to A_{1g} vibration mode of SnO_2. The weak bands around 478 cm^{-1} and 772 cm^{-1} correspond to E_g and B_{2g} vibration modes of SnO_2 [30]. After doping F ions, the dominant bands of SnO_2 broaden and have a slight shift, which might be evidence that F ions have substituted for a portion of O^{2-} in the SnO_2. Compared with FTO, the intensity of Cu/FTO samples become stronger, indicating that there may be an intense interaction between Cu NPs and FTO [22]. Especially, the 19.2% Cu/FTO sample exhibits the strongest intensity and this SERS enhancement may be due to surface plasmon resonance [31]. Moreover, Figure S1 shows the Raman spectra at low wavelength region. Compared with FTO, there is no other band appearing for 19.2% Cu/FTO, which may exclude the presence of oxidized Cu [32].

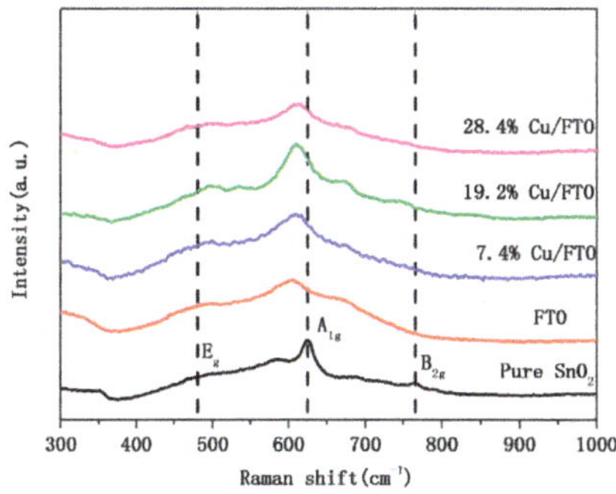

Figure 3. Raman spectra of pure SnO_2, FTO and Cu/FTO samples.

Morphologies and microstructures of the samples were studied by scanning electron microscopy (SEM), transmission electron microscopy (TEM), and high resolution TEM (HRTEM). Figure 4a shows the SEM image of the FTO sample. The morphology looks like irregular particles and many aggregations exist. Figure 4b–e shows the SEM images of Cu/FTO (7.4%, 19.2%, and 28.4%) samples, respectively. After the introduction of Cu, the particle size of Cu/FTO samples seems to be larger and there appear to be more aggregations. Especially for the 28.4% Cu/FTO sample, bulk agglomeration appears in the image (Figure 4e). The EDX spectrum (Figure 4f) and the elemental mapping patterns (Figure 4g) confirm that C, O, Sn, Cu, and F exist, which is consistent with the result of XPS below. Meanwhile, we can see that the main elements are uniformly distributed. The TEM images (Figure S2a–c) shows that the size distribution of 19.2% Cu/FTO sample is about 40–80 nm and there are some particles joining together. The HRTEM image of 19.2% Cu/FTO is presented in Figure S2d. The lattice spacing of d = 0.209 nm corresponds to the (111) plane of Cu [15]. The lattice spacing of d = 0.237 nm and 0.335 nm are attributed to the (200) plane and the (110) plane of SnO$_2$ (JCPDS No. 41-1445), respectively [33]. It can be found that Cu NPs and FTO have a close contact for developing the heterogeneous interface and this can promote the migration efficiency of photoinduced electrons.

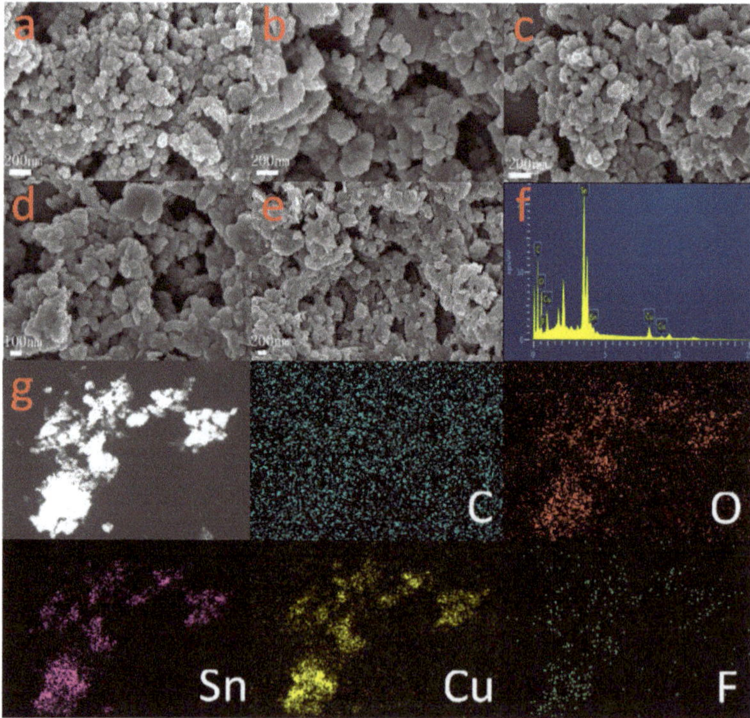

Figure 4. SEM (scanning electron microscopy) images (**a**) FTO; (**b**) 7.4% Cu/FTO; (**c**,**d**) 19.2% Cu/FTO; (**e**) 28.4% Cu/FTO; EDX (energy dispersive X-ray) spectrum (**f**); and elemental mapping patterns (**g**) of the 19.2% Cu/FTO sample.

The XPS spectra of the as-prepared 19.2% Cu/FTO sample were analyzed to further determine the chemical composition and the chemical state of elements. The results are shown in Figure 5. The standard C 1s peak at 284.8 eV was used as a reference to correct the peak shifts. The survey spectra of Cu/FTO sample (Figure 5a) suggest the presence of C, O, F, Sn, and Cu elements. Figure 5b shows the high-resolution XPS spectra of C 1s, the peaks at 284.8 eV, 285.9 eV, and 288.6 eV correspond to C–C, C–O, and O–C=O, respectively [34]. The XPS spectrum of O 1s (Figure 5c) exhibits two fitted peaks located at 531.1 eV and 532.2 eV, which are assigned to Sn–O and C–O. The XPS spectrum of Sn 3d (Figure 5d) shows two peaks attributed to Sn $3d_{5/2}$ and Sn $3d_{3/2}$. The first signal peak can be fitted into two peaks at 486.9 eV and 487.4 eV, which can be ascribed to Sn–O and Sn–F, respectively. The F 1s spectrum (Figure S3) exhibits only one major peak, which is assigned to F–Sn bond [14,35]. As shown in Figure 5f, the characteristic peaks of Cu 2p at 932.5 eV and 952.6 eV correspond to the binding energy of Cu $2p_{3/2}$ and Cu $2p_{1/2}$, respectively. The characteristic satellite peaks are not observed, suggesting that the high value state of Cu^{2+} is inexistence. Giving that the XPS spectra of Cu 2p is not able to distinguishing Cu^+ or Cu^0, Cu 2p Auger electron spectroscopy (AES) was carried out by XPS. As illustrated in Figure 5f, the only peak at 918.6 eV is assigned to Cu^0 rather than Cu^+, which is accordance with the result of the XRD analysis [15,25,27]. The abovementioned results show that the Cu^{2+} species are fully reduced to Cu^0 rather than Cu^+, which is consistent with the results of the XRD and Raman analysis.

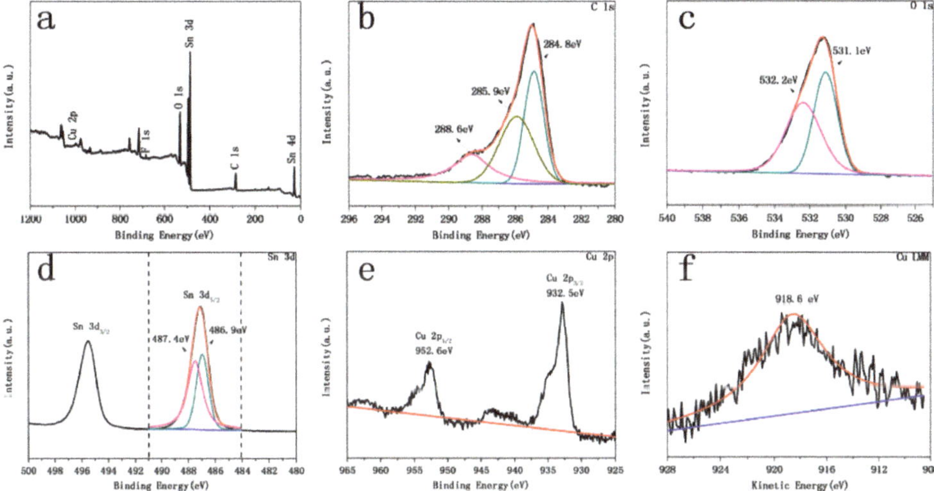

Figure 5. XPS (X-ray photoelectron spectroscopy) spectra of 19.2% Cu/FTO sample. (**a**) Survey spectra; (**b**) C 1s; (**c**) O 1s; (**d**) Sn 3d; (**e**) Cu 2p; and (**f**) Cu 2p AES (Auger electron spectroscopy).

2.2. Optical and Photoelectrochemical Properties

To investigate the optical property of pure SnO_2, FTO, and Cu/FTO samples, UV-VIS diffuse reflectance spectra was carried out. As displayed in Figure S4a, pure SnO_2 has a limited visible light absorption for itself because of its intrinsic band gap. However, after doping F ions, FTO exhibits strong absorption in the visible light range, which is beneficial for enhancing the H_2 evolution under visible light irradiation.

The optical band gap of semiconductors was determined by the Tauc equation [36]:

$$(\alpha h v)^n = A(h v - E_g)$$

where α is the absorption coefficient, $h v$ is the photon energy, A and n are constant, E_g is the optical band energy. The value of n is 0.5 and 2 for the indirect and direct band gap, respectively [37]. The E_g was calculated by the intercept of the extrapolated linear part of the curve with the energy axis. As presented in Figure S4b, the value of n is 2 and the band gap values of pure SnO_2 and FTO are 3.51 eV and 2.26 eV, respectively. As depicted in Figure 6, after the introduction of Cu, the absorption intensity of Cu/FTO samples is improved greatly in the visible light region. Furthermore, the 19.2% Cu/FTO sample exhibits the best absorption in the visible light region.

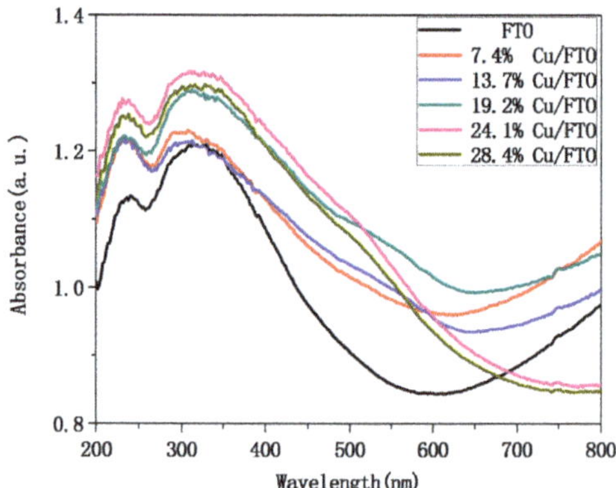

Figure 6. UV-VIS DRS (UV-visible diffuse-reflectance spectrum) of FTO and Cu/FTO samples.

To study the interaction between Cu NPs and FTO, we performed photoluminescence (PL) spectra and the results are presented in Figure 7. As displayed, all the samples of FTO and Cu/FTO exhibit a similar emission spectra shape with different intensities. The FTO shows the highest emission peak attributed to its higher recombination of the photogenerated electron-hole pairs. The emission intensity decreases with the introduction of Cu, indicating that the recombination of photogenerated electron-hole pairs is restrained in the Cu/FTO. This phenomenon could be due to the fast transfer of electrons from the FTO to the Cu and this could enhance the photocatalytic activity. Especially, the 19.2% Cu/FTO sample has the lowest emission intensity, suggesting that it may have the optimal content of Cu to inhibit carrier recombination, and this is consistent with the result of the photocatalytic H_2 evolution experiment. Due to the lower charge's recombination, more electrons will participate in photacatalytic reaction and, thus, enhance photocatalytic H_2 evolution. However, when further increasing the content of Cu, the Cu/FTO samples exhibit an increase tendency in emission intensity. The reason may be that the introduction of excessive Cu results in serious agglomeration, which could inhibit the fast migration of electrons [11].

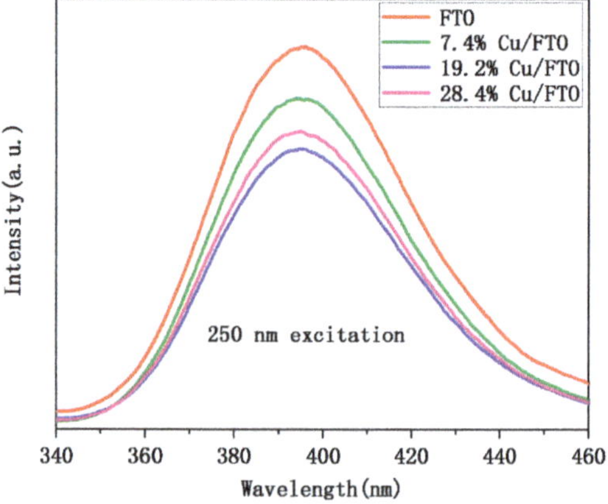

Figure 7. PL (photoluminescence) spectra of FTO and Cu/FTO samples with different contents of Cu.

As we know, the photocurrent test of semiconductors can be used to evaluate the separation and transfer efficiency of photogenerated electron–hole pairs. Figure 8a shows the transient photocurrent response of FTO and Cu/FTO film electrodes for six on-off cycles under visible light irradiation. Upon light irradiation, the photocurrent response appears in all FTO and Cu/FTO samples, but when the light is turned off, the value of the photocurrent decreases quickly to zero, which is reproducible in the six on-off cycles. The Cu/FTO samples show higher photocurrent density than that of FTO, indicating that there is less carrier recombination and much faster transfer of charges due to the intense surface interaction between Cu and FTO. Moreover, the 19.2% Cu/FTO sample exhibits the highest photocurrent density and the value is about 4.5 times as high as that of bare FTO. All these results evidently indicate the separation efficiency and transfer of the photogenerated electron–hole pairs are enhanced significantly by the introduction of Cu.

In order to further investigate the charge transfer capability, the EIS (electrochemical impedance spectra) Nyquist plots of both FTO and Cu/FTO samples were obtained and presented in Figure 8b. Generally, the smaller the diameter is, the lower charge transfer resistance and the faster interface charge transfer are. Compared with FTO, the Cu/FTO samples show smaller diameter, indicating that the introduction of Cu can diminish the impedance. Furthermore, the 19.2% Cu/FTO sample shows the smallest diameter due to its lower impedance. This result is consistent with that received by the photocurrent analysis.

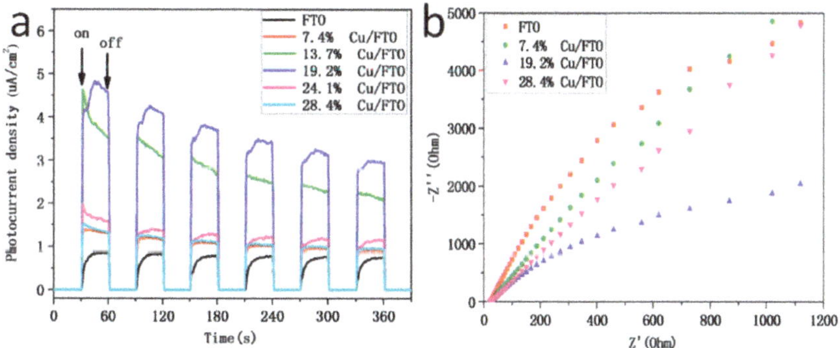

Figure 8. (a) The transient photocurrent of FTO and Cu/FTO samples with different contents of Cu and (b) EIS (electrochemical impedance spectra) Nyquist plots of FTO and Cu/FTO samples with different contents of Cu under visible light irradiation.

2.3. Photocatalytic Hydrogen Evolution

The photocatalytic hydrogen evolution of samples is evaluated under visible light irradiation with lactic acid as the sacrificial agent. There is no H_2 evolution of all samples to be detected without light irradiation, suggesting that H_2 evolution arises from photocatalytic reactions.

On one hand, when the content of Cu is 19.2%, we performed a series of photocatalytic H_2 evolution experiments under visible light irradiation to determine the optimal content of F. Figure 9 shows photocatalytic H_2 evolution rates of 19.2% Cu/SnO$_2$ and 19.2% Cu/FTO with different Sn/F molar ratios under visible light irradiation. Compared with 19.2% Cu/SnO$_2$, all the 19.2% Cu/FTO samples with different Sn/F molar ratios exhibit higher photocatalytic activity of H_2 evolution. Moreover, when the Sn/F molar ratios is 10:5, the 19.2% Cu/FTO sample achieves the highest photocatalytic H_2 evolution rate (11.22 μmol h^{-1}), which could be ascribed to the appropriate band gap and fewer defect sites of FTO (Sn/F = 10:5) than other FTO samples [37].

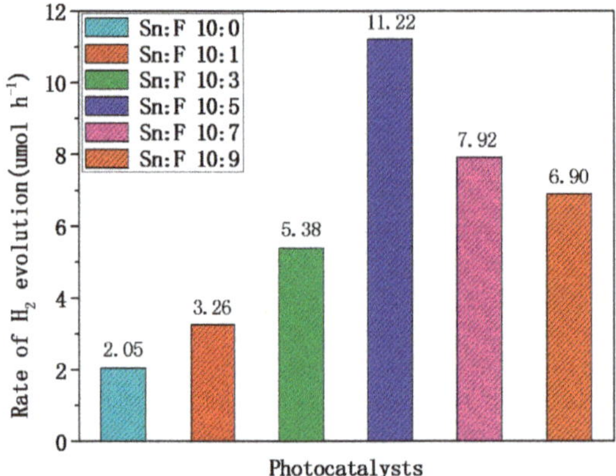

Figure 9. Photocatalytic H_2 evolution rates of 19.2% Cu/SnO_2 and 19.2% Cu/FTO samples with different Sn/F molar ratios under visible light irradiation.

On the other hand, when the Sn/F molar ratio is 10:5, we performed a series of photocatalytic H_2 evolution experiments under visible light irradiation to determine the optimal content of Cu in the Cu/FTO samples and the results were shown in Figure 10. The bare FTO sample does not show any photocatalytic activity of H_2 evolution, which may be due to the fast recombination of photoinduced electron-hole pairs. However, after the introduction of Cu, the Cu/FTO samples exhibit excellent photocatalytic activity of H_2 evolution. The H_2 evolution rate of Cu/FTO samples increases with the content of Cu and the highest photocatalytic H_2 evolution rate is up to 11.22 $\mu mol \cdot h^{-1}$ when the content of Cu is 19.2 wt % for FTO. However, when the content of Cu reach 19.2%, the H_2 evolution rate of Cu/FTO decreases with further increasing content of Cu. The results suggest the appropriate content of Cu can enhance photocatalytic activity because of the high absorption of light and the faster transfer of electrons. However, the introduction of excessive Cu may aggregate together and become new recombination centers of photogenerated electron-hole pairs, which could inhibit the photocatalytic activity of H_2 evolution [38]. This result is consistent with the result of the PL spectra (Figure 7).

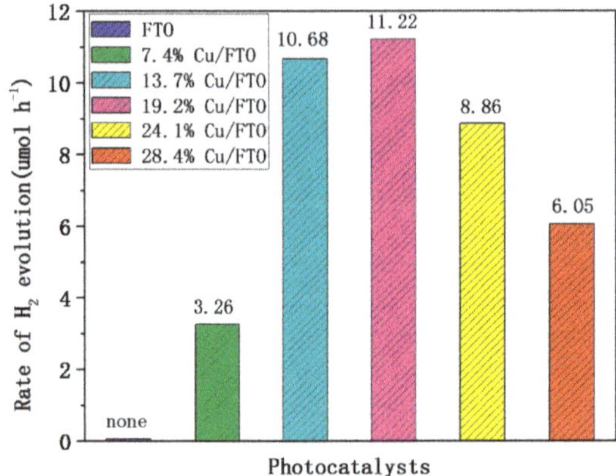

Figure 10. Photocatalytic H_2 evolution rate of FTO (Sn/F = 10/5) and Cu/FTO (Sn/F = 10/5) with different contents of Cu.

Considering practical applications, the representative 19.2% Cu/FTO sample was selected to investigate the recycling and stability of Cu/FTO samples. As shown in Figure 11, the recycling photocatalytic test was carried out for six runs under the same condition. The photocatalytic H_2 evolution of the sample is steady for six repeated runs under visible light irradiation, suggesting that the as-prepared Cu/FTO samples have excellent stability and repeatability. Furthermore, no obvious change of XRD patterns (Figure S5) before and after the stability test is observed, indicating the photocatalysts have good stability during photocatalytic H_2 evolution [34].

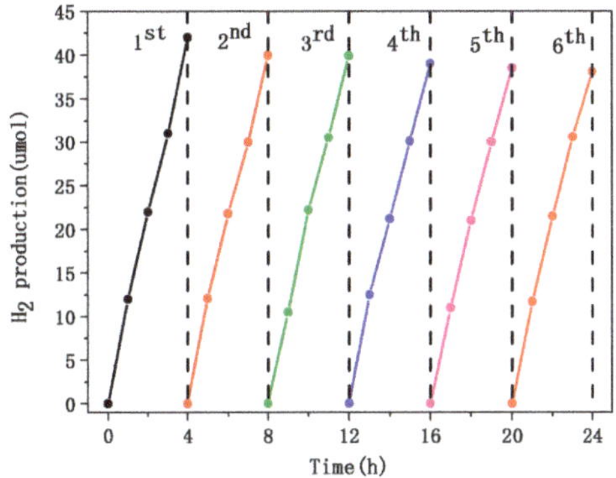

Figure 11. Recycling test of as-prepared 19.2% Cu/FTO (Sn/F = 10/5) under visible light irradiation for H_2 evolution (irradiation time = 24 h).

2.4. Mechanism

On the basis of all the above results and discussions, the proposed schematic mechanism of photocatalytic H_2 evolution of Cu/FTO photocatalyst is illustrated in Figure 12. Upon visible light irradiation, the electrons (e^-) of the valence band (VB) can be excited to the conduction band (CB) and the holes (h^+) remain in the valence band (VB). Due to the fast recombination of photogenerated electron-hole pairs, the photocatalytic activity of naked FTO is very low. However, after the introduction of Cu NPs, the recombination of photogenerated electron-hole pairs is greatly inhibited. Due to the existence of trapping centers (Cu NPs), the electrons (e^-) from the conduction band (CB) are tend to transfer quickly to Cu NPs and then react with H^+ for H_2 more easily. Meanwhile, the holes (h^+) are consumed by the sacrificial reagents (lactic acid). Thereby, the as-prepared Cu/FTO photocatalyst is beneficial for photocatalytic H_2 evolution.

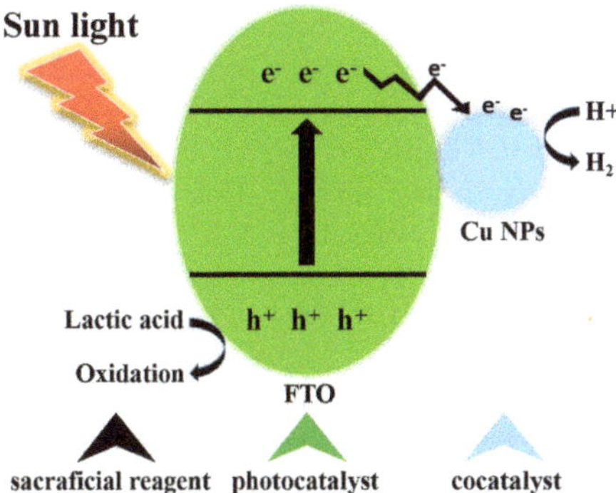

Figure 12. Schematic mechanism for photocatalytic H_2 evolution on the Cu/FTO system.

3. Materials and Methods

3.1. Materials

All the chemical reagents and solvents were analytical grade and used without further purification. Deionized water was supplied by a Millipore Milli-Q system (Millipore, Suzhou, China). Copper acetate hydrate ($Cu(CH_3COO)_2 \cdot H_2O$, $\geq 99.0\%$), tin chloride dehydrate ($SnCl_2 \cdot 2H_2O$, $\geq 98.0\%$), hydrofluoric acid (HF, $\geq 40\%$), ammonium hydroxide ($NH_3 \cdot H_2O$, $\geq 25\%$), lactic acid ($C_3H_6O_3$, $\geq 97.0\%$) and glucose ($C_6H_{12}O_6$, $\geq 98.0\%$) were purchased from Sinopharm Chemical Reagent Co. Ltd. (Dalian, China).

3.2. Synthesis of FTO Nanopowders

A typical synthesis process is as following. Firstly, 15 g of tin chloride dehydrate ($SnCl_2 \cdot 2H_2O$) was dissolved in 10 mL of deionized water and then ammonium hydroxide ($NH_3 \cdot H_2O$) was added dropwise into the solution under vigorous stirring until the pH value of the solution was up to 7. The solution was centrifuged at 9000 rpm to obtain the precipitation and the precipitation was washed three times with deionized water. Secondly, a certain amount of hydrofluoric acid (HF) was slowly dropped into the precipitation under vigorous stirring and obtained the sol mixture. Then the sol mixture was dried in a drying oven at 200 °C for 4 h and the gel mixture obtained. Finally, the gel mixture was calcined at 450 °C for 2 h under air in a muffle furnace and subsequently ground into fine nanopowders. Similarly, The FTO nanopowders with different Sn/F molar ratios (10:0, 10:1, 10:3, 10:5, 10:7, 10:9) were synthesized by following the same method. The Sn/F molar ratios (10:0, 10:1, 10:3, 10:5, 10:7, 10:9) are theoretical values calculated by the precursor ratios. If there is no special instructions in the paper, the Sn/F molar ratio of FTO is 10:5.

3.3. Synthesis of Cu/FTO Nanocomposites

In a typical synthesis process, 10 mL of deionized water, 15 mL of copper acetate (0.2 M $Cu(CH_3COO)_2$), 1.2 g of glucose ($C_6H_{12}O_6$) and 800 mg of FTO nanopowders were mixed together to form a homogeneous suspension by sonicating for 30 min. Then the suspension was transferred to a Teflon-lined autoclave (SX2-12-10Q, Suzhou Jiangdong Precision Instrument Co. Ltd., Shanghai, China) and heated at 120 °C for 12 h. Finally, the resulting solution was centrifuged to obtain the solid product. The solid product was washed three times by deionized water and ethanol, respectively. The obtained product was dried in a vacuum drying oven at 50 °C for 5 h. The Cu/FTO nanocomposites with different contents of Cu (0%, 7.4%, 13.7%, 19.2%, 24.1%, 28.4%) were prepared by following the same method. Cu wt % (0%, 7.4%, 13.7%, 19.2%, 24.1%, 28.4%) is the theoretical value calculated by the precursor ratios.

3.4. Characterization

The crystalline structure of the samples was characterized by using D8 Advantage X-ray diffraction (XRD, Bruker, Beijing, China) with Cu Kα radiation. Fourier transform infrared spectroscopy (FTIR) was recorded in the wave number range from 400 cm^{-1} to 4000 cm^{-1} by a FTIR spectrometer (Nicolet 670, Bruker, Beijing, China). The morphology of the samples was characterized by scanning electron microscopy (Merlin, Zeiss, Beijing, China). Transmission electron microscopy (TEM) test was conducted by an electron microscope (JEOL, JEM-2100, Bruker, Beijing, China) equipped with energy dispersive X-ray (EDX) spectrometer. X-ray photoelectron spectroscopy (XPS) was recorded on a Kratos Axis-Ultra DLD (Axis Ultra DLD, Bruker, Beijing, China). UV-VIS diffuse reflectance absorption spectra were obtained by a Hitachi U-3010 spectrophotometer (Shimadzu, Beijing, China). The photoluminescence spectra (PL) was conducted on a Hitachi F-4500 fluorescence spectrophotometer (Shimadzu, Beijing, China). Raman spectra were obtained by Laser Confocal Raman Microscopy system (LabRAM Aramis, H.J.Y, Beijing, China). The photocurrents and electrochemical impedance spectra (EIS) were recorded on an electrochemical workstation (CHI660C, CH Instruments, Inc., Beijing, China).

3.5. Photocatalytic Hydrogen Evolution Experiment

The photocatalytic experiment was performed by using a Xe lamp with an ultraviolet cut-off filter (PLS-SXE300CUV, Perfect light Co. Ltd., Beijing, China) ($\lambda > 420$ nm) as a light source in a quartz reaction cell, which was connected to a closed gas circulation. Based on our previous research, the as-synthesized photocatalyst (50 mg), lactic acid (10 mL), and water (60 mL) were mixed together by constant magnetic stirring (85-2, Gongyi yuhua Instrument Co. Ltd., Zhengzhou, China) to form a homogeneous suspension. The solution was evacuated with N_2 before every photocatalytic reaction. The evolved H_2 under irradiation was analyzed by online gas chromatography (GC-7900, Tianmei Co. Ltd., Beijing, China) with a thermal conductivity detector (TCD).

3.6. Photoelectrochemical Measurements

Electrochemical measurements were performed by using an electrochemical workstation (CHI660C Instruments, Shanghai, China). FTO and Cu/FTO electrodes served as the working electrode, platinum wire was used as the counter electrode, and Ag/AgCl as the reference electrode. The 0.5 M Na_2SO_4 aqueous solution served as the electrolyte and a 300 W Xe lamp was used as the light source. The working electrode was prepared as follows: 0.1 g of as-synthesized catalyst was dispersed in 1 mL of ethanol solution by sonication to get slurry. Then the slurry was spread onto the cleaned ITO glass and the as-prepared electrolyte was dried at 323 K for 4 h under vacuum.

4. Conclusions

In summary, we prepared a series of FTO nanopowders with different Sn/F molar ratios by a typical sol-gel method and then they were modified with different contents of Cu via a facile hydrothermal method. Compared with pure SnO_2, the fluorine-doped tin oxide (FTO) exhibits stronger absorption in the visible light range, but it does not show any photocatalytic activity for H_2 evolution because of the fast recombination of photogenerated electron-hole pairs. However, after the introduction of Cu, the Cu/FTO nanocomposites show excellent photocatalytic activity for H_2 evolution. When the loading content of Cu is 19.2 wt % for FTO, the highest H_2 evolution rate is up to 11.22 μmol h^{-1} under visible light irradiation. The photocatalysts exhibit excellent stability and repeatability for H_2 evolution. Moreover, the introduction of Cu could greatly promote the separation and transfer efficiency of the photogenerated electron-hole pairs and, thus, can enhance photocatalytic H_2 evolution. This work provides an insight for developing effective photocatalysts, which could be used for the conversion of solar energy to new energy sources.

Supplementary Materials: The following are available online at www.mdpi.com/2073-4344/7/12/385/s1, Figure S1. Raman spectra of FTO and 19.2% Cu/FTO. Figure S2. TEM (a) and HRTEM (b) of as-prepared 19.2% Cu/FTO sample. Figure S3. F 1s spectra of 19.2% Cu/FTO with XPS characterization. Figure S4. UV-VIS diffuse reflectance spectra (a) and optical band gap spectra (b) of pure SnO_2 and FTO. Figure S5. XRD patterns of 19.2% Cu/FTO before and after the stability test of hydrogen evolution.

Acknowledgments: We gratefully acknowledge financial support from the National Natural Science Foundation of China (Nos. 21571064, 21371060) and the Fundamental Research Funds for the Central Universities (No. 2015ZM162).

Author Contributions: Heping Zeng and Hui Liu conceived and designed the experiments; Hui Liu performed the experiments; Hui Liu, An Wang, Quan Sun and Tingting Wang analyzed the data; Hui Liu wrote the paper.

Conflicts of Interest: The authors declare no conflict of interest.

References

1. Wang, Y.; Wang, Q.; Zhan, X.; Wang, F.; Safdar, M.; He, J. Visible light driven type II heterostructures and their enhanced photocatalysis properties: A review. *Nanoscale* **2013**, *5*, 8326–8339. [CrossRef] [PubMed]
2. Jafari, T.; Moharreri, E.; Amin, A.; Miao, R.; Song, W.Q.; Suib, S. Photocatalytic Water Splitting—The Untamed Dream: A Review of Recent Advances. *Molecules* **2016**, *21*, 900. [CrossRef] [PubMed]

3. Huo, J.P.; Zeng, H.P. A novel triphenylamine functionalized bithiazole–metal complex with C_{60} for photocatalytic hydrogen production under visible light irradiation. *J. Mater. Chem. A* **2015**, *3*, 6258–6264. [CrossRef]

4. Fujishima, A.; Honda, K. Photolysis-decomposition of water at the surface of an irradiated semiconductor. *Nature* **1972**, *238*, 37–38. [CrossRef] [PubMed]

5. Huo, J.P.; Fang, L.T.; Lei, Y.L.; Zeng, G.C.; Zeng, H.P. Facile preparation of yttrium and aluminum co-doped ZnO via a sol–gel route for photocatalytic hydrogen production. *J. Mater. Chem. A* **2014**, *2*, 11040–11044. [CrossRef]

6. Qin, J.Y.; Huo, J.P.; Zhang, P.Y.; Zeng, J.; Wang, T.T.; Zeng, H.P. Improving the photocatalytic hydrogen production of $Ag/g-C_3N_4$ nanocomposites by dye-sensitization under visible light irradiation. *Nanoscale* **2016**, *8*, 2249–2259. [CrossRef] [PubMed]

7. Chen, Z.W.; Pan, D.Y.; Li, Z.; Jiao, Z.; Wu, M.H.; Shek, C.; Wu, C.M.L.; Lai, J.K.L. Recent Advances in Tin Dioxide Materials: Some Developments in Thin Films, Nanowires, and Nanorods. *Chem. Rev.* **2014**, *114*, 7442–7486. [CrossRef] [PubMed]

8. Khan, M.M.; Ansari, S.A.; Khan, M.E.; Ansari, M.O.; Min, B.K. Visible light-induced enhanced photoelectrochemical and photocatalytic studies of gold decorated SnO_2 nanostructures. *New J. Chem.* **2015**, *39*, 2758–2766. [CrossRef]

9. Babu, B.; Kadam, A.N.; Ravikumar, R.; Chan, B. Enhanced visible light photocatalytic activity of Cu-doped SnO_2 quantum dots by solution combustion synthesis. *J. Alloys Compd.* **2017**, *703*, 330–336. [CrossRef]

10. Ghimbeu, C.M.; Lumbreras, M.; Siadat, M.; van Landschoot, R.C.; Schoonman, J. Electrostatic sprayed SnO_2 and Cu-doped SnO_2 films for H_2S detection. *Sens. Actuators B Chem.* **2008**, *133*, 694–698. [CrossRef]

11. Chandra, S.; George, G.; Ravichandran, K.; Thirumurugan, K. Influence of simultaneous cationic (Mn) and anionic (F) doping on the magnetic and certain other properties of SnO_2 thin films. *Surf. Interface* **2017**, *7*, 39–46. [CrossRef]

12. Seema, H.; Christian, K.K.; Chandra, V.; Kim, K.S. Graphene-SnO_2 composites for highly efficient photocatalytic degradation of methylene blue under sunlight. *Nanotechnology* **2012**, *23*, 355705. [CrossRef] [PubMed]

13. Kent, C.A.; Concepcion, J.J.; Dares, C.J.; Torelli, D.A.; Rieth, A.J.; Miller, A.S.; Meyer, T.J. Water Oxidation and Oxygen Monitoring by Cobalt-Modified Fluorine-Doped Tin Oxide Electrodes. *J. Am. Chem. Soc.* **2013**, *135*, 8432–8435. [CrossRef] [PubMed]

14. Samad, W.Z.; Goto, M.; Kanda, H.; Nordin, N.; Liew, K.H.; Yarmo, M.A.; Yusop, M.R. Fluorine-doped tin oxide catalyst for glycerol conversion to methanol in sub-critical water. *J. Supercrit. Fluids* **2017**, *120*, 366–378. [CrossRef]

15. Zhang, P.Y.; Song, T.; Wang, T.T.; Zeng, H.P. In-situ synthesis of Cu nanoparticles hybridized with carbon quantum dots as a broad spectrum photocatalyst for improvement of photocatalytic H_2 evolution. *Appl. Catal. B Environ.* **2017**, *206*, 328–335. [CrossRef]

16. He, W.; Kim, H.K.; Wamer, W.G.; Melka, D.; Callahan, J.H. Photogenerated Charge Carriers and Reactive Oxygen Species in ZnO/Au Hybrid Nanostructures with Enhanced Photocatalytic and Antibacterial Activity. *J. Am. Chem. Soc.* **2014**, *136*, 750–757. [CrossRef] [PubMed]

17. Logar, M.; Jancar, B.; Sturm, S.; Suvorov, D. Weak polyion multilayer-assisted in situ synthesis as a route toward a plasmonic Ag/TiO_2 photocatalyst. *Langmuir* **2010**, *26*, 12215–12224. [CrossRef] [PubMed]

18. Shiraishi, Y.; Kofuji, Y.; Kanazawa, S.; Sakamoto, H.; Ichikawa, S.; Tanaka, S.; Hirai, T. Platinum nanoparticles strongly associated with graphitic carbon nitride as efficient co-catalysts for photocatalytic hydrogen evolution under visible light. *Chem. Commun.* **2014**, *50*, 15255–15258. [CrossRef] [PubMed]

19. Zou, Y.L.; Kang, S.; Li, X.Q.; Qin, L.X.; Mu, J. TiO_2 nanosheets loaded with Cu: A low-cost efficient photocatalytic system for hydrogen evolution from water. *Int. J. Hydrog. Energy* **2014**, *39*, 15403–15410. [CrossRef]

20. Han, C.; Yang, M.Q.; Zhang, N.; Xu, Y.J. Enhancing the visible light photocatalytic performance of ternary CdS–(graphene–Pd) nanocomposites via a facile interfacial mediator and co-catalyst strategy. *J. Mater. Chem. A* **2014**, *2*, 19156–19166. [CrossRef]

21. Liu, E.Z.; Qi, L.L.; Bian, J.J.; Chen, Y.H.; Hu, X.Y.; Fan, J.; Liu, H.C.; Zhu, C.J.; Wang, Q.P. A facile strategy to fabricate plasmonic Cu modified TiO_2 nano-flower films for photocatalytic reduction of CO_2 to methanol. *Mater. Res. Bull.* **2015**, *68*, 203–209. [CrossRef]

22. Huo, J.P.; Zeng, H.P. Copper nanoparticles embedded in the triphenylamine functionalized bithiazole–metal complex as active photocatalysts for visible light-driven hydrogen evolution. *J. Mater. Chem. A* **2015**, *3*, 17201–17208. [CrossRef]

23. Yang, Y.; Xu, D.; Wu, Q.Y.; Diao, P. Cu_2O/CuO Bilayered Composite as a High-Efficiency Photocathode for Photoelectrochemical Hydrogen Evolution Reaction. *Sci. Rep.* **2016**, *6*, 35158. [CrossRef] [PubMed]

24. Liu, X.; Cui, S.S.; Sun, Z.J.; Du, P.W. Copper oxide nanomaterials synthesized from simple copper salts as active catalysts for electrocatalytic water oxidation. *Electrochim. Acta* **2015**, *160*, 202–208. [CrossRef]

25. Zhang, P.Y.; Wang, T.T.; Zeng, H.P. Design of $Cu\text{-}Cu_2O$/g-C_3N_4 nanocomponent photocatalysts for hydrogen evolution under visible light irradiation using water-soluble Erythrosin B dye sensitization. *Appl. Surf. Sci.* **2017**, *391*, 404–414. [CrossRef]

26. Liu, H.Y.; Wang, T.T.; Zeng, H.P. CuNPs for Efficient Photocatalytic Hydrogen Evolution. *Part. Part. Syst. Charact.* **2015**, *32*, 869–873. [CrossRef]

27. Zheng, L.R.; Zheng, Y.H.; Chen, C.Q.; Zhan, Y.Y.; Lin, X.Y.; Zheng, Q.; Wei, K.M.; Zhu, J.F. Network Structured SnO_2/ZnO Heterojunction Nanocatalyst with High Photocatalytic Activity. *Inorg. Chem.* **2009**, *48*, 1819–1825. [CrossRef] [PubMed]

28. Zhu, S.J.; Meng, Q.N.; Wang, L.; Zhang, J.H.; Song, Y.B.; Jin, H.; Zhang, K.; Sun, H.C.; Wang, H.Y.; Yang, B. Highly Photoluminescent Carbon Dots for Multicolor Patterning, Sensors, and Bioimaging. *Angew. Chem. Int. Ed.* **2013**, *52*, 3953–3957. [CrossRef] [PubMed]

29. Zhang, B.; Tian, Y.; Zhang, J.X.; Cai, W. The characterization of fluorine doped tin oxide films by Fourier Transformation Infrared spectrum. *Appl. Phys. Lett.* **2011**, *98*, 021906. [CrossRef]

30. Noor, N.; Parkin, I.P. Enhanced transparent-conducting fluorine-doped tin oxide films formed by Aerosol-Assisted Chemical Vapour Deposition. *J. Mater. Chem. C* **2013**, *1*, 984–996. [CrossRef]

31. Liu, E.; Kang, L.; Yang, Y.; Sun, T.; Hu, X. Plasmonic Ag deposited TiO_2 nano-sheet film for enhanced photocatalytic hydrogen production by water splitting. *Nanotechnology* **2014**, *25*, 165401–165410. [CrossRef] [PubMed]

32. Deng, Y.L.; Handoko, A.D.; Do, Y.H.; Xi, S.B.; Yeo, B.S. In Situ Raman Spectroscopy of Copper and Copper Oxide Surfaces during Electrochemical Oxygen Evolution Reaction: Identification of Cu^{III} Oxides as Catalytically Active Species. *ACS Catal.* **2016**, *6*, 2473–2481. [CrossRef]

33. Chen, X.; Zhou, B.; Yang, S.; Wu, H.; Wu, Y. In situ construction of an SnO_2/g-C_3N_4 heterojunction for enhanced visible-light photocatalytic activity. *RSC Adv.* **2015**, *5*, 68953–68963. [CrossRef]

34. Song, T.; Huo, J.P.; Liao, T.; Zeng, J.; Qin, J.Y.; Zeng, H.P. Fullerene $[C_{60}]$ modified $Cr_{2-x}Fe_xO_3$ nanocomposites for enhanced photocatalytic activity under visible light irradiation. *Chem. Eng. J.* **2016**, *287*, 359–366. [CrossRef]

35. Wu, S.S.; Yuan, S.; Shi, L.Y.; Zhao, Y.; Fang, J.H. Preparation, characterization and electrical properties of fluorine-doped tin dioxide nanocrystals. *J. Colloid Interface Sci.* **2010**, *346*, 12–16. [CrossRef] [PubMed]

36. Chu, D.M.; Zhang, C.Y.; Yang, P.; Du, Y.K.; Lu, C. WS_2 as an Effective Noble-Metal Free Cocatalyst Modified $TiSi_2$ for Enhanced Photocatalytic Hydrogen Evolution under Visible Light Irradiation. *Catalysts* **2016**, *6*, 136. [CrossRef]

37. Yu, J.C.; Yu, J.G.; Ho, W.K.; Jiang, Z.T.; Zhang, L.Z. Effects of F^- Doping on the Photocatalytic Activity and Microstructures of Nanocrystalline TiO_2 Powders. *Chem. Mater.* **2002**, *14*, 3808–3816. [CrossRef]

38. Chao, Y.G.; Zheng, J.F.; Chen, J.Z.; Wang, Z.J.; Jia, S.P.; Zhang, H.X.; Zhu, Z.P. Highly efficient visible light-driven hydrogen production of precious metal-free hybrid photocatalyst: CdS@NiMoS core–shell nanorods. *Catal. Sci. Technol.* **2017**, *7*, 2798–2804. [CrossRef]

Article

Synthesis, Characterization of Nanosized ZnCr$_2$O$_4$ and Its Photocatalytic Performance in the Degradation of Humic Acid from Drinking Water

Raluca Dumitru [1,*], **Florica Manea** [1], **Cornelia Păcurariu** [1], **Lavinia Lupa** [1], **Aniela Pop** [1], **Adrian Cioablă** [2,*], **Adrian Surdu** [3] **and Adelina Ianculescu** [3]

[1] Faculty for Industrial Chemistry and Environmental Engineering, Politehnica University Timisoara, Piata Victoriei No. 2, Timisoara 300006, Romania; florica.manea@upt.ro (F.M.); cornelia.pacurariu@upt.ro (C.P.); lavinia.lupa@upt.ro (L.L.); aniela.pop@upt.ro (A.P.)

[2] Faculty of Mechanical Engineering, Politehnica University Timisoara, 1 Mihai Viteazu Blv., Timisoara 300222, Romania

[3] Department of Oxide Materials Science and Engineering, Faculty of Applied Chemistry and Materials Science, Polytehnic University of Bucharest, Gh. Polizu Street No.1-7, Bucharest 011061, Romania; adrian.surdu@live.com (A.S.); a_ianculescu@yahoo.com (A.I.)

* Correspondence: ralucadumitru2002@yahoo.com (R.D.); adrian.cioabla@upt.ro (A.C.); Tel.: +40-256-404-188 (R.D.)

Received: 31 March 2018; Accepted: 12 May 2018; Published: 15 May 2018

Abstract: Zinc chromite (ZnCr$_2$O$_4$) has been synthesized by the thermolysis of a new Zn(II)-Cr(III) oxalate coordination compound, namely [Cr$_2$Zn(C$_2$O$_4$)$_4$(OH$_2$)$_6$]·4H$_2$O. The coordination compound has been characterized by chemical analysis, infrared spectroscopy (IR), and thermal analysis. The zinc chromite obtained after a heating treatment of the coordination compound at 450 °C for 1 h has been investigated by XRD, FE-SEM, TEM/HR-TEM coupled with selected area electron diffraction (SAED) measurements. The photocatalytic performance of nanosized zinc chromite was assessed for the degradation and mineralization of humic acid (HA) from a drinking water source, envisaging the development of the advanced oxidation process for drinking water treatment technology. A mineralization efficiency of 60% was achieved after 180 min of 50 mg L^{-1} HA photocatalysis using zinc chromite under UV irradiation, in comparison with 7% efficiency reached by photolysis.

Keywords: zinc chromite; photocatalysis activity; oxalate; humic acid

---

1. Introduction

Zinc chromite (ZnCr$_2$O$_4$), a mixed oxide that exhibits a normal spinel structure and crystallizes in the cubic system, has attracted considerable interest in material science, due to its physical-chemical properties suitable for various applications. Many applications of the nanocrystalline ZnCr$_2$O$_4$ spinel have been reported, e.g., as catalytic material for a variety of reactions [1–3], a photocatalyst [4–7], a sensor for toxic gases [8], and for humidity [9,10].

The solid state reactions that consist of the mixing of oxides or carbonates, followed by calcination and grinding, are considered the most general method for preparing spinel [11–13]. A higher temperature of calcination (>1000 °C) is needed for the reactions' completion, which leads to the obtaining of spinel powder with a small surface area [14,15]. Various synthesis methods, such as co-precipitation [16,17], sol-gel [6,18], thermolysis of polymer-metal complex [19], microwave [14,20], and hydrothermal methods [15,21,22] have been investigated in order to obtain spinel powders with higher specific surface areas. Among all these methods, the most efficient one, which considers the lack of stoichiometry control and allows the formation of homogeneous nanoparticles with very good properties for catalytic and photocatalytic usage, is considered to be synthesis at a low temperature, starting from different precursors [23–25].

In recent years, photocatalytic oxidation as an advanced oxidation process with semiconducting materials has been studied for both water and wastewater treatment. The most studied research has considered TiO_2-based materials [26–28], and recently, cromite-type materials have raised a high interest due to their certain peculiarities for photocatalysis applications, e.g., small band-gap energy and chemical stability [29,30]. $ZnCr_2O_4$ is considered a very important spinel compound, due to its potential application as an efficient catalyst, and its photocatalytic activity has been reported for the removal of organic dyes from wastewaters [6,31].

The goal of this research is to obtain nanosized $ZnCr_2O_4$ through this new and efficient method, based on the thermal conversion at 450 °C of the $[Cr_2Zn(C_2O_4)_4(OH_2)_6]\cdot4H_2O$ precursor, and to study the compound's photocatalytic activity during the removal process of humic acid from drinking waters. This proposed method of zinc chromite spinel synthesis exhibits the advantages of a short reaction time, low reaction temperatures, no by-product generation, environmentally friendliness, and the ability to be easily scaled up due to its simplicity and low costs, compared to the conventional methods. The catalytic activity of synthesized $ZnCr_2O_4$ was studied under UV irradiation for colour removal, degradation, and the mineralization of humic acids from water. Humic acids are considered to be the main component of natural organic matters (NOMs) from drinking water sources, which should be eliminated due to their potential toxicity and carcinogeneous character in particular, during the disinfection step. The photocatalytic study was conducted in comparison with the photolysis process, and considered a prior sorption step before photocatalysis to assess the contribution of each process.

2. Results and Discussion

2.1. Synthesis and Characterization of the Cr^{3+}-Zn^{2+} Coordination Compound

The synthetic route developed for the synthesis of the coordination compound precursor is based on the redox reaction between 1,2-ethanediol and a nitrate ion in the presence of nitric acid (2M):

$$12C_2H_4(OH)_2 + 6([Cr(OH_2)_6]^{3+} + 3NO_3^-) + 3([Zn(OH_2)_6]^{2+} + 2NO_3^-) + 8(H^+ + NO_3^-)$$
$$\xrightarrow{\Delta t^\circ} 3Cr_2Zn(C_2O_4)_4(OH_2)_6]\cdot4H_2O_{(s)} + 32NO_{(g)} + 76H_2O_{(g)}$$

The IR spectrum of the oxalate precursor (Figure 1a and Table S1 Electronic Supplementary Material) displays a strong band at 1622 cm^{-1}, attributed to $\nu_{asym(OCO)}$ vibration, and the bands at 1391 cm^{-1} and 910 cm^{-1} attributed to $\nu_{sym(OCO)}$ confirm that the ligand acts as a bidentate [32,33].

The strong band at 1719 cm^{-1}, assigned to the $\nu_{asym(O=C-O)}$ vibration, as well as the band at 1268 cm^{-1} attributed to $\nu_{sym(O=C-O)}$, confirms that the ligand acts as a tetradentate [34,35].

The broad and intense band with the maximum at 3565 cm^{-1} assigned to vibration $\nu_{(OH)}$ confirms the presence of water in the Cr(III)-Zn(II) oxalate [34–36].

The IR spectrum of the ligand (Figure 1b and Table S2 Electronic Supplementary Material) is similar to the one reported by the literature for oxalic acid [33].

The thermal decomposition of the coordination compound (Figure 2) occurs in the 25–450 °C temperature range, and is characterized by a four-stage mass loss. During the first endothermic step (25–150 °C, mass loss (found/calc.) = 9.75/10.26%) is eliminated the lattice water. The second decomposition stage (150–321 °C, mass loss (found/calc.) = 14.83/15.39%) corresponds to the removal of the six coordinated molecules of water. The oxidative degradation of oxalate occurs in the third step (321–406 °C), and the mass loss ((found/calc.) = 40.69/41.06%) indicates the obtaining of a mixture of oxides: zinc oxide and nonstoichiometric oxide Cr_2O_{3+x}, x = 0.25. This assumption is supported by the existence of the fourth decomposition stage (406–450 °C). In this decomposition process, the continuous mass loss assigned to oxygen evolving is in agreement with the data from literature [35,37], and represents the formation of Cr_2O_3 followed by obtaining the zinc chromite.

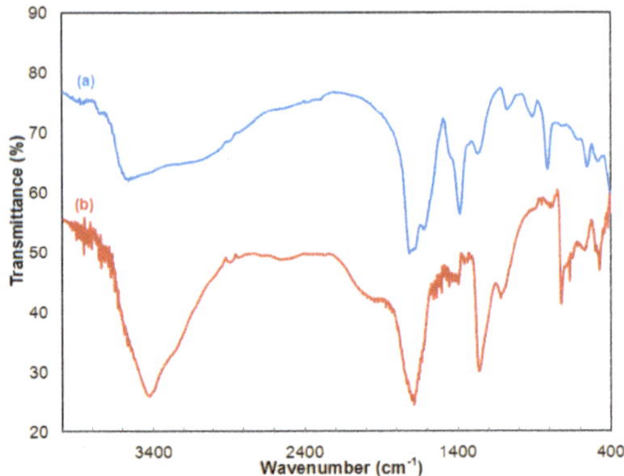

Figure 1. IR vibrational spectra of (**a**) the $[Cr_2Zn(C_2O_4)_4(OH_2)_6]\cdot 4H_2O$ compound and (**b**) isolated oxalic acid.

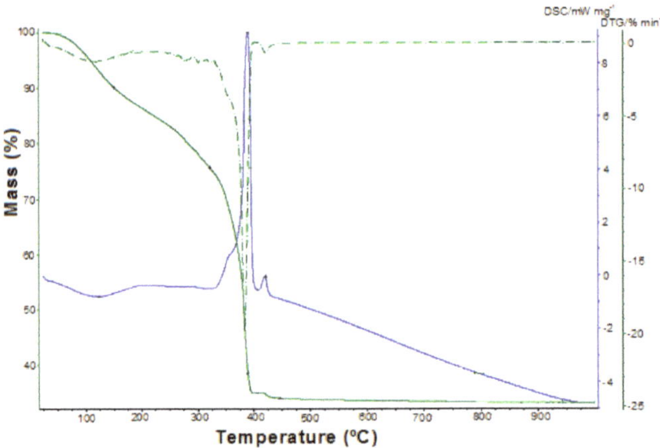

Figure 2. Thermal curves (TG, DTG and DSC) of $[Cr_2Zn(C_2O_4)_4(OH_2)_6]\cdot 4H_2O$ in air atmosphere.

Based on TG, DTG and DSC curves analysis, the thermal decomposition of $[Cr_2Zn(C_2O_4)_4(OH_2)_6]\cdot 4H_2O$ in air atmosphere can be represented by the following sequence:

$$[Cr_2Zn(C_2O_4)_4(OH_2)_6]\cdot 4H_2O_{(s)} \xrightarrow{(I)} 4H_2O_{(g)} + [Cr_2Zn(C_2O_4)_4(OH_2)_6]_{(s)}$$

$$[Cr_2Zn(C_2O_4)_4(OH_2)_6]_{(s)} \xrightarrow{(II)} 6H_2O_{(g)} + Cr_2Zn(C_2O_4)_{4(s)}$$

$$Cr_2Zn(C_2O_4)_{4(s)} \xrightarrow[\;\;\;\;\;\;\;]{+\,(1/2+1/8)O_2\;;\quad (III)} 4CO_{2(g)} + 4CO_{(g)} + ZnO_{(s)} + Cr_2O_{3,25(s)}$$

$$Cr_2O_{3,25(s)} \xrightarrow{(IV)} 1/8O_{2(g)} + Cr_2O_{3(s)}$$

$$ZnO + Cr_2O_3 \rightarrow ZnCr_2O_4$$

The obtained zinc chromite at 480 °C was confirmed through elemental analysis ($ZnCr_2O_4$, calc./found: Zn% = 28.01/28.24; Cr% = 44.55/44.48).

2.2. Characterization of $ZnCr_2O_4$ Powders

XRD investigations at room temperature performed on the powders resulting after calcination at various temperatures indicate that the crystallization process is already in progress starting with temperatures above 450 °C (Figure 3). Thus, the powder obtained after calcination at 450 °C is already crystallized, consisting of $ZnCr_2O_4$ with the typical cubic spinel structure (space group Fd-3m), identified as a unique phase with the ICCD card no. 01-079-5291. The diffraction peaks of $ZnCr_2O_4$ become sharper as the calcination temperature increases (Figure 3). This evolution is very similar to that reported by Gingasu et al. for their zinc chromite powders, synthesized by using the precursor method via thermal decomposition of tartarte/gluconate compounds [24].

From the structural point of view, an obvious increase of lattice parameter *a* and unit cell volume *V* takes place with the increase of the thermal treatment temperature from 450 to 600 °C, when the crystallization process is still in progress. A further rise in calcination temperature seems to no longer significantly affect the network, so that only a slight variation of the unit cell parameters and volume was observed (Table 1, Figure 3).

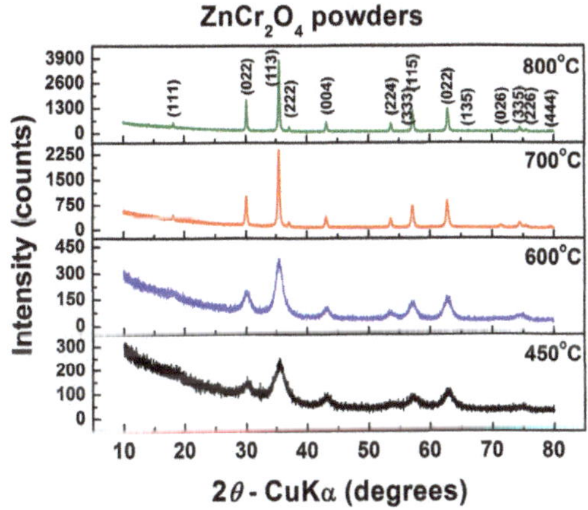

Figure 3. XRD patterns at room temperature for $ZnCr_2O_4$ powders annealed at various temperatures.

Table 1. Structural parameters obtained from XRD data by Rietveld refinement (ICCD no. 01-079-5291).

Calcination Temperature	450 °C	600 °C	700 °C	800 °C
Structure/Space group	Cubic/Fd-3m			
a (Å)	8.3181 ± 0.0028	8.3296 ± 0.0065	8.3258 ± 0.0009	8.3295 ± 0.0005
V (Å³)	575.54	577.92	577.14	577.90
$<D>$ (nm)	3.54 ± 0.34	5.20 ± 0.39	23.4 ± 2.26	45.79 ± 11.6
$<S>$ (%)	2.85 ± 1.65	1.91 ± 0.64	0.41 ± 0.20	0.20 ± 0.04
R_e	17.43	17.60	12.88	12.44
R_p	8.02	8.83	6.50	6.46
R_{wp}	11.30	11.98	8.65	8.73
Goodness of fit (GOF)	0.42	0.46	3.15	0.49

As expected, the increase of the calcination temperature influenced the crystallite size *D* and lattice strain *S*. Thus, the average crystallite size increased from 3.5 nm to 45.8 nm for the $ZnCr_2O_4$ powders that were thermally treated in the temperature range of 450–800 °C (Table 1). The coarsening process evolves faster at annealing temperatures above 600 °C. However, the value of the average crystallite size was maintained in the nanometric range even for the powder calcined at 800 °C. Concurrently, the increase of the crystallite sizes induced by higher annealing temperatures led to a structural relaxation, pointed out by the decreasing trend of the values of the lattice strains (Table 1).

The morphology of the synthesized powders calcined at various temperatures is presented in the FE-SEM images of Figure 4a–d. Figure 4a,b show that nanosized particles were obtained for the powders thermally treated at temperatures ranging between 450–600 °C. In this case, the estimation of the shape and average size of particles is almost impossible, because they are very small and not well-defined, exhibiting a high agglomeration tendency. The $ZnCr_2O_4$ powder annealed at 700 °C exhibits a duplex morphology, consisting of a large fraction of rounded, small particles of a few tens of nanometers (10–30 nm), as well as a lower proportion of polyhedral particles of 70–120 nm with well-defined edges and corners, most with an almost octahedral shape (Figure 4c). This kind of morphology was maintained in the powder calcined at 800 °C, when the coarsening process occurs for both smaller and larger particles. One can notice that, in spite of their size (of 40–50 nm), the smaller particles seem also to evolve toward a well-faceted, polyhedral shape (Figure 4d).

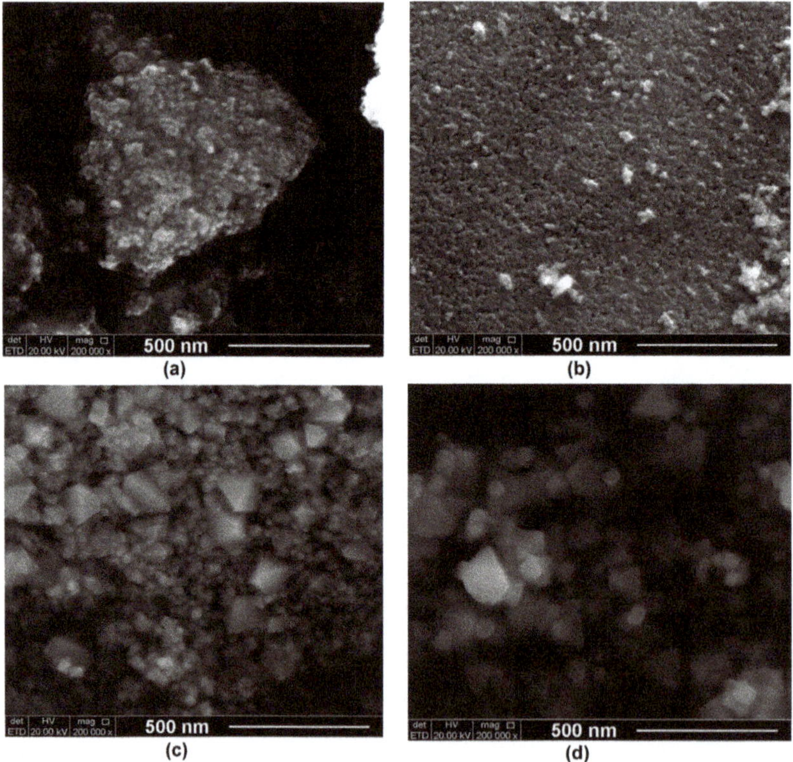

Figure 4. FE-SEM images showing some morphological features and the agglomeration tendency of $ZnCr_2O_4$ powders calcined at different temperatures: (**a**) 450 °C; (**b**) 600 °C; (**c**) 700 °C; and (**d**) 800 °C.

In order to have a more realistic view of the size and morphology of the $ZnCr_2O_4$ particles, TEM investigations are required. Thus, the TEM images of the powders annealed at lower temperatures indicate that, in both cases, the individual particles exhibit sizes of only a few nanometers. Average particle size values $<d_{TEM}>$ of 4.7 nm and 6.9 nm were estimated for the powders obtained after annealing at 400 and 600 °C, respectively (Figures 5a,b and 6a). These $<d_{TEM}>$ values fit quite well the average crystallite size values $<D>$ calculated from the XRD data (3.54 nm and 5.20 nm, respectively), proving the single-crystalline nature of the $ZnCr_2O_4$ particles, which each consist of 6–8 spinel unit cells. For the powder annealed at 700 °C, the TEM image of Figure 7a reveals the two types of particles also emphasized by the FE-SEM investigations. One can observe that some of the smaller particles already exhibit well-faceted morphology, while the larger particles seem also to be single crystals. An overall $<d_{TEM}>$ value of 32.5 nm was estimated for the average particle size, taking into account the two types of particles. The same tendency toward a bimodal particle size distribution was also observed in the case of the powder thermally treated at 800 °C, where a few larger octahedral particles of 120–200 nm coexist with the major fraction of smaller particles of 30–50 nm (Figure 8a,b). The overall value of the average particle size is 66.3 nm.

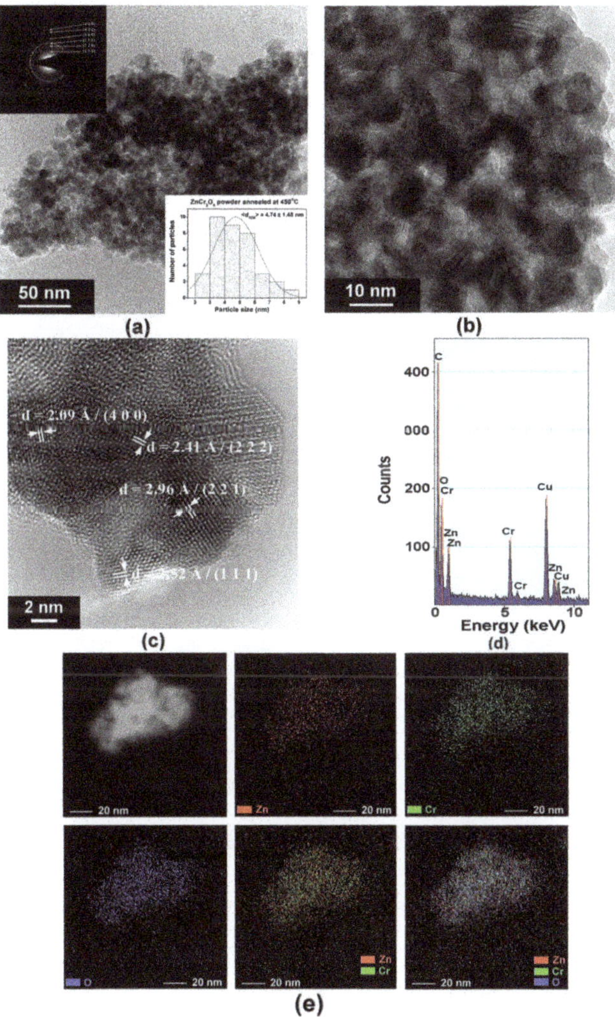

Figure 5. Morphology, crystallinity, and compositional uniformity of $ZnCr_2O_4$ powder calcined at 450 °C: (**a**) TEM image showing a general view of a large aggregate of nanosized particles—insets show the SAED pattern and histogram indicating the particle size distribution; (**b**) TEM image showing a detail inside an agglomerate; (**c**) HR-TEM image; (**d**) EDX spectrum; (**e**) STEM image and elemental (at. %) mapping performed on a small agglomerate of a few $ZnCr_2O_4$ particles.

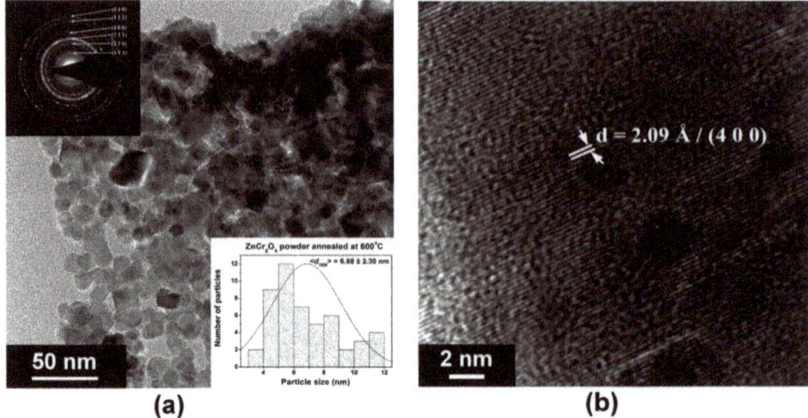

Figure 6. Morphology and crystallinity of ZnCr$_2$O$_4$ powder calcined at 600 °C: (**a**) TEM image showing a general view of a large aggregate of nanosized particles—insets show the SAED pattern and histogram indicating the particle size distribution; (**b**) HR-TEM image.

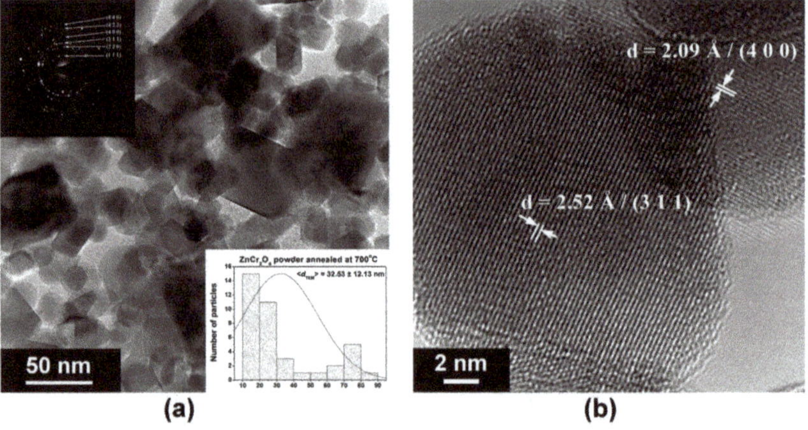

Figure 7. Morphology and crystallinity of ZnCr$_2$O$_4$ powder calcined at 700 °C: (**a**) TEM image showing a duplex morphology consisting of a large fraction of nanosized, equiaxial particles and a small fraction of submicron, polyhedral, well-faceted particles—insets show the SAED pattern and histogram indicating the particle size distribution; (**b**) HR-TEM image.

Despite the small size, especially in the case of the powders thermally treated at lower temperatures (450 and 600 °C), the particles exhibit a high crystallinity degree, proved by the HR-TEM images of Figures 5c, 6b, 7b and 8c, which clearly show highly-ordered fringes spaced at specific distances corresponding to various crystalline planes of the spinel structure, as well as the bright spots that form well-defined diffraction rings in the SAED patterns (insets of Figures 5a, 6a, 7a and 8a).

EDX spectra of Figures 5d and 8d reveal only the presence of the Zn, Cr, and O species. No other foreign cations were identified, which proves the lack of any contamination during the synthesis process and the high purity degree of the as-prepared powdered samples.

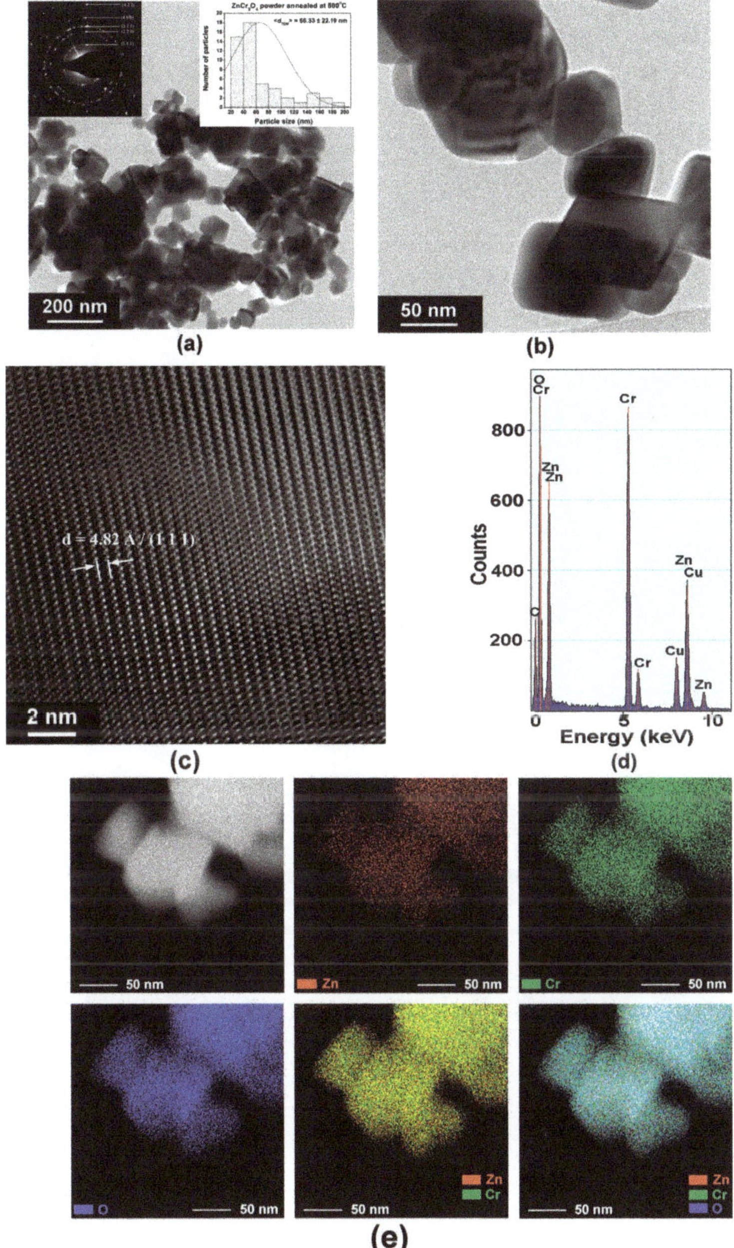

Figure 8. Morphology, crystallinity, and compositional uniformity of $ZnCr_2O_4$ powder calcined at 800 °C: (**a**) TEM image showing a general view of several particles of different size and shape—insets show the SAED pattern and histogram indicating the particle size distribution; (**b**) TEM image showing details of a few polyhedral particles; (**c**) HR-TEM image; (**d**) EDX spectrum; (**e**) STEM image and elemental (at. %) mapping performed on five $ZnCr_2O_4$ particles.

EDX mapping was carried out in order to investigate the compositional homogeneity of a small agglomerate of particles of the powder annealed at 450 °C, as well as of five larger $ZnCr_2O_4$ particles of the powder annealed at 800 °C. The superposition of the maps corresponding to all three elemental species of the $ZnCr_2O_4$ compound indicates a uniform distribution of Zn and Cr cations and oxygen anions, with respect to their atomic ratio in the spinel structure. This suggests that some compositional gradients or segregation of some potential secondary phases, undetected by XRD investigations, appear in the powders under investigation (Figures 5e and 8e).

Figure 9 represents, as an example, the UV-VIS diffuse reflectance spectrum of $ZnCr_2O_4$ (450 °C), and a broad absorption band within a 200–650 nm wavelength can be seen. The band gap energy was determined using the Kubelka–Munk equation [38]:

$$F(R) = (1 - R)^2/2R, \tag{1}$$

where R is reflectance.

To determine the band gap energy, the $(F(R)h\nu)^2$-$h\nu$ dependence is presented in the inset of Figure 6, where $h\nu = 1240/\lambda$, which determines the band gap energy value of 2.45 eV, lower than the 2.9 eV reported by Abbasi et al., [17] for $ZnCr_2O_4$ and $ZnCr_2O_4/Ag$. Therefore, the results of the band gap energy values for the $ZnCr_2O_4$ powder calcinated at 600, 700, and 800 °C are gathered in Table 2, and the values are larger in comparison to $ZnCr_2O_4$ (450 °C). The particle size affects the band gap energy value in relation to composition, crystallinity, and crystallite size. The band gap energy decreasing as crystallite size decreases has been also reported by Irfan et al. [39]. This result showed the high photocatalytic potential of the material under both UV and VIS irradiation up to the 650 nm wavelength.

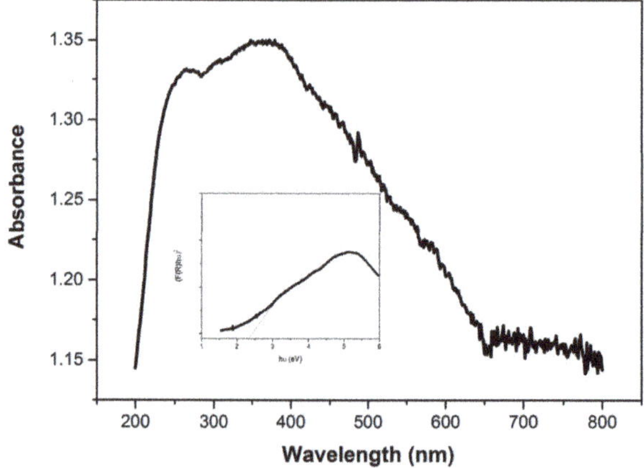

Figure 9. DRUV-VIS diffuse reflectance spectrum of $ZnCr_2O_4$. The inset shows the $(F(R)h\nu)^2$-$h\nu$ plot.

Table 2. The band gap energy values of $ZnCr_2O_4$.

Calcination Temperature/°C	The Band Gap Energy Values/eV
450	2.90
600	3.19
700	3.22
800	3.25

2.3. Photocatalysis Activity

The initial UV-VIS spectrum profile for 50 mg/L HA and its evolution during the photocatalysis using $ZnCr_2O_4$ (450 °C) is shown in Figure 10a, in comparison to the photolysis process (inset of Figure 10a). Absorbance recorded at wavelengths 254 nm and 365 nm is considered to be a suitable parameter for the characterization of aquatic humic substances, selected due to the fact that they indicate the presence of aromatic compounds. The aromatic compounds derived from an aquatic heterotrophic metabolism consist of nitrogen-based functions characterized by the chromophore characteristics. The comparative results of photocatalysis involving the prior sorption step for 30 min, photolysis under UV irradiation, and sorption under dark conditions, as well as the application results applied for HA removal from water expressed as RE (%), are shown in Figure 10b.

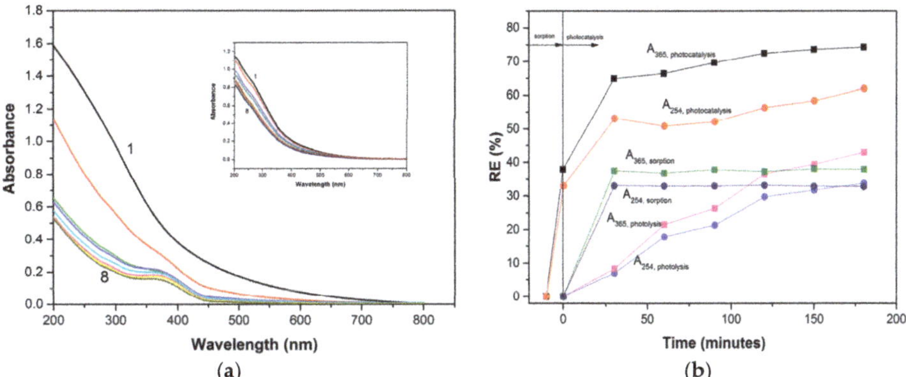

Figure 10. Evolution of UV-VIS spectrum profile of 50 mg/L humic acid (HA) during photocatalysis using $ZnCr_2O_4$. Inset shows the evolution of UV-VIS spectrum profile of 50 mg/L HA during the photocatalysis (**a**) and the removal efficiency (RE) evolution for 50 mg/L HA removal from water by application of sorption-based photocatalysis (solid line) and photolysis (dash line), assessed in terms of A_{254} and A_{365} (**b**).

It is obvious that $ZnCr_2O_4$-based photocatalysis enhanced the humic acid (HA) degradation in comparison with photolysis. Also, it can be seen that after adsorption for 30 min, the HA removal efficiency of about 30% was reached. Once the UV irradiation process started, a very significant decrease in the HA spectrum intensity was observed. It is noteworthy that during photocatalysis application, two distinct rate steps were delimited, except the sorption stage. The first one was characterised by a fast kinetics rate until 30 min of photocatalysis, followed by slower kinetics rate. Thus, for 30 min of the photocatalysis, a synergic effect regarding removal efficiency was found that was related to the sorption and the photolysis effect. This aspect denoted a good photocatalytic capacity of $ZnCr_2O_4$ for HA degradation that also considers the sorption effect as a compulsory stage in the photocatalytic mechanism. A different behaviour is observed for the photolysis application, which is characterised by a constant slow kinetics rate. The sorption characteristics of the $ZnCr_2O_4$ photocatalyst have been proved by kinetics modelling. The pseudo-first kinetics model was used to fit the experimental results for both photolysis and the photocatalysis, but the low correlation coefficient obtained for the photocatalysis showed that this model did not describe precisely. The second-order kinetics model better fit the experimental results for the photocatalysis application, which corresponded mainly to the sorption process, thus indicating a major contribution of sorption within the photocatalysis. A longer irradiation time in the presence of $ZnCr_2O_4$ indicated the slow kinetics of the photocatalysis process, and at its peak the cumulating effect of sorption and photolysis processes was noticed. The first-order kinetics model described the photocatalysis very well, and the results related to the rate constants are given in Table 3.

Table 3. Apparent rate constants calculated by fitting experimental data through the pseudo-second-order kinetics model for photocatalysis and the pseudo-first-order kinetics model for photolysis.

Process/Catalyst	Parameters	Value/Absorbance	
		A_{254}	A_{365}
Photocatalysis/ZnCr$_2$O$_4$	K_{app} (g mg^{-1} min^{-1})	$4 \cdot 10^{-3}$	$16.2 \cdot 10^{-3}$
	R^2	0.997	0.988
Photolysis	K_{app} (min^{-1})	$2.1 \cdot 10^{-3}$	$2.9 \cdot 10^{-3}$
	R^2	0.923	0.937

As we mentioned already, based on the RE the kinetics results sustained ZnCr$_2$O$_4$ as a photocatalyst for UV irradiation for humic acid removal. Moreover, to support this statement, the mineralization degree was assessed by the total organic carbon (TOC) parameter for both photocatalysis and photolysis. The mineralization degree was about 60% for photocatalysis, compared to 7% for photolysis and 30% for sorption, which showed a net superiority of ZnCr$_2$O$_4$ for HA mineralization. Also, the mineralization degree is very close to the degradation efficiency that confirmed the effectiveness of the HA mineralization through photocatalysis, using the ZnCr$_2$O$_4$ photocatalyst. Based on these results, it can be concluded that a complex mechanism of HA degradation is based on sorption in the first stage and on mineralization through hydroxyl radicals generated under UV irradiation [26].

3. Materials and Methods

For the synthesis of the Zn(II)-Cr(III) oxalate coordination compound, Cr(NO$_3$)$_3$·9H$_2$O, Zn(NO$_3$)$_2$·6H$_2$O, 1,2-ethanediol, and 2M nitric acid (Merck) were employed.

A water solution containing 1,2-ethanediol, chromium nitrate, zinc nitrate, and nitric acid (2M) in a molar ratio 2:1:0.5:1.3 was used. This mixture was heated in a water bath for about 40 min. The reaction was finished when no more gas evolved. The resulting solid product was purified by washing with acetone and dried under a room temperature environment. The metal content was determined by the atomic absorption spectrophotometry (VARIAN Spectra 110 spectrophotometer). Carbon and hydrogen were analyzed using a Carlo Erba 1108 elemental analyzer. The data of the elemental analysis for the oxalate precursor is indicated in the Table 4.

Table 4. Composition and elemental analysis data.

Compound	Cr(III) %		Zn(II) %		C %		H %	
(composition formula)	calc.	found	calc.	found	calc.	found	calc.	found
Cr$_2$Zn(C$_2$O$_4$)$_4$·10H$_2$O	14.82	14.90	9.32	9.25	13.68	13.70	2.85	2.91

In order to identify the ligand, the Cr^{3+}-Zn^{2+} oxalate was treated with R-H cationite (Purolite C-100). After the metallic cations were retained and the remained solution evaporated, the solid oxalic acid (H$_2$C$_2$O$_4$·2H$_2$O) was obtained. The isolated ligand was identified through three procedures: elemental analysis (H$_2$C$_2$O$_4$·2H$_2$O, calc./found: C% = 19.04/19.22; H% = 4.76/4.98), FTIR spectroscopy (Figure 1b and Table S2 Electronic Supplementary Material), and specific reactions [32].

FTIR spectra (KBr pellets) of the compound and of the decomposition product were recorded on a Jasco FT-IR spectrophotometer, in the range of 4000–400 cm^{-1}. Thermal measurements (TG, DTG, and DSC) were performed using a NETZSCH-STA 449C instrument, in the range of 25–1000 °C, using alumina crucibles. The experiments were carried out in artificial air flow of 20 mL min^{-1} and a heating rate of 10 K min^{-1}. The phase purity and crystal structure of $ZnCr_2O_4$ powders were determined by X-ray diffraction (XRD) investigations, performed at room temperature by means of a Rigaku Ultima IV multipurpose diffraction system (Rigaku Co., Tokyo, Japan). The diffractometer was set in a parallel beam geometry, using Ni-filtered CuKα radiation (λ = 1.5418 Å), CBO optics, and a graphite monochromator, operated at 40 kV and 40 mA. The measurements were performed in θ–2θ mode, with a scan step increment of 0.01°, in the 2θ range of (10°–80°). Phase identification was performed using HighScore Plus 3.0e software, connected to the ICDD PDF-4+ 2017 database. Lattice parameters were refined by the Rietveld method. After removing the instrumental contribution, the full-width at half-maximum (FWHM) of the diffraction peaks can be interpreted in terms of crystallite size and lattice strain. A pseudo-Voigt function was used to refine the shapes of the $ZnCr_2O_4$ peaks, and a Caglioti function was used for FWHM approximation. The size and the agglomeration tendency of the $ZnCr_2O_4$ particles was assessed by scanning electron microscopy (FE-SEM), using a high-resolution FEI QUANTA INSPECT F microscope with field emission gun (FEI Co., Eindhoven, The Netherlands). For a high-accuracy estimation of the morphology and crystallinity degree of the $ZnCr_2O_4$ particles, additional transmission electron microscopy (TEM/HR-TEM) and selected area electron diffraction (SAED) investigations were performed. The bright-field and high resolution images, as well as the SAED patterns, were collected by means of a TecnaiTM G^2 F30 S-TWIN transmission electron microscope (FEI Co., Eindhoven, The Netherlands). An image-corrected 80–200 kV Titan Themis transmission and scanning transmission electron microscope (S/TEM), equipped with a high brightness Schottky field emission gun (X-FEG) tip and a four diode Super X-ray energy dispersive spectroscopy (X EDS) detector (Thermo Fisher Scientific, former FEI Co., Eindhoven, Netherlands) was used for rapid compositional analyses (acquisition of STEM images and EDX mappings). For these purposes, small amounts of powdered samples were suspended in ethanol by 15 min ultrasonication. For FE-SEM analyses, a drop of the suspension was put on a carbon tape stuck on a stub and dried under an IR lamp for 5 min. Finally, the sample was sputtered with a gold film. For TEM observations, a drop of suspension was put onto a 400 mesh holey carbon-coated film Cu grid and dried. The average particle size for the $ZnCr_2O_4$ powders was determined using the OriginPro 9.0 software, by taking into account size measurements on 50–60 particles (from TEM images of appropriate magnifications obtained from various microscopic fields), performed by means of the microscope software DigitalMicrograph 1.8.0. The UV-Vis diffuse reflectance spectrum of the zinc chromite was obtained on a UV-VIS Carry 100 Varian spectrophotometer.

The photocatalytic experiments were carried out under magnetic stirring at 20 °C into a RS-1 photocatalytic reactor (Heraeus, Hanau, Germany), which consisted of a submerged UV lamp surrounded by a quartz shield. For each experiment, an adsorption step of 30 min was assured under the same hydrodynamic conditions without UV irradiation. At a certain running time, the suspension was sampled and filtered through a 0.4 μm membrane filter. The concentration of humic acid was measured in terms of absorbance at 254 nm (A_{254}) and at 345 nm (A_{345}), in order to evaluate the decolorizing and respective degradation degrees, using a Carry 100 Varian spectrophotometer. Humic acid has a very complex structure, containing heterogeneous mixtures of small molecules, which usually result from the biological transformations of dead cells associated with a supramolecular structure that can be separated into smaller molecules by chemical fractionation.

To assess the mineralization degree, the total organic carbon (TOC) parameter was measured for initial and final HA concentration, using the TOC analyzer from Shimadzu (Tokyo, Japan). The HA removal efficiency was calculated using the following equation:

$$\text{Removal efficiency (RE)} = \frac{C_0 - C_t}{C_0} \times 100 \ (\%)\tag{2}$$

where C_0 and C_t are the concentrations of HA in aqueous solution, in terms of the A_{254}, A_{345}, and TOC parameters at initial time and certain time t, respectively (mg L^{-1}). Kinetics data were fitted with the pseudo-second-order kinetic model [28], and expressed as:

$$\frac{t}{q_t} = \frac{1}{k_2 \cdot q_e^2} + \frac{t}{q_e}, \tag{3}$$

where k_2 is the rate constant of the pseudo-second-order adsorption kinetics (g mg^{-1} min^{-1}) and q_e is the equilibrium adsorption capacity (mg g^{-1}). The simplified pseudo-first-order equation was used also, to fit the kinetics data:

$$\ln\left(C_0/C_t\right) = kKt = k_{app}t, \tag{4}$$

where r is the rate of humic acid degradation and colour removal (mg L^{-1} min^{-1}), C_0 is the initial humic acid concentration (mg L^{-1}), C_t is the concentration of the humic acid at time t (mg L^{-1}), t is the irradiation time (min), and k is the reaction rate constant (min^{-1}).

4. Conclusions

Nanospinel zinc chromite was successfully synthesized using a new method, via thermal decomposition of the $[Cr_2Zn(C_2O_4)_4(OH_2)_6]\cdot4H_2O$ oxalate compound at the temperature of 450 °C. XRD investigations, FE-SEM, and TEM/HR-TEM analyses indicated that phase-pure $ZnCr_2O_4$ nanoparticles were obtained after calcination at temperatures ranging between 450–800 °C. The average particle size values increased with increasing temperature, from 4.7 nm (450 °C) to 66.3 nm (800 °C), which affected the band gap energy value. The lowest band gap energy value of 2.45 eV was determined for the $ZnCr_2O_4$ powder after calcination at 450 °C, which exhibited photocatalytic activity towards the degradation and the mineralization of humic acids (HAs) from water. The sorption efficiency of this spinel for HAs was about 30% after 30 min, and the photocatalytic efficiency for HA degradation was about 60%, which was similar to the mineralization degree, thus confirming the effectiveness of photocatalysis for mineralization while avoiding the generation of by-products. The sorption step was confirmed by the pseudo-second-order kinetics model to fit the experimental results best, which is characteristic to sorption in comparison with the photocalysis process, which was described as the best by the pseudo-first kinetics model. Two steps of the photocatalysis process using nanosized $ZnCr_2O_4$ spinel were found: the first step was characterized by the fast rate, while the second step occurred at the slow rate. In comparison with the photolysis and sorption process, a synergic effect was found for the photocatalysis process during the first 30 min. Then, a maximum cumulative effect was found until 180 min. Taking into account the energy consumption that corroborates these study results, the optimum irradiation time of the photocatalysis with $ZnCr_2O_4$ is 30 min. This study informed about the great potential of this spinel to be used for drinking water treatment via its photocatalytic application.

Supplementary Materials: The following are available online at http://www.mdpi.com/2073-4344/8/5/210/s1, Table S1: Characteristic IR absorption bands of the compound $[Cr_2Zn(C_2O_4)_4(OH_2)_6]\cdot4H_2O$, Table S2: Characteristic IR absorption bands of the oxalic acid.

Author Contributions: In this paper, R.D. and F.M. conceived and designed the experiments; C.P., L.L., A.P. and A.S. performed the experiments; R.D., F.M., A.P., A.C. and A.I. analyzed the data; A.C. contributed reagents/materials/analysis tools; R.D., F.M. and A.I. wrote the paper.

Acknowledgments: This work was partially supported by a grant of the Romanian National Authority for Scientific Research and Innovation, CNCS-UEFISCDI, project number PN-II-RU-TE-2104-4-1043. The TEM characterization was possible due to EU-funding project POSCCE-A2-O2.2.1-2013-1/Priority Axe 2, Project No. 638/12.03.2014, ID 1970, SMIS-CSNR code 48652. The contribution of Bogdan Vasile in performing some TEM/EDX investigations is highly acknowledged.

Conflicts of Interest: The authors declare no conflict of interest.

References

1. Gabr, R.M.; Girgis, M.M.; El-Awad, A.M. Formation, conductivity and activity of zinc chromite catalyst. *Mater. Chem. Phys.* **1992**, *30*, 6619–6623. [CrossRef]
2. Epling, W.S.; Hoflund, G.B.; Minahan, D.M. Reaction and surface characterization study of higher alcohol synthesis catalysts: VII Cs- and Pd-promoted 1:1 Zn/Cr spinel. *J. Catal.* **1998**, *175*, 175–184. [CrossRef]
3. El-Sharkawy, E.A. Textural, structural and catalytic properties of $ZnCr_2O_4/Al_2O_3$ catalysts. *Adsorpt. Sci. Technol.* **1998**, *6*, 193–216. [CrossRef]
4. Peng, C.; Gao, L. Optical and photocatalytic properties of spinel $ZnCr_2O_4$ nanoparticles synthesized by a hydrothermal route. *J. Am. Soc.* **2008**, *91*, 2388–2390. [CrossRef]
5. Thennarasu, G.; Sivasamy, A. Synthesis and characterization of nanolayered $ZnO/ZnCr_2O_4$ metal oxide composites and its photocatalytic activity under visible light irradiation. *J. Chem. Technol. Biotechnol.* **2015**, *90*, 514–524. [CrossRef]
6. Yazdanbakhsh, M.; Khosravi, I.; Goharshadi, E.K.; Youssefi, A. Fabrication of nanospinel $ZnCr_2O_4$ using sol-gel method and its application on removal of azo-dye from aqueous solutions. *J. Hazard. Mater.* **2010**, *184*, 684–689. [CrossRef] [PubMed]
7. He, H.Y. Catalytic and photocatalytic activity of $ZnCr_2O_4$ particle synthesized using metallo-organic precursor. *Mater. Technol.* **2008**, *23*, 110–113. [CrossRef]
8. Niu, X.; Du, W.; Du, W. Preparation and gas sensing properties of ZnM_2O_4 (M = Fe, Co, Cr). *Sens. Actuators B* **2004**, *99*, 405–409. [CrossRef]
9. Pokhrel, S.; Jeyaraj, B.; Nagaraja, K.S. Humidity-sensing properties of $ZnCr_2O_4$-ZnO composites. *Mater. Lett.* **2003**, *57*, 3534–3548. [CrossRef]
10. Bayahn, M.; Hashemi, T.; Brinkman, A.W. Sintering and humidity-sensitive behavior of the $ZnCr_2O_4$-K_2CrO_4 ceramic system. *J. Mater. Sci.* **1997**, *32*, 6619–6623. [CrossRef]
11. Marinkovic Stanojevic, Z.V.; Mancic, L.; Marcic, L.; Milosevici, O. Microstructural characterization of mechanically activated ZnO-Cr_2O_3 sysstem. *J. Eur. Ceram. Soc.* **2005**, *25*, 2081–2084. [CrossRef]
12. Marinkovic Stanojevic, Z.V.; Romcevic, N.; Stoganovic, B. Spectroscopic study of spinel $ZnCr_2O_4$ obtained from mechanically activated ZnO-Cr_2O_3 mixtures. *J. Eur. Ceram. Soc.* **2007**, *27*, 903–907. [CrossRef]
13. Imelda, E.; Myriam, P.; Roberto, M.; Adriana, G.C.; Guadalupe, S.L.; Luisa, M.F.V.; Octavia, D. Solid state reaction in Cr_2O_3-ZnO nanoparticles synthesized by triethanolamine chemical precipitation. *Mater. Sci. Appl.* **2011**, *2*, 1584–1592. [CrossRef]
14. Mousavi, Z.; Esmaeili-Zare, M.; Salavati-Niasari, M. Magnetic and optical properties of zinc chromite nanostructures prepared by microwave method. *Trans. Nonferr. Met. Soc. China* **2015**, *25*, 3980–3986. [CrossRef]
15. Naz, S.; Durrani, S.K.; Mehmood, M.; Nadeem, M. Hydrothermal synthesis, structural and impedance studies of nanocrystalline zinc chromite spinel oxide material. *J. Saudi. Chem. Soc.* **2016**, *20*, 585–593. [CrossRef]
16. Mousavi, Z.; Soofivand, F.; Esmaeili-Zare, M.; Salavati-Niasari, M.; Bagheri, S. $ZnCr_2O_4$ nanoparticles: Facile synthesis, characterization and photocatalytic properties. *Sci. Rep.* **2016**, *6*, 20071. [CrossRef] [PubMed]
17. Babar, A.R.; Kumbhar, S.B.; Shinde, S.S.; Moholkar, A.V.; Kim, J.H.; Rajpure, K.Y. Structural, compositional and electrical properties of co-precipitated zinc stannate. *J. Alloy. Comp.* **2011**, *509*, 7508–7514. [CrossRef]
18. Cui, H.; Zayat, M.; Levy, D. Sol-gel synthesis of nanoscaled spinel using propylene oxide as a gelation agent. *Sol-Gel Sci. Technol.* **2005**, *35*, 175–181. [CrossRef]
19. Gene, S.A.; Saion, E.; Shaari, A.H.; Kamarudin, M.A.; Al-Hada, N.M.; Kharazmi, A. Structural, optical and magnetic characterization of spinel zinc chromite nanocrystallines synthesized by thermal treatment methods. *J. Nanomater.* **2014**, *2014*, 416765. [CrossRef]
20. Hailiang, L.; Wenling, M.; Xiaohua, Y.; Jungi, L. Preparation and characterization of $MgCr_2O_4$ nanocrystals by microwave method. *Adv. Mater. Res.* **2011**, *152*, 1000–1003. [CrossRef]
21. Durani, S.K.; Hussain, S.Z.; Saeed, K.; Khan, Y.; Arif, M.; Ahmed, H. Hydrothermal synthesis and characterization of nanosized transition metal chromite spinels. *Turk. J. Chem.* **2012**, *36*, 111–120. [CrossRef]
22. Durani, S.K.; Naz, S.; Hayat, K. Thermal analysis and phase evolution of nanocrystalline perovskite oxide materials synthesized via hydrothermal and self-combustion methods. *J. Therm. Anal. Calorim.* **2014**, *115*, 1371–1380. [CrossRef]

23. Stefanescu, M.; Barbu, M.; Vlase, T.; Barvinschi, P.; Barbu-Tudoran, L.; Stoia, M. Novel low temperature synthesis method for nanocrystalline zinc and magnesium chromites. *Thermochim. Acta* **2011**, *526*, 130–136. [CrossRef]

24. Gingasu, D.; Mindru, I.; Patron, L.; Culita, D.C.; Calderon-Moreno, J.M.; Diamandescu, L.; Feder, M.; Oprea, O. Precursor method—A nonconventional route for the synthesis of $ZnCr_2O_4$ spinel. *J. Phys. Chem. Solid* **2013**, *74*, 1295–12302. [CrossRef]

25. Gingasu, D.; Mindru, I.; Culita, D.C.; Patron, L.; Calderon-Moreno, J.M.; Preda, S.; Oprea, O.; Osiceanu, P.; Pineda, E.M. Investigation of nanocrystalline zinc chromite obtained by two soft chemical routes. *Mater. Res. Bull.* **2014**, *49*, 151–159. [CrossRef]

26. Fujishima, A.; Rao, T.N.; Tryk, D.A. Titanium Dioxide Photocatalysis. *J. Photochem. Photobiol. C Photochem. Rev.* **2000**, *1*, 1–21. [CrossRef]

27. Orha, C.; Pode, R.; Manea, F.; Lazau, C.; Bandas, C. Titanium dioxide-modified activated carbon for advanced drinking water treatment. *Process Saf. Environ. Prot.* **2017**, *108*, 26–33. [CrossRef]

28. Orha, C.; Manea, F.; Pop, A.; Bandas, C.; Lazau, C. TiO_2-nanostructured carbon composite sorbent/photocatalyst for humic acid removal from water. *Desalin. Water Treat.* **2016**, *57*, 14178–14187. [CrossRef]

29. Yuan, W.; Liu, X.; Li, L. Synthesis, characterization and photocatalytic activity of cubic-like $CuCr_2O_4$ for dye degradation under visible light irradiation. *Appl. Surf. Sci.* **2014**, *319*, 350–357. [CrossRef]

30. Vader, V.T. Ni and Co substituted zinc ferri-chromite: A study of their influence in photocatalytic performance. *Mater. Res. Bull.* **2017**, *85*, 18–22. [CrossRef]

31. Abbasi, A.; Hamadanian, M.; Salavati-Niasari, M.; Mortazavi-Derazkola, S. Facile size-controlled preparation of highly photocatalytically active $ZnCr_2O_4$ and $ZnCr_2O_4$/Ag nanostructures for removal of organic contaminants. *J. Colloid Interface Sci.* **2017**, *500*, 276–284. [CrossRef] [PubMed]

32. Dumitru, R.; Papa, F.; Balint, I.; Culita, D.; Munteanu, C.; Stanica, N.; Ianculescu, A.; Diamandescu, L.; Carp, O. Mesoporous cobalt ferrite: A rival of platinum catalyst in methane combustion reaction. *Appl. Catal. A Gen.* **2013**, *467*, 178–186. [CrossRef]

33. Nakamoto, K. *Infrared and Raman Spectra of Inorganic and Coordination Compounds*; Wiley: New York, NY, USA, 1986; ISBN 0471010669.

34. Fujita, J.; Nakamoto, K.; Kobayshi, M. Infrared Spectra of Metallic Complexes. II. The Absorption Bands of Coördinated Water in Aquo Complexes. *J. Am. Chem. Soc.* **1956**, *78*, 3963–3965. [CrossRef]

35. Bîrzescu, M.; Niculescu, M.; Dumitru, R.; Carp, O.; Segal, E. Synthesis, structural characterization and thermal analysis of the cobalt(II) oxalate obtained through the reaction of 1,2-ethanediol with $Co(NO_3)_2 \cdot 6H_2O$. *J. Therm. Anal. Calorim.* **2009**, *96*, 979–986. [CrossRef]

36. Bîrzescu, M. Combinations with Ethyleneglycol and Their Oxidation Products. Ph.D. Thesis, University of Bucharest, Bucharest, Romania, 1998.

37. Stefănescu, M.; Sasca, V.; Bîrzescu, M. Thermal behaviour of the homopolynuclear glyoxylate complex combinations with Cu(II) and Cr(III). *J. Therm. Anal. Calorim.* **2003**, *72*, 515–524. [CrossRef]

38. Kubelka, P.; Munk, F. Ein Beitrag Zur Optik Der Farbanstriche. *Z. Tech. Phys.* **1931**, *12*, 593–601.

39. Irfan, S.; Shen, Y.; Rizwan, S.; Wang, H.C.; Khan, S.B.; Nan, C.W. Band-gap engineering and enhanced photocatalytic activity of Sm and Mn doped $BiFeO_3$ nanoparticles. *J. Am. Ceram. Soc.* **2016**, *100*, 31–40. [CrossRef]

 *catalysts*

Article

Synthesis of NaOH-Modified TiOF$_2$ and Its Enhanced Visible Light Photocatalytic Performance on RhB

Chentao Hou *, Wenli Liu and Jiaming Zhu

College of Geology and Environment, Xi'an University of Science and Technology, Xi'an 710054, China; 13572955186@163.com (W.L.); XRD610@126.com (J.Z.)
* Correspondence: houct@xust.edu.cn; Tel.: +86-029-8558-3188

Received: 16 July 2017; Accepted: 16 August 2017; Published: 22 August 2017

Abstract: NaOH-modified TiOF$_2$ was successfully prepared using a modified low-temperature hydrothermal method. Scanning electron microscopy shows that NaOH-modified TiOF$_2$ displayed a complex network shape with network units of about 100 nm. The structures of NaOH-modified TiOF$_2$ have not been reported elsewhere. The network shape permits the NaOH-modified TiOF$_2$ a S$_{BET}$ of 36 m^2·g^{-1} and a pore diameter around 49 nm. X-ray diffraction characterization shows that TiOF$_2$ and NaOH-modified TiOF$_2$ are crystallized with a pure changed cubic phase which accords with the SEM results. Fourier transform infrared spectroscopy characterization shows that NaOH-modified TiOF$_2$ has more O–H groups to supply more lone electron pairs to transfer from O of O–H to Ti and O of TiOF$_2$. UV–vis diffuse reflectance spectroscopy (DRS) shows that the NaOH-modified TiOF$_2$ sample has an adsorption plateau rising from 400 to 600 nm in comparison with TiOF$_2$, and its band gap is 2.62 eV, lower than that of TiOF$_2$. Due to the lower band gap, more O–H groups adsorption, network morphologies with larger surface area, and sensitization progress, the NaOH-modified TiOF$_2$ exhibited much higher photocatalytic activity for Rhodamine B (RhB) degradation. In addition, considering the sensitization progress, O–H groups on TiOF$_2$ not only accelerated the degradation rate of RhB, but also changed its degradation path. As a result, the NaOH-modified TiOF$_2$ exhibited much higher photocatalytic activity for RhB degradation than the TiOF$_2$ in references under visible light. This finding provides a new idea to enhance the photocatalytic performance by NaOH modification of the surface of TiOF$_2$.

Keywords: TiOF$_2$; NaOH-modified TiOF$_2$; network shape; photocatalysis; RhB

---

1. Introduction

Nowadays, environmental pollution is affecting human survival and development. Photocatalysis is considered an efficient, stable, and environmentally friendly method for controlling environmental pollution [1]. In the past, TiO$_2$ has been widely used as a photocatalyst in the photo-degradation of organic pollutants. However, it has a wide energy band gap (3.1–3.2 eV) which only permit its UV light response and can easily cause electron–hole recombination [1–4]. Thus, studies on changing morphology [1–3], modification [1,4,5], and other methods were conducted to decrease its band gap or inhibit its electron–hole recombination. The discovery of non-titanium semiconductor photocatalysts with a narrow intrinsic energy band gap, efficiently driven by visible light, may also attract much attention [5–13].

Recently, Li's research group found that TiOF$_2$ cubes—considered a promising anode material for lithium ion batteries (LIBs) [14–19]—showed visible-light driven property and exhibited excellent performance in photodegradation of Rhodamine B (RhB) and 4-chlorophenol (4-CP) [7]. TiOF$_2$ is also proven to be more active and durable at room temperature due to the covalent bonds of F species with Ti [8,9]. Only a few studies focused on the photocatalytic activity of TiOF$_2$ have been reported [7–9].

As usual, $TiOF_2$ nanoparticles were synthesized via hydrothermal [10,14–18] and solvothermal [7,11,12] methods from titanium (IV) isopropoxide (TIP), and have a cubic shape [7–18]. The size of $TiOF_2$ nanocubes could be affected by alcoholysis time, alcohol kind, solvothermal temperature, different H_2O production rate and amount [7,17]. While the photocatalytic activity of $TiOF_2$ is still unsatisfactory, it is necessary to explore novel approaches to improve its photocatalytic performance.

Alkali modification is proven to be an effective method to enhance the catalytic performance for α-pinene isomerization, formaldehyde oxidation, and benzene hydroxylation [20–23]. Thus, it stimulated us to modify the $TiOF_2$ catalyst obtained from our earlier studies. In this study, we firstly reported a network-shaped NaOH-modified $TiOF_2$ treated by a hydrothermal process under low temperature. The FTIR measurement showed that more associated O–H exists on the surface of $TiOF_2$, which can remarkably enhance the catalytic activity of $TiOF_2$ toward RhB oxidation under visible light. The NaOH-modified $TiOF_2$ had better photocatalytic performance than $TiOF_2$ in Li's research [7].

2. Results and Discussion

2.1. Phase Structures and Morphology

The phase and crystallinity of the as-prepared $TiOF_2$ and NaOH-modified $TiOF_2$ samples were tested by XRD analysis. It can be seen from Figure 1, the patterns of as-prepared $TiOF_2$ and NaOH-modified $TiOF_2$ samples all have sharp peaks at $2\theta = 23.6°$, $48.1°$, and $54.2°$, corresponding to the (100), (200), and (210) planes of the cubic $TiOF_2$ phase (JCPDS no. 08-0060) [7,13] and no peak of any anatase TiO_2 (JCPDS no. 21-1272) [24,25] crystal appears, indicating that the as-prepared samples have high crystallinity and pure phase of cubic $TiOF_2$. It also indicates that the height of the (100) crystal planes of the $TiOF_2$ and NaOH-modified $TiOF_2$ are higher, and the (200) and (210) crystal planes are lower than that of the standard cubic $TiOF_2$, indicating a new shape of the as-prepared $TiOF_2$ and NaOH-modified $TiOF_2$. The Scherrer formula was used to calculate the normal distance of certain crystal surfaces of $TiOF_2$ and NaOH-modified $TiOF_2$

$$\tau = \kappa\lambda / (\beta \cos\theta) \tag{1}$$

where τ, κ, λ, β, and θ are the mean normal distance of certain crystal surfaces, the shape constant with a value of 0.89 when β is the half width of the diffraction peak (FWHM), the diffracted ray wavelength (0.15418 nm for Cu-Ka), and the diffraction angle in radians, respectively [26].

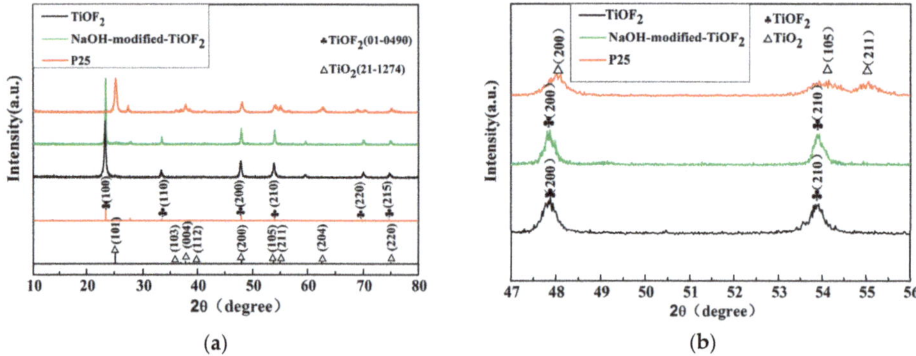

Figure 1. XRD pattern with 2θ from $10°$ to $80°$ (**a**) and partial enlarged detail of 2θ from $47°$ to $56°$ (**b**) of as-prepared $TiOF_2$, NaOH-modified $TiOF_2$, commercial P25 and a standard card of $TiOF_2$ and TiO_2.

The normal distances of $TiOF_2$ are 7.26, 17.81, 11.67, 10.77, 17.38, and 16.98 nm along the (100), (110), (200), (210), (220), and (215) planes [7,13], respectively, while the distanced of crystal

NaOH-modified $TiOF_2$ change to 15.99, 42.18, 26.26, 30.78, 40.80, and 48.39 nm along the corresponding planes, respectively. It can be seen that the normal distance of the crystalline phase of NaOH-modified $TiOF_2$ shrunk 8.88% and 3.72% along (100) and (110) planes. However, an increase was observed along (200), (210), and (215) planes. This can be explained in that NaOH-modifying induces more O–H adsorbed onto (100) and (110) planes of $TiOF_2$ and, thus, induces the planes' exposure.

Morphologies and microstructure of original $TiOF_2$ and NaOH-modified $TiOF_2$ were checked by SEM characterization. In Figure 2a,b, $TiOF_2$ crystals displayed a mixture of the cubic image which is accords with the cubic image in [7–18]. Each individual particle crystal is about 50–300 nm and tends to aggregate, forming larger particles while—in Figure 2c,d—the NaOH-modified $TiOF_2$ displayed a more complex network shape with network units in about 100 nm. The NaOH-modified $TiOF_2$ shows phases assembling along certain directions. This accords with the XRD results. The network shape permits much more surface area for photocatalysis. These structures of NaOH-modified $TiOF_2$ have not been reported elsewhere. The Barrett-Joyner-Halenda (BJH) method was used to analyze the pore size distribution and pore volume and the surface area (S_{BET}) was calculated using the BET method. The Figure 3a demonstrated that the NaOH-modified $TiOF_2$ showed a typical IV type N_2 adsorption–desorption isotherm and mesoporous structure with an average pore diameter of about 49 nm. Thus, its S_{BET} can reach as high as 36 $m^2 \cdot g^{-1}$, while the average pore diameter and S_{BET} of $TiOF_2$ are only 3 nm and 2.7 $m^2 \cdot g^{-1}$, which is much lower than that of NaOH-modified $TiOF_2$. The larger surface area permits more O–H and pollutant adsorption and the formation of additional mesopores affects the improvement of mass transfer, enhancing photocatalytic performance accordingly [21,22].

Figure 2. SEM of as-synthesized samples: (**a**,**b**) $TiOF_2$; and (**c**,**d**) NaOH-modified $TiOF_2$.

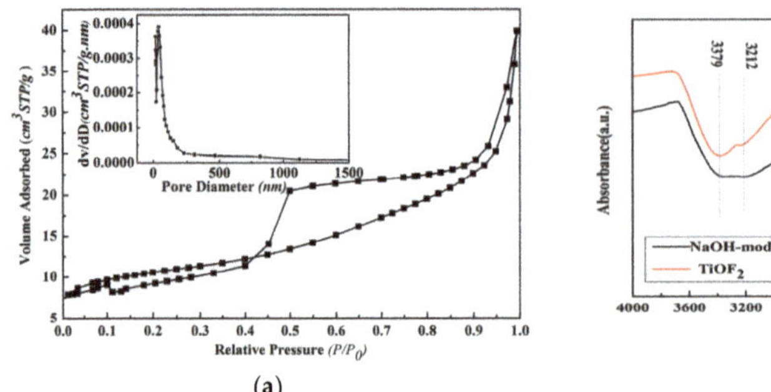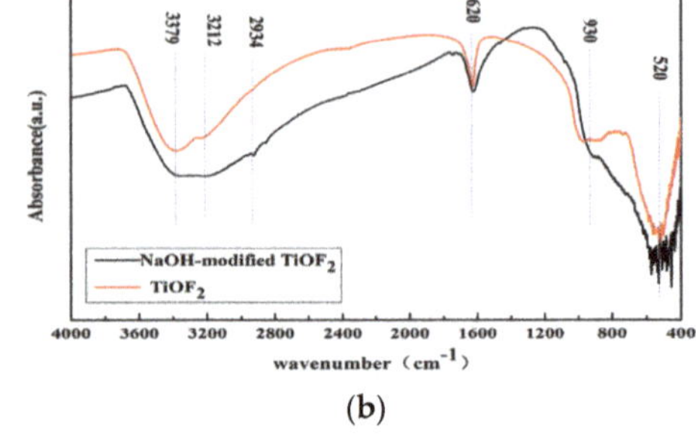

(a)   (b)

Figure 3. N_2 adsorption-desorption isotherm of the NaOH-modified $TiOF_2$ (a) and the FTIR spectra for $TiOF_2$ and NaOH-modified TiOF (b).

2.2. FTIR Analysis

Figure 3b shows the FTIR spectra of $TiOF_2$ and NaOH-modified $TiOF_2$. The strong band around 700–500 cm^{-1} could contribute to the Ti–O–Ti stretching vibration [22–24]. The peak around 3379 cm^{-1} and the broad band centered around 3212 cm^{-1} were due to the free and bonding O–H stretching vibration of Ti−OH, respectively [27,28]. The peak at 1620 cm^{-1} was due to the O–H bending vibration of Ti−OH [16,22,29–33]. The broad band centered around 3212 cm^{-1} in NaOH-modified $TiOF_2$ becomes broader than that in $TiOF_2$, meaning that more O–H bonds or associated O–H appeared in NaOH-modified $TiOF_2$. According to previous work, the free O–H stretching vibration used to appear at about 3600 cm^{-1} without bonding O–H [27,28]. It can be seen that the O–H frequency for $TiOF_2$ and NaOH-modified $TiOF_2$ is 221 cm^{-1} and 388 cm^{-1} lower than 3600 cm^{-1}, indicating a strong hydrogen bond impact [27,28]. The O–H on the TiO_2 surface can enhance the transference of photo-generated electrons and then enhance photocatalytic performance [29]. The peaks around 930 cm^{-1} were due to the Ti–F vibrations in the $TiOF_2$ [16]. The peak intensity decreased from $TiOF_2$ to NaOH-modified $TiOF_2$, indicating F was exchanged by O–H after NaOH modification. All of these show that the NaOH-modified $TiOF_2$ samples contain more O–H groups than $TiOF_2$. It can be explained that $TiOF_2$ was modified in NaOH solution, thus, more O–H would be chemisorbed onto $TiOF_2$, and further exchanged with F. Then, more lone pair electrons in the O–H groups transferring from the O of O–H to Ti and the O of $TiOF_2$, the performance of $TiOF_2$ can be enhanced accordingly [21]. In addition, because RhB is a cationic dye, NaOH brings more O–H onto the surface of $TiOF_2$ to hold more RhB and accelerate its degradation rate [34,35].

2.3. UV–Vis Analysis

Figure 4 shows that the UV–vis absorption spectroscopy and band gap of as-prepared $TiOF_2$ and NaOH-modified $TiOF_2$ samples. The NaOH-modified $TiOF_2$ has a raised adsorption plateau from 400 to 600 nm, which indicates stronger visible light absorption than that of $TiOF_2$ (Figure 4a). Band gap estimation can be seen in Figure 4b showing that the band gap of NaOH-modified $TiOF_2$ is 2.62 eV, which is lower than that of $TiOF_2$ (2.80 eV) and lower than anatase TiO_2 (3.2 eV) [1–4], NiO (4.0 eV) [6], and other oxides, indicating easier excitation by visible light. This can be explained in that NaOH treatment causes certain facet exposure and network morphologies of $TiOF_2$, changing its light absorption properties. Thus, the NaOH treatment lowered the band gap of $TiOF_2$, enhanced its visible light absorption, and further enhanced its visible light photocatalytic properties.

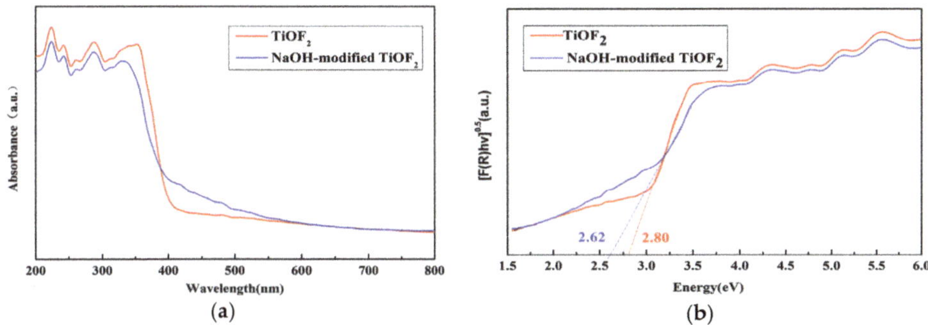

Figure 4. UV–vis DRS spectra (**a**) and band gap (**b**) of $TiOF_2$ and NaOH-modified $TiOF_2$.

2.4. Catalytic Activity

Figure 5 shows the visible light photcatalytic properties of $TiOF_2$, NaOH-modified $TiOF_2$, $TiOF_2$ in reference ($TiOF_2$-Ref) [7,13], $TiOF_2$-crushed in reference ($TiOF_2$-crushed-Ref) [7], and P25. It can be seen in Figure 5a that, in the adsorption test in dark and in light on the process without the catalyst for RhB, the decrease of RhB is very small. It can be concluded that the adsorption and sensitization mechanisms can be negligible in the degradation process. Thus, the degradation of RhB was a photocatalytic process. The concentration of RhB decreased under the same conditions, which means that all samples are visible-light active. It also shows that NaOH-modified $TiOF_2$ can cause almost complete decomposition of RhB in 3 h, having better photocatalytic performance than all of the $TiOF_2$ in reference [7,13]. While P25 and $TiOF_2$ performed poorly compared to NaOH-modified $TiOF_2$ and $TiOF_2$-Ref. The reaction rate of all of the samples are shown in Figure 5b. It can be seen that the data was fitted with the first-order reaction equation as

$$\ln(C_0/C) = kt \tag{2}$$

where t is the reaction time, C_0 is concentration of RhB at time 0, C is the concentration of RhB at time t, and k is the reaction rate constant.

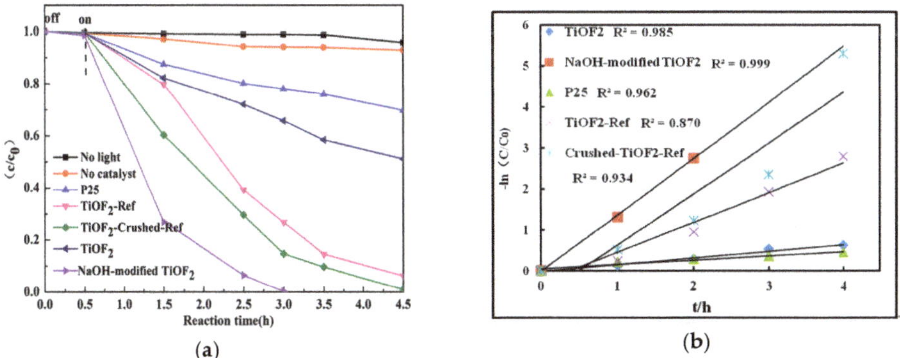

Figure 5. Catalytic activity of RhB under visible light: (**a**) concentration dependent on time and (**b**) kinetic fit for the degradation of RhB.

It can be seen that P25 and $TiOF_2$ had rate constants of only 0.10 and 0.16 h^{-1}, indicating poor photocatalytic performance. The result is consistent with previous work [8,20]. The calculated rate constants are 1.37, 0.73, and 1.24 h^{-1} for NaOH-modified $TiOF_2$, $TiOF_2$-Ref, and $TiOF_2$-crushed-Ref, respectively. The NaOH-modified $TiOF_2$ sample shows the best performance among all the photocatalysts, whose degradation rates are much higher than that of P25 and $TiOF_2$ in our samples and 10.4% higher than that of $TiOF_2$-crushed-Ref The excellent performance could be mainly attributed to its larger S_{BET} (32 $m^2 \cdot g^{-1}$ for $TiOF_2$-crushed-Ref) and more bonding O–H [7,21].

2.5. Effect of the Sensitization Mechanism

According to previous studies, dyes can be degraded on TiO_2 through a sensitized process under visible light [34,35]. In order to know whether there is a similar path on $TiOF_2$ and NaOH-modified $TiOF_2$, UV–vis absorption spectral changes of RhB with visible light irradiation time in the suspension of $TiOF_2$ and NaOH-modified $TiOF_2$ were tested. The results are shown in Figure 6. It can be seen in Figure 6 that the spectral change of RhB with the irradiation time on NaOH-modified $TiOF_2$ is quite different from that on $TiOF_2$. There is a blue-shift from 558 to 498 nm in the absorption maximum with irradiation time for NaOH-modified $TiOF_2$, while none in that of $TiOF_2$. This is attributed to the N-deethylation products of RhB, which confirms the possibility of the sensitization mechanism [34,35]. Thus, NaOH treatment can induce the sensitization process and change the degradation path of $TiOF_2$.

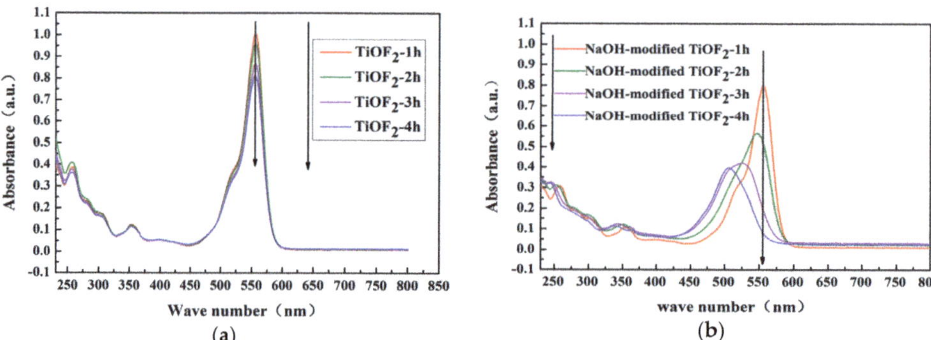

Figure 6. UV–vis absorption spectra of RhB in (**a**) $TiOF_2$ and (**b**) NaOH-modified $TiOF_2$ suspension under visible light.

3. Materials and Methods

Tetrabutyl titanate (TBOT, A.R. grade) was purchased from Fu Chen Chemical Reagent Factory, Tianjin, China. Absolute ethyl alcohol (C_2H_5OH, A.R. grade) and sodium hydroxide (NaOH, A.R. grade) was purchased from Fuyu Fine Chemical Co., Ltd., Tianjin, China. Hydrofluoric acid (HF, A.R. grade) was purchased from Xilong Chemical Industry Co., Ltd., Chengdu, China. All reagents are used without further purification. Ultrapure water was used as the experimental water.

NaOH-modified $TiOF_2$ was synthesized via a modified low-temperature hydrothermal method. In a typical synthesis, 30.4 mL absolute ethyl alcohol was added into 35.2 mL TBOT, which was named solution A. Absolute ethyl alcohol (30.4 mL) and 20.2 mL HF were added into 180 mL ultrapure water, which was named solution B. Solution A was dropped into solution B under medium-speed magnetic stirring at 20 °C for 1.5 h to obtain a faint yellow sol. The sol was aged at room temperature for 2 days to change to a gel. The gel was then transferred into a 50-mL Teflon-lined stainless steel autoclave. When sealed, the autoclave was placed at 100 °C for 2 h in a drying box, then was naturally cooled to room temperature. Ultra-pure water and absolute ethanol were used to wash the obtained white precipitates several times to reach a pH of 7, and then the precipitates were dried at 100 °C. The as-prepared sample was $TiOF_2$. One gram of the $TiOF_2$ precursor was dispersed in 100 mL 5 $mol \cdot L^{-1}$ NaOH solution under magnetic stirring with a speed of 4000 $r \cdot min^{-1}$ for 1 h, then the suspension was also washed with ultra-pure water and absolute ethanol to reach a pH of 7. The product was dried at 100 °C for 12 h. The sample was denoted as NaOH-modified $TiOF_2$. The crystal structure as analyzed by a XD-2 X-ray diffractometer (Beijing Purkinje, Beijing, China) with Cu Kαradiation with a scan rate of 4.0000° $\cdot$ min^{-1}. The morphology was examined by field emission scanning electron microscopy (FESEM, JEOL JSM6700, Tokyo, Japan). Fourier transform infrared (FTIR) spectra were recorded using a Bruker TENSOR27 (Karlsurhe, Germeny) using the KBr method. The optical properties were determined by UV–vis diffuse reflectance spectroscopy (UV–vis DRS: (Shimadzu 2600, Beijing, China). N_2 adsorption-desorption isotherms were measured at 77 K and the BET method was used to calculate the surface area (S_{BET}) by a JW-BK122F (Beijing, China).

The degradation of RhB was conducted at room temperature in a 150 mL double-layered quartz reactor containing 50 mg catalyst and 50 mL 5.0 mg$\cdot$L^{-1} RhB solution. A 300 W Xe lamp (Jiguang-300, Shanghai, China) was located at a distance of 15 cm from the RhB solution to simulate solar light. A cutoff filter (JB-420, Shanghai, China) was chosen to filter off the light whose wavelength was less than 420 nm to simulate visible light. The solution was magnetically stirred for 30 min to ensure the adsorption–desorption equilibrium, then the xenon lamp was turned on to start the photocatalytic degradation. At 30 min time intervals, about 5.0 mL RhB solution was extracted and centrifuged at high-speed (11,000 r$\cdot$min^{-1}) to remove catalysts. Then the concentration of the remaining RhB solutions were analyzed with a Purkinje UV1901 UV–vis spectrophotometer at 554 nm. The photocatalyst was separated from the RhB solution and another run of the reaction was started to investigate the durability of the catalysts.

4. Conclusions

NaOH-modified $TiOF_2$ was successfully prepared via a modified low-temperature solvothermal method. It exhibited much better photocatalytic performance for RhB degradation. XRD characterization shows that $TiOF_2$ and NaOH-modified $TiOF_2$ are crystallized with a pure changed cubic phase which is accord with the SEM results. SEM shows that $TiOF_2$ crystals displayed a mixture of the cubic images, while the NaOH-modified $TiOF_2$ displayed a more complex network shape with network units in about 100 nm. These structures of NaOH-modified $TiOF_2$ have not been reported elsewhere. The network shape permits the NaOH-modified $TiOF_2$ a surface area of 36 m$^2\cdot$g^{-1} and a pore diameter about 49 nm, which will enhance the adsorption of O–H groups and pollutants. FTIR characterization shows that NaOH-modified $TiOF_2$ has more O–H groups to supply more lone electron pairs transferring from O of the O–H groups to Ti and O of $TiOF_2$, in accordance with the BET analysis. UV–vis absorption spectroscopy shows that the NaOH-modified $TiOF_2$ samples have an adsorption plateau rising from 400 to 600 nm in comparison with $TiOF_2$ and its band gap is 2.62 eV, lower than that of $TiOF_2$. Due to the lower band gap, more O–H groups adsorption, network morphologies with larger surface area, and sensitization process, the NaOH-modified $TiOF_2$ exhibited much higher photocatalytic activity for RhB degradation. In addition, considering the sensitization process, O–H on $TiOF_2$ not only accelerated the degradation rate of RhB, but also changed its degradation path. This finding provides a new idea to enhance the photocatalytic performance by NaOH modification of the surface of $TiOF_2$. Considering its synthesizing process, the NaOH-modified $TiOF_2$ needs much lower temperature and shorter time than $TiOF_2$-crushed-Ref, but has much better photocatalytic performance, which provides a more economic choice.

Acknowledgments: Financial support was provided by Shaanxi Key Industrial Projects (2014GY2-07) and the Shaanxi Province Education Department Science and Technology Research Plan (15JK1460).

Author Contributions: In this paper, Chentao Hou and Wenli Liu designed the experiments; Wenli Liu and Jiaming Zhu conducted the experiments; Chentao Hou and Wenli Liu analyzed the data; and Chentao Hou wrote the article.

Conflicts of Interest: There is no conflict of interest existing in the manuscript submission, and it is approved by all of the authors for publication. All the authors listed have approved the manuscript to be enclosed.

References

1. Chen, X.; Mao, S.S. Titanium dioxide nanomaterials: Synthesis, properties, modifications, and applications. *Chem. Rev.* **2007**, *107*, 2891–2959. [CrossRef] [PubMed]
2. Nunes, D.; Pimentel, A.; Santos, L.; Barquinha, P.; Fortunato, E.; Martins, R. Photocatalytic TiO_2 Nanorod Spheres and Arrays Compatible with Flexible Applications. *Catalysts* **2017**, *7*, 60. [CrossRef]
3. Zuo, S.; Jiang, Z.; Liu, W.; Yao, C.; Chen, Q.; Liu, X. Synthesis and Characterization of Urchin-like Mischcrystal TiO_2 and Its Photocatalysis. *Mater. Charact.* **2014**, *96*, 177–182. [CrossRef]
4. Xie, C.; Yang, S.; Shi, J.; Niu, C. Highly Crystallized C-Doped MesoporousAnatase TiO_2 with Visible Light Photocatalytic Activity. *Catalysts* **2016**, *6*, 117. [CrossRef]
5. Cherian, C.T.; Reddy, M.V.; Magdaleno, T.; Sow, C.-H.; Ramanujachary, K.V.; Rao, G.V.S.; Chowdari, B.V.R. (N,F)-Co-doped TiO_2: Synthesis, anatase-rutile conversion and Li-cycling properties. *CrystEngComm* **2012**, *14*, 978–986. [CrossRef]
6. Ramasami, A.K.; Reddy, M.V.; Balakrishna, G.R. Combustion synthesis and characterization of NiO nanoparticles. *Mater. Sci. Semicond. Process.* **2015**, *40*, 194–202. [CrossRef]
7. Wang, J.; Cao, F.; Bian, Z.; Leung, M.K.H.; Li, H. Ultrafine single-crystal $TiOF_2$ nanocubes with mesoporous structure, high activity and durability in visible light driven photocatalysis. *Nanoscale* **2014**, *6*, 897–902. [CrossRef] [PubMed]
8. Wang, Z.; Ana, C.; Zhang, J. Synthesis of heterostructured $Pd@TiO_2/TiOF_2$ nanohybrids with enhanced photocatalytic performance. *Mater. Res. Bull.* **2016**, *80*, 337–343.
9. Dongn, P.; Cui, E.; Hou, G.; Guan, R.; Zhang, Q. Synthesis and photocatalytic activity of $Ag_3PO_4/TiOF_2$ composites with enhanced stability. *Mater. Lett.* **2015**, *143*, 20–23. [CrossRef]
10. Lv, K.; Yu, J.; Cui, L.; Chen, S.; Li, M. Preparation of thermally stable anatase TiO_2 photocatalyst from $TiOF_2$ precursor and its photocatalytic activity. *J. Alloys Compd.* **2011**, *509*, 4557–4562. [CrossRef]
11. Wang, Z.; Huang, B.; Dai, Y.; Zhang, X.; Qin, X.; Li, Z.; Zheng, Z.; Cheng, H.; Guo, L. Topotactic transformation of single-crystalline $TiOF_2$ nanocubes to ordered arranged 3D hierarchical TiO_2 nanoboxes. *CrystEngComm* **2012**, *14*, 4578–4581. [CrossRef]
12. Huang, Z.; Wang, Z.; Lv, K.; Zheng, Y.; Deng, K. Transformation of $TiOF_2$ cube to a hollow nanobox assembly from anatase TiO_2 nanosheets with exposed {001} facets via solvothermal strategy. *ACS Appl. Mater. Interfaces* **2013**, *5*, 8663–8669. [CrossRef] [PubMed]
13. Wen, C.Z.; Hu, Q.H.; Guo, Y.N.; Gong, X.Q.; Qiao, S.Z.; Yang, H.G. From titanium oxydifluoride ($TiOF_2$) to titania (TiO_2): Phase transition and non-metal doping with enhanced photocatalytic hydrogen (H_2) evolution properties. *Chem. Commun.* **2011**, *47*, 6138–6140. [CrossRef] [PubMed]
14. Jung, M.; Kim, Y.; Lee, Y. Enhancement of the electrochemical capacitance of $TiOF_2$ obtained via control of the crystal structure. *J. Ind. Eng. Chem.* **2017**, *47*, 187–193. [CrossRef]
15. Louvain, N.; Karkar, Z.; El-Ghozzi, M.; Bonnet, P.; Gúerin, K.; Willmann, P. Fluorinationof anatase TiO_2 towards titanium oxyfluoride $TiOF_2$: A novel synthesis approach and proof of the Li-insertion mechanism. *J. Mater. Chem. A* **2014**, *2*, 15308–15315. [CrossRef]
16. Li, W.; Body, M.; Legein, C.; Dambournet, D. Identify OH groups in $TiOF_2$ and their impact on the lithium intercalation properties. *J. Solid State Chem.* **2017**, *246*, 113–118. [CrossRef]
17. Chen, L.; Shen, L.; Nie, P.; Zhang, X.; Li, H. Facile hydrothermal synthesis of single crystalline $TiOF_2$ nanocubes and their phase transitions to TiO_2 hollow nanocages as anode materials for lithium-ion battery. *Electrochim. Acta* **2012**, *62*, 408–415. [CrossRef]
18. Reddy, M.V.; Madhavi, S.; Rao, G.V.S.; Chowdari, B.V.R. Metal oxyfluorides $TiOF_2$ and NbO_2F as anodes for Li-ion batteries. *J. Power Sources* **2006**, *162*, 1312–1321. [CrossRef]
19. Reddy, M.V.; Rao, G.V.S.; Chowdari, B.V.R. Metal Oxides and Oxysalts as Anode Materials for Li Ion Batteries. *Chem. Rev.* **2013**, *113*, 5364–5457. [CrossRef] [PubMed]
20. Zhou, P.; Yu, J.; Nie, L.; Jaroniec, M. Dual-dehydrogenation-promoted catalytic oxidation of formaldehyde on alkali-treated Pt clusters at room temperature. *J. Mater. Chem. A* **2015**, *3*, 10432–10438. [CrossRef]
21. Nie, L.; Yu, J.; Li, X.; Cheng, B.; Liu, G.; Jaroniec, M. Enhanced Performance of NaOH-Modified Pt/TiO_2 toward Room Temperature Selective Oxidation of Formaldehyde. *Environ. Sci. Technol.* **2013**, *47*, 2777–2783. [CrossRef] [PubMed]

22. Gopalakrishnan, S.; Zampieri, A.; Schwieger, W. Mesoporous ZSM-5 zeolites via alkali treatment for the direct hydroxylation of benzene to phenol with N_2O. *J. Catal.* **2008**, *260*, 193–197. [CrossRef]

23. Wu, Y.; Tian, F.; Liu, J.; Song, D.; Jia, C.; Chen, Y. Enhanced catalytic isomerization of α-pinene over mesoporous zeolite beta of low Si/Al ratio by NaOH treatment. *Microporous Mesoporous Mater.* **2012**, *162*, 168–174. [CrossRef]

24. Wang, Z.; Lv, K.; Wang, G.; Deng, K.; Tang, D. Study on the shape control and photocatalytic activity of high-energy anatasetitania. *Appl. Catal. B Environ.* **2010**, *100*, 378–385. [CrossRef]

25. Niu, L.; Zhang, Q.; Liu, J.; Qian, J.; Zhou, X. TiO_2 nanoparticles embedded in hollow cube with highly exposed {001} facets: Facile synthesis and photovoltaic applications. *J. Alloys Compd.* **2016**, *656*, 863–870. [CrossRef]

26. He, M.; Wang, Z.; Yan, X.; Tian, L.; Liu, G.; Chen, X. Hydrogenation effects on the lithium ion battery performance of $TiOF_2$. *J. Power Sources* **2016**, *306*, 309–316. [CrossRef]

27. Iwata, T.; Watanabe, A.; Iseki, M.; Watanabe, M.; Kandor, H. Strong Donation of the Hydrogen Bond of Tyrosine during Photoactivation of the BLUF Domain. *J. Phys. Chem. Lett.* **2011**, *2*, 1015–1019. [CrossRef]

28. Ning, Y. *Structural Identification of Organic Compounds and Organic Spectroscopy*; Science Press: Beijing, China, 2000; ISBN 978-7-03-0074-126.

29. Zhang, Y.; Zhang, Q.; Xia, T.; Zhu, D.; Chen, Y.; Chen, X. The influence of reaction temperature on the formation and photocatalytic hydrogen generation of (001) faceted TiO_2 nanosheets. *ChemNanoMat* **2015**, *1*, 270–275. [CrossRef]

30. Vaiano, V.; Sannino, D.; Almeida, A.R.; Mul, G.; Ciambelli, P. Investigation of the Deactivation Phenomena Occurring in the Cyclohexane Photocatalytic Oxidative Dehydrogenation on $MoOx/TiO_2$ through Gas Phase and in situ DRIFTS Analyses. *Catalysts* **2013**, *3*, 978–997. [CrossRef]

31. Samsudin, E.M.; Hamid, S.B.A.; Basirun, W.J.; Centi, G. Synergetic effects in novel hydrogenated F-doped TiO_2 photocatalysts. *Appl. Surf. Sci.* **2016**, *370*, 380–393. [CrossRef]

32. Ding, X.; Hong, Z.; Wang, Y.; Lai, R.; Wei, M. Synthesis of square-like anatase TiO_2 nanocrystals based on $TiOF_2$ quantum dots. *J. Alloys Compd.* **2013**, *550*, 475–478. [CrossRef]

33. Nishikiori, H.; Hayashibe, M.; Fujii, T. Visible Light-Photocatalytic Activity of Sulfate-Doped Titanium Dioxide Prepared by the Sol−Gel Method. *Catalysts* **2013**, *3*, 363–377. [CrossRef]

34. Kowalska, E.; Remita, H.; Colbeau-Justin, C.; Hupka, J.; Belloni, J. Modification of Titanium Dioxide with Platinum Ions and Clusters: Application in Photocatalysis. *J. Phys. Chem. C* **2008**, *112*, 1124–1131. [CrossRef]

35. Park, H.; Choi, W. Photocatalytic Reactivities of Nafion-Coated TiO_2 for the Degradation of Charged Organic Compounds under UV or Visible Light. *J. Phys. Chem. B* **2005**, *109*, 11667–11674. [CrossRef] [PubMed]

MDPI

St. Alban-Anlage 66

4052 Basel

Switzerland

Tel. +41 61 683 77 34

Fax +41 61 302 89 18

www.mdpi.com

Catalysts Editorial Office

E-mail: catalysts@mdpi.com

www.mdpi.com/journal/catalysts